The concept of normativity spans a series of interrelated dichotomies that lie at the heart of philosophical inquiry: fact and value, is and ought, the objective and the subjective, causes and reasons, the natural world and human sensibilities. Much philosophical effort has been devoted to accentuating the gaps between the concepts juxtaposed by each of these pairs, and the fallacies involved in their conflation. This volume, however, seeks to bridge these gaps. The papers collected here—all written expressly for this volume—set out to show that normative discourse must be sensitive to the facts, and that reasoning about facts is inherently value-laden. They demonstrate that the descriptive and the normative meet in language, in expressions that are both descriptive and normative. And they highlight the objective aspects of moral reasoning, and the normative aspects of objectivity. These challenges to the traditional view are as relevant to social and political discourse as they are to philosophy.

Yemima Ben-Menahem, The Hebrew University of Jerusalem

This distinguished volume of essays draws creatively on several rich traditions in philosophy—including Wittgenstein, Murdoch, philosophy of law, critical social theory, and Deweyan and Peircean pragmatism—to bring together an important variety of new challenges to the supposed "fact/value gap" and its alleged consequences for philosophy. For all those who want think harder and deeper about "fact and value", it will be essential reading.

Sophie-Grace Chappell, Open University

Marchetti and Marchetti have gathered here a comprehensive collection of positions critical of the coherence of the fact/value dichotomy; each coming at the problem from a different perspective; each offering a different (dis)solution. Their introductory overview of the origins of the dichotomy in the Western philosophic tradition is valuable, tying together the various strands, from Hume through Russell, that have provided support for the distinction in its current emotivist and non-cognitivist forms. The collection is also noteworthy for its attention to the critical voices found in the American pragmatist tradition and taken up by a number of contemporary thinkers whose work is represented here, such as Hilary Putnam.

Sharyn Clough, Oregon State University

Explicit interest in questions about the relation between facts and values has waxed and waned, inside and outside the academy, over the last several decades. But the questions themselves, which often turn up under different labels, remain immediately relevant both to our efforts to do justice to the world and to ethical challenges we confront within it. Marchetti and Marchetti have given us a collection that clearly brings out the importance of fact-value debates while also stressing the debates' multifaceted character. Taken together, these essays – from a group of distinguished thinkers – offer not only a helpful tour of the complexity of the issues but also a forceful impression of how and why they matter today.

Alice Crary, The New School

That there is a clear and unequivocal distinction between facts and values is something all too often assumed and only very seldom actively interrogated. *Facts and Values* takes up this question from a range of philosophical perspectives (including but not restricted to the 'analytic') that nevertheless converge in their rejection of the idea that the distinction can be given any unqualified application. This has important consequences, allowing us to recognise, for instance, the interconnection between ethics and ontology, and forcing us also to acknowledge the way in which evaluative commitments are inextricably bound up with all of our engagements in the world. The volume provides both an excellent point of entry into the topic at the same time as it also sets out important new insights and approaches. It contains contributions from many key figures in the area, but is especially notable for including one of the last pieces of writing by a seminal thinker of the last fifty years, Hilary Putnam, who together with Ruth Anna Putnam, has been central in bringing philosophical attention back to this important question.

Jeff Malpas, Univerity of Tasmania

Throughout the history of analytic philosophy, the fact-value distinction has been baked into nearly every research program within the tradition. Hilary Putnam has shown that we had better not just assume that the distinction—or its sophisticated variants, e.g., the contrast between science and ethics—can be sustained. If it cannot, there is a great deal of rethinking to be done, both across the many philosophical subspecializations and in the broader intellectual culture. This collection keeps this very important ball rolling, advancing an agenda with the potential to reshape philosophy.

Elijah Millgram, University of Utah

Facts and Values

This collection offers a synoptic view of current philosophical debates concerning the relationship between facts and values, bringing together a wide spectrum of contributors committed to testing the validity of this dichotomy, exploring alternatives, and assessing their implications. The assumption that facts and values inhabit distinct, unbridgeable conceptual and experiential domains has long dominated scientific and philosophical discourse, but this separation has been seriously called into question from a number of corners. The original essays here collected offer a diversity of responses to fact-value dichotomy, including contributions from Hilary Putnam and Ruth Anna Putnam who are rightly credited with revitalizing philosophical interest in this alleged opposition. Both they, and many of our contributors, are in agreement that the relationship between epistemic developments and evaluative attitudes cannot be framed as a conflict between descriptive and normative understanding. Each chapter demonstrates how and why contrapositions between science and ethics, between facts and values, and between objective and subjective are false dichotomies. Values cannot simply be separated from reason. *Facts and Values* will therefore prove essential reading for analytic and continental philosophers alike, for theorists of ethics and metaethics, and for philosophers of economics and law.

Giancarlo Marchetti is Associate Professor at the Università degli studi di Perugia, Italy. He is the author of *Il neo-pragmatismo* (1999) and *Verità e valori. Tra pragmatismo e filosofia analitica* (2008), and the co-editor of *Il Pragmatismo. Dalle origini agli sviluppi contemporanei* (with Giovanni Maddalena and Rosa Maria Calcaterra, 2015).

Sarin Marchetti is Postdoctoral Fellow at the Università degli Studi di Milano, Italy. He is the author of *Ethics and Philosophical Critique in William James* (2015) and *James* (2016), and the co-editor of *Pragmatism and the European Traditions: Encounters with Analytic Philosophy and Phenomenology Before the Great Divide* (with Maria Baghramian, 2017).

Routledge Studies in Contemporary Philosophy

For a full list of titles in this series, please visit www.routledge.com

79 **A Social Theory of Freedom**
Mariam Thalos

80 **The Cognitive Basis of Aesthetics**
Cassirer, Crowther, and the Future
Elena Fell and Ioanna Kopsiafti

81 **Interactive Justice**
A Proceduralist Approach to Value Conflict in Politics
Emanuela Ceva

82 **The Epistemological Skyhook**
Determinism, Naturalism, and Self-Defeat
Jim Slagle

83 **Time and the Philosophy of Action**
Edited by Roman Altshuler and Michael J. Sigrist

84 **McTaggart's Paradox**
R. D. Ingthorsson

85 **Perspectives on Ignorance from Moral and Social Philosophy**
Edited by Rik Peels

86 **Self-Reflection for the Opaque Mind**
An Essay in Neo-Sellarsian Philosophy
T. Parent

87 **Facts and Values**
The Ethics and Metaphysics of Normativity
Edited by Giancarlo Marchetti and Sarin Marchetti

Facts and Values

The Ethics and Metaphysics of Normativity

Edited by Giancarlo Marchetti and Sarin Marchetti

Routledge
Taylor & Francis Group
NEW YORK AND LONDON

First published 2017
by Routledge
605 Third Avenue, New York, NY 10017

and by Routledge
4 Park Square, Milton Park, Abingdon, Oxon OX14 4RN

Routledge is an imprint of the Taylor & Francis Group, an informa business

Library of Congress Cataloging in Publication Data
Names: Marchetti, Giancarlo, editor.
Title: Facts and values : the ethics and metaphysics of normativity / edited by Giancarlo Marchetti and Sarin Marchetti.
Description: 1 [edition]. | New York : Routledge, 2016. | Series: Routledge studies in contemporary philosophy ; 87 | Includes bibliographical references and index.
Identifiers: LCCN 2016036040 | ISBN 9781138955516 (hardback : alk. paper)
Subjects: LCSH: Normativity (Ethics) | Facts (Philosophy) | Values.
Classification: LCC BJ1458.3 .F33 2016 | DDC 121/.8—dc23
LC record available at https://lccn.loc.gov/2016036040

ISBN: 978-1-138-95551-6 (hbk)
ISBN: 978-1-315-66629-7 (ebk)

Typeset in Sabon
by Apex CoVantage, LLC

In memory of Hilary Putnam (1926–2016)

Contents

Preface

According to a deep-seated assumption, facts and values belong to different—indeed opposite—epistemic fields. Statements about facts deal exclusively with objective justifications when it seems our very rationality is at stake. In contrast, value statements are relegated to the sphere of subjective and contingent impressions, residing outside the scope of rationality altogether—or, at best, inhabiting a watered-down, second-order variety of practical reason. When expressing a certain value, we are therefore not committed to a fully-fledged rational norm, as our *merely* practical normativity remains parochially tied to the activity under consideration. This, surprisingly widespread, view has crystallized into an unbridgeable opposition between the descriptive and the evaluative, casting doubts upon the reality of values themselves. And, underlying this division lies the further assumption that facts alone reference a genuine λόγος, that is, a normative standard capable of determining what is the case, and hence what can justifiably be thought or said. What is defined as "factual" comes to define all that is "rational," thereby denigrating affective or emotional commitments as sub-rational.

Although the dogma of a fact-value dichotomy has long held sway in scientific and philosophical discourse and practice, it has been called into question by some of our most eminent thinkers: to name but a few, Charles S. Peirce, William James, F. C. S. Schiller, John Dewey, C. I. Lewis, Iris Murdoch, Philippa Foot, Donald Davidson, Hilary Putnam, Ruth Anna Putnam, Richard Rorty, and John McDowell, all of whom helped show how the assumed heterogeneity of descriptive and evaluative considerations and judgments seriously misrepresents the nature and operation of both.

Building upon the critical tradition launched by such thinkers, the authors in this volume lend further credence to their claims that the relationship between epistemic activities and evaluative attitudes cannot be understood as an opposition between the descriptive and the normative. According to them, the contrapositions between facts and values—and hence between science and ethics, as well as between objectivity and subjectivity—are indeed obfuscatory pseudo-dualisms. The authors of this volume invite us to rethink the relationship between these two reciprocally indispensable and intertwined epistemic fields, questioning the alleged fact/value dichotomy with the aim of restoring a more inclusive and integrated notion of rationality and normativity—one capable of putting us back in touch with the world and with ordinary life, and illuminating the essential connection between what we understand reality to be and our variegated ways of coping with it.

G. M. and S. M.

Acknowledgements

We would like to thank our editors at Routledge, Margo Irvin, Andrew Weckenman, and Alexandra Simmons for having followed the book with great care in all its phases. Working together has been indeed a genuine pleasure. We are also grateful to the two Routledge referees for their precious criticisms and suggestions, which greatly improved the volume and spurred us to put together a better and stronger collection. Finally and most importantly, we would like to warmly thank the authors for having taken time to write, rework, and polish their chapters time and again: we have learned a great deal from them and feel honored to have our names associated with theirs. May all collaborative projects be as smooth, congenial, and rewarding as this one.

Behind and Beyond the Fact/ Value Dichotomy

Giancarlo Marchetti and Sarin Marchetti

1 The Wider Stakes of the Dichotomy

Among the central philosophical riddles—both those *investigated* by philosophers and those *fabricated* by them anew—there is the thorny issue of the relationship between matters of fact and matters of evaluative judgment. Despite the many different philosophical clothes it has worn, the so called "fact/value dichotomy" has been a constant, and rather uncomfortable, companion of modernity since the scientific revolution. As the finalistic and teleological conception of nature unraveled, the world came to be seen as an inhospitable place for the old ontological furniture of metaphysics: once stripped of its intrinsic meanings and powers, reality became a cold plane of causality upon which we project our values and in the light of which we sort out our preferences. This picture continues to define the modern mindset, and is indeed one of the most lasting inheritances of Enlightenment, despite three centuries of political and cultural upheaval. From a historical perspective, the fact/value dichotomy acts as a kind of cultural-historical landmark, alongside epochal themes of disenchantment, naturalism, and normativity.[1]

Nowadays, hardly anyone would defend an absolute dichotomy between the descriptive and the evaluative when debating the rational and normative foundations of our thinking and acting—although there remains a small cluster of philosophers stubborn enough to do so. Nevertheless, we still feel compelled to rigorously maintain the nominal distinction between the two, as if out of fear that objectivity or sincerity will themselves lose integrity if we no longer take empirical science as the supreme measure of all human inquiry. We are told that, in failing to distinguish the two, we run the risk of both overlooking factual evidence *and* obscuring our evaluative attitudes. In everyday conversation, as well as in more formal environments like philosophy seminars, it is not uncommon to hear someone question whether a claim is based on facts or is merely the expression of a value. What lies behind this question is the idea that our most considered judgments are those arrived at through the dispassionate assessment of evidence and the careful weighting of the various values involved in the critical handling of the facts so distilled. This way of setting things up has a number of consequences for the way in which we conceive of what counts as a fact and what counts as a value, which is in turn motivated by such conceptions. Little wonder then that a significant slice of the debate focuses on the possibility that values refer to different phenomena than what is described by facts; that values embody a distinct linguistic system, following their own epistemological pathway; or that values are themselves constitutive of a particular subset of facts. This thesis has been cashed out in semantic, linguistic, and metaphysical terms, hence differing slightly in its rationale and critical target—more on this in a moment.

That there is a sharp dichotomy—rather than, say, a functional distinction—between facts and values in the first place is not uncontroversial, and has been forcefully debated by philosophers and non-philosophers alike. One of the most interesting and pressing intellectual concerns for ourselves, the great-grandchildren of the Enlightenment, is how to revise its more problematic conceits without falling back upon pre-modern outlooks with their problematic metaphysical sureties. If this is too grand a question for any single generation—let alone a single thinker—to resolve, the recent past has hosted a wealth of insightful attempts to think anew the relationship between facts and values. The authors of the present collection arriving from the three distinct (yet at times converging) realms of pragmatism, analytic philosophy, and critical theory, each offer analyzes aiming to steer the debate towards a new picture of descriptive and evaluative understanding. As against the fact/value dichotomy, each author promotes some degree of fact-value entanglement while also tracking down diverse points of exchange, transition, and convergence. In the light of this revision of one of the key assumptions of modernity, which consists in nothing short of a re-categorization of some of its central claims, this volume as a whole represents an attempt at realigning the ethical and metaphysical dimensions of our thinking and acting and their relation to the collective horizon of our social practices. This amounts, among other things, to the portraying of the normativity of our thoughts and conducts as an integral component of our *practices of knowledge and action*, rather than superimposed on them from without.

The philosophical scene in which the various chapters of this work move is complex and variegated, and hence in what follows we shall reconstruct a portion of the events and some of the most salient steps that brought the discussion where it currently stands, working in all effects as an integration of the four historical chapters comprising part I. In conclusion, a few words on the more contemporary scene and state of the art will be offered, so to engage more directly with part II and III of the collection.

2 The Anatomy of the Dichotomy

According to the advocates of the dichotomy, facts and values occupy two separate epistemic fields. Factual statements, as empirically verifiable or falsifiable by means of neutral and intersubjective methods, refer to "true" gnoseological problems and trade in explanations, scientific prevision, and objectivity. They are associated with those areas of "homogeneous rationality" and are naturalistically rooted to reality (Schwartz 2007, 69). Value statements, being empirically unverifiable and objectively unjustifiable, seem to be relegated to what is simply subjective, relative, emotional, and contingent—that is, to those "non-homogenous", sub-rational areas lying outside the domain of reason.

According to a definite empiricist line in epistemology, which had in Hume its godfather and in Russell, Ayer, and Stevenson its most notable paladins, the fact/value dichotomy shows that nothing is objective outside the sphere of natural science. According to this line of argument, knowledge is grounded on the presupposition that science, dealing solely in facts, remains neutral, objective, and hence *value-free* (cf. Kincaid, Dupré, and Wylie 2007); values are instead biased as well as unstable, and hence cannot be considered bearers of truth. But this picture of knowledge is shaped by an "intellectualist myth" according to which our cognitive or epistemic functions refer to a field of thought whose primary goal is the accurate and objective representation of "reality" (Corradi Fiumara 2001, 76). Our reigning representationalist epistemologies appear therefore vitiated by the exclusion of value components, whose theoretical relevance is silenced and undermined. Such epistemologies hinge upon an extremely narrow conception of rationality, by which values figure as emotive and non-cognitive obstacles on the road towards truth. By aiming at a unitary conception of science and knowledge, these epistemologies are vitiated by a metaphysical attitude of "exclusion" and "denial" that, as Corradi Fiumara writes, "tacitly transform illuminating distinctions into innumerable divisive and oppositional paradigms. These are so pervasive as to become almost inconspicuous—and, ironically, even canonic" (Corradi Fiumara 2001, 80). Derived from the Humean categories of "matters of fact" and "relation of ideas", those dualisms opposing facts and values, the descriptive and the normative, rationality and the emotion, science and ethics, imprint upon our culture so as to compartmentalize modes of experience and thinking that are actually formed in continuous interaction which each other.

This bifurcated conception of rationality, which subordinates values to facts, and the normative to the descriptive, has been undermined by recent, alternative epistemologies questioning the simplistic opposition of "inside" to "outside": think of the dichotomical categories of, e.g., mind/body, emotion/reason, or subject/object. Such a "spatial" construal of rationality is seen as restrictive precisely because it confers an empty placeholder status to values, despite their clear, commonsensical entwinement with facts in our effort to understand and navigate the world.[2] According to these alternative epistemologies, our philosophical thinking about these matters has been corrupted by an intellectualistic presupposition blinding us from acknowledging the intersections and synergies between cognitive process and evaluative factors. This alternative conception of rationality does not commit us to any premodern model of cognition or normativity, but rather recalibrates our self-understanding *within* the boundaries of a secularized yet re-enchanted worldview.

3 Hume and the Fact/Value Dichotomy

Historically speaking, the expulsion of values from the natural world is quite a recent event. Dewey, for example, writes:

> For centuries, until, say, the sixteenth and seventeenth centuries, nature was supposed to be what it is because of the presence within it of ends. In their very capacity as ends they represented complete or *perfect* Being. All natural changes were believed to be striving to actualize these ends as the goals towards which they moved by their own nature. Classic philosophy identified *ens, verum*, and *bonum*, and the identification was taken to be an expression of the constitution of nature as the object of natural science. In such a context there was no call and no place for any *separate* problem of valuation and values, since what are now termed values were taken to be integrally incorporated in the very structure of the world. But when teleological considerations were eliminated from one natural sciences after another, and finally from the sciences of physiology and biology, the problem of value arose as a separate problem.
>
> (Dewey 1939, 192)

It is in fact with Hume that modern thought sees the affirmation of what Hilary Putnam has called "the last dogma of empiricism" (Putnam 2002, 145), the projection of an insurmountable heterogeneity between descriptive and prescriptive judgments.

The *locus classicus* of the non-derivability of values from facts is Hume's *A Treatise of Human Nature*. The precise content of this thesis, better known as "Hume's Law", is itself the subject of considerable debate, as a result of diverse and often conflicting readings advanced by his interpreters. The standard interpretation is the one offered by various classical and contemporary empiricists (cf. Hudson 1969), who hold that Hume's Law represents an essentially logical question, and thereby precludes any attempt to derive values from facts, or, more generally, any normative conclusions from purely empirical considerations—as Hume himself famously claimed, "one cannot infer an 'ought' from an 'is'" (Hume 1978, III, I, I). According to Hume, it would be a fallacy to derive evaluative conclusions from descriptive premises, as there would be an irreducible division between propositions whose proper function is that of describing what is the case, and propositions whose proper function is prescribing what ought to be thought or done.

The idea that Hume's Law is first and foremost a logical concern operating along formal inferences has been questioned by interpreters—most notably, Hilary Putnam, who holds that Hume's Law represents a *metaphysical* dichotomy, between "matters of fact" and "relations of ideas."[3] According to Putnam, Hume's "matters of fact" are grounded in a "pictorial" semantics, according to which the only way to conceptualize "matters of fact" is by visual "resemblance."[4] For Hume, ideas always derive from sensible impressions, of which they remain but faint images (Hume 1978, I, I, I), whose persistence depends on the sensible quality of their reproduction. The Scottish philosopher described impressions and ideas—which are not just visual but also tactile, auditory, olfactory, and so on—in primarily visual terms, as though our ideas were best judged according to their descriptive accuracy. From this assumption Putnam concludes that there can be no "matters of fact" for Hume concerning what can be considered virtuous, right, good, and so on. Hume's pictorial semantics cannot in fact accommodate discussion of virtue, righteousness, and goodness because it lacks the ability to depict virtue in the way we might depict the shape, color, and texture of a cherry (Putnam 2002, 15). Ideas are not simply pictorial; they are also tied to emotional states and sentiments. Given that, for Putnam, the latter are the only properties of ideas Hume used to explain why it seems that there are such "matters of facts" in the first place, it seems evident that those constitutive elements of our ideas corresponding to judgments of virtue cannot but be sentiments aroused in us by the " 'contemplation' of the relevant actions owing to the 'particular structure and fabric' of our minds" (Putnam 2002, 15).

For Hume, reason and ethics, science and value, proof and volition are separate spheres which cannot be bridged by any valid logical inference. However, according to Putnam, Hume's Law is not based on any rules of formal inference governing statements, but rather on deeper ontological assumptions about their content. As such, Hume's Law amounts to a hypostatization of metaphysical entities ("facts" and "values"), whose gnoseological architecture chains rationality to sense certainty, to a depleted, naturalistic environment, bereft of the virtues and passions that make our world come alive.

4 The Dichotomy after Hume

4.1 Emotivism and the Legacy of Non-Cognitivism

In the first decades of the twentieth century, an inflated version of Hume's Law found a most congenial soil in metaethical emotivism. Emotivism, as it is well known, is a form of non-cognitivism rooted in a methodological monism that recalls the logical positivist's image of a value-free science built upon the strict semantics of logical demonstration and empirical verification.[5] On the one hand, *synthetic* statements derive their truth-value and meaning from the observable predicates of phenomena encountered in direct experience. On the other hand, the truth-value of *analytic* (or tautological) statements depends entirely upon their internal logical form or on the definitional content of their component parts. This classification scheme banishes ethical, aesthetic, and evaluative statements from the sphere of reason—because they resist dissolution into simple analytic or synthetic statements, they are treated as cognitively meaningless pseudo-propositions on the par with other "metaphysical" statements.[6]

Emotivism is perhaps the most recognizable and contentious form of non-cognitivism, which came to dominate the metaethical agenda of Anglophone philosophy during the first half of the twentieth century, and which has maintained its influence over the way values are conceptualized well into our present. If we accept that the 1950s marks a dramatic turning point for the development of metaethics, we can begin to see the wider seismic effects of this development for moral philosophy as a whole—even if non-cognitivism's proponents themselves have proven fickle in fashioning a preferred label for their approach (with prescriptivism, error theory, expressivism, and quasi-realism representing the latest philosophical trends today). Emotivism, specifically as an issue for analytic metaethics, is our present focus, because it is here that we find the clearest articulation of fact/value dichotomy in all its semantic and metaphysical finery, coupled with the most explicit meta-philosophical machinery necessary for its implementation. Emotivism represents in fact only the most radical outcome of the non-cognitivist approach to moral discourse and practice, which equates the cognitively assessable with the descriptive and relegates the spurious evaluative dimension to what falls outside the proper scope of objectivity. Virtually all the above-mentioned subsequent versions of non-cognitivism hold in fact the fact/value dichotomy in one form or another as, despite their significant differences, they all variously subscribe to a representationalist epistemology pretending to carve up the world at its joints, and hence separating the proper constituents of reality (facts and objects) from those inhabiting our minds or language only (values and norms).

As we have already observed, this dichotomy not only exerts influence over the way we think about facts and about values, it also reflects our deeper assumptions about the aim and scope of philosophical and ethical investigations alike. In fact, by depicting values as belonging to a domain outside the sphere of rational argumentation, moral philosophy itself came to be considered an exercise in ethical persuasion rather than critical assessment. What indeed often lurks behind this divide, and behind representationalism as the preferred epistemological standpoint, is a doctrinal interest in promoting certain other philosophical approaches more akin to the natural sciences, at the expense (if not wholesale dismissal) of approaches committed to non-empirical evaluation (which are treated pejoratively as a mode of aesthetic contemplation).

4.2 Bertrand Russell

Early in his philosophical career, Russell held fast to an intuitionist or realist conception of ethical judgments, analogous to the one advanced by Moore in his *Principia Ethica*. Later on, under the influence of Wittgenstein's *Tractatus* and Santayana's criticism of ethical realism (cf. Santayana 1913), Russell would abandon his objectivist intuitionism and embrace, first, a variation of subjectivist emotivism (Cf. Russell 1914), and, secondly, his "enlightened emotivism" (Potter 2006, 1–6). According to the latter view, value judgments are nothing but the expression of sentiments and emotions:

> questions as to "values" . . . lie outside the domain of science . . . [indeed they] lie wholly outside the domain of science. That is to say when we assert that this or that has "value," we are giving expression to our own emotions, not to a fact which would still be true if our personal feelings were different.
>
> (Russell 1997, 230–1)[7]

According to Russell, the only valid scientific propositions are analytic statements (which are rationally justifiable), and synthetic statements, (which are empirically verifiable). By contrast, value judgments are neither empirically verifiable nor deducible from empirical observations.

In fact, their declarative grammatical form leads us to a mistaken conclusion about their truth-aptness, due in part to a superficial linguistic resemblance to the logic of scientific assertions, which serves to mask the subjective, sentimental character of their expression:

> when a man says 'this is good in itself', he seem to be making a statement, just as much as if he said 'this is square' or 'this is sweet . . . I think that what the man really means is: 'I wish everybody to desire this' or rather 'Would that everybody desired this.' If what he says is interpreted as a statement, it is merely an affirmation of his own personal wish; if, on the other hand, it is interpreted in a general way, it states nothing, but merely desires something.
>
> (Russell 1997, 235–6)

For Russell, it is a mistake to think that ethical judgments manifest the same empirical qualities as factual statements. According to him, ethical and value statements are not declarative or descriptive statements at all—they offer no factual propositions, causal explanations or clarifying descriptions of our worldviews, and as such remain expressions of subjective desires, hopes, and preferences. Ethical and evaluative disagreement can be resolved only by means of emotional appeals or persuasive strategies, so that it "is not a disagreement as to any kind of truth, but a difference of taste" (Russell 1997, 237).

If it is hard to find a bolder statement of the divide, Russell complicates the picture by claiming that, whether or not wishes are personal or subjective, what is being desired is itself universal. This more conciliatory tone moves in the direction of granting at least *some* objective weight to values, even if this facticity is still considered a different species of truth. Ethical and value judgments are now said to comprise the "attempt to give universal, and not merely personal, importance to certain of our desires" (Russell 1997, 232). Distinct from ethical judgments, scientific observation looks to "the causes of desires, and the means for realizing them, but it cannot contain any genuinely ethical sentences, because it is concerned with what is true or false." Ethical judgments have no bearing on what is actually true or false, but rather "consist of desires of a certain general kind, namely such as are concerned with the desires of mankind in general" (Russell 1997, 237).

According to Russell, a desire becomes universal when it is not in conflict with the general desires of the majority of human beings. This consideration brings him to relate ethics and politics and put them under a single banner:

> ethics is . . . closely related to politics: it is an attempt to bring the collective desires of group to bear upon individuals; or . . . it is an attempt by an individual to cause his desires to become those of his

> group. This latter is, of course, only possible if his desires are not too obviously opposed to the general interest.
>
> (Russell 1997, 232)

Russell's emotivism is unlike the other forms of emotivism in its "universal" aspiration and "political" dimension. According to him, there are a great number of common values, reflecting our shared humanity, and this makes him a "universalistic emotivist" (Aiken 1963, 109) rather than a subjectivist one.

4.3 Alfred Jules Ayer

A much more radical form of emotivism is found in Ayer's *Language, Truth and Logic*. The main targets of this work are ethical naturalism and intuitionism, specifically their claims regarding the possibility of our apprehending moral facts through intuition. According to Ayer, the objectivism of the naturalistic and "absolutistic" (intuitionist) conception of ethics is unsustainable, as it is impossible to find an objective criterion for assessing ethical and value judgments. This is not because "they have an 'absolute' validity which is mysteriously independent of ordinary sense-experience, but because they have no objective validity whatsoever" (Ayer 1952, 108).

For Ayer, there are only two classes of meaningful propositions: logical and factual ones. Value or normative judgments belong to neither, and therefore cannot be treated as objects of knowledge and of scientific investigation:

> there cannot be such a thing as ethical science, if by ethical science one means the elaboration of a "true" system of morals. For we have seen that, as ethical judgments are mere expressions of feeling, there can be no way of determining the validity of any ethical system, and, indeed, no sense in asking whether any such system is true.
>
> (Ayer 1952, 112)

Ethical propositions are normative and, as such, pseudo-concepts, as they do not describe facts or states of affairs. As a consequence, no "science of ethics" is possible. If, however, we consider ethical judgment as a branch of psychology, then it can be said to constitute a form of knowledge. According to Ayer, it is the proper task of psychology to account for the "different feelings that the different ethical terms are used to express" as well as "the different reactions that they customarily provoke" (Ayer 1952, 112). By contrast, ethical and evaluative propositions are sheer expressions of emotion, and are therefore pure and simple *nonsense* reflecting only the subjective mindset of a person seeking to persuade or otherwise influence the actions of another. Their conative force has the express purpose of moving others toward a desired outcome: "the ethical terms do not serve only to express feeling. They are calculated also to arouse feeling, and so to stimulate action" (Ayer 1952, 108). For Ayer, values and normative judgments carry no rational weight, express merely personal preferences, and thus lie outside the domain of meaningful expression, being neither factually verifiable nor logically compelling (cf. Ayer 1954).

4.4 Charles Leslie Stevenson

Russell's and Ayer's emotivist assertions were given a somewhat more subtle defense by Stevenson. In a series of articles published in *Mind*, as well as in his *Ethics and Language* (1944), Stevenson argued for a "liberalized" emotivist metaethics, wholly distinct from Ayer's high-handed dismissal of ethical, aesthetical, and value judgments.[8] Stevenson does not evoke the image of a "linguistic malady" that must be expunged from the field of rational knowledge, but rather suggests that value judgments are a legitimate field for critical investigation. He suggests that the very term *emotive* should "be kept as a tool for use in careful study, not as a device for relegating the nondescriptive aspects of language to limbo" (Stevenson 1944, 79).

By deploying a much wider array of tools than could be found in the machinery of the linguistic analysis of the time, Stevenson granted value judgment a dual role: equal parts emotive and descriptive. Emotive meaning refers to the manifold of attitudes—understood as "any psychological disposition of being *for* or *against* something" (Stevenson 1963, 1–2)—or sentiments deployed more or less consciously by the speaker in order to influence or persuade her audience. This first component of meaning is grounded in the capacity to condition one's listeners, to influence their behavior by means of sincere expression, rather than by descriptively detailing one's attitudes. Descriptive meaning, by contrast, consists in the "disposition to affect cognition" (Stevenson 1944, 70), which takes place whenever the stimulus or response is a cognitive mental state—such as believing, thinking, supposing, presuming (Stevenson 1944, 62)—or as part of a comprehensive system of beliefs.

Stevenson rejects the pictorial theory of meaning and identifies the meaning of an assertion with the tendency of a sign to trigger particular mental responses—that is, when our very words move our interlocutors to act in accordance with our intentions or plans. The presence of this psychological component is what allows Stevenson to distinguish emotive from descriptive meaning: "ethical judgments alter attitudes, not by an appeal to self-conscious efforts (as in the case with imperatives), but by the more flexible mechanism of *suggestion*. Emotive terms present the subject of which they are predicated in a bright and dim light, so to speak, and thereby *lead* people, rather than command them, to alter their attitudes" (Stevenson 1944, 33).

The bipolar character of ethical statements is revealed in the kind of *agreement* and *disagreement* they generate. Stevenson questions whether the nature of ethical agreement or disagreement is "parallel to that found in the natural sciences, differing only with regard to the relevant subject matter" or whether it is "of some broadly different sort" (Stevenson 1944, 2), to which he suggests the latter is the case. The disagreements that emerge in science, history, and biography are in fact "disagreements in belief." They concern "how matters are truthfully to be described and explained" (Stevenson 1944, 4), and refer to the *matters of fact* involved in the situation under investigation, to which we refer when engaged in problem-solving. When we disagree, we do so with regards to the truth or falsity of a particular assertion or description held in dispute, where someone claims the truth about something that another explicitly denies. Suppose Peter and Lisa disagree about the issue of an immigrant's right to vote: their disagreement stems from divergent beliefs about what each "knows" to be the facts of the case. Peter believes this is a question of social progress, which Lisa disputes. Their disagreement can only be resolved through further analyzing the facts of the matter, including deliberation on what should count as social progress and what a good example of this might be. We should note too how, for Stevenson, the fact that they are discussing how a substantive social/political issue can be defined does not make their disagreement one of conflicting tastes or preferences—it is still a matter of establishing the *factual* and *evidential* conditions of the matter under dispute.[9]

Ethical problems, on the other hand, have to do with explicit disagreements in attitudes, including our desires, preferences, aspirations, needs, and evaluations. In this case, when two parties disagree they do so on the basis of antithetical orientations towards a common object, one approving, the other disapproving. Unlike scientific and factual disagreements, ethical disputes are much harder to resolve. In the above example, the disagreement about voting rights is informed by conflicting perceptions of immigrant communities: Peter prefers to think of immigrants as equal citizens, whereas Lisa believes the distinction between citizens and non-citizens is important—their conflict is not over an interpretation of the facts but rather the values they have committed themselves to. This kind of disagreement will only be resolved by means of "persuasive, not empirical or rational" argumentations, with a view to bringing all parties to endorse the same conclusion (Stevenson 1963, 2). While it is possible to advance reasons both in favor and against one's preferred position, "these 'reasons' are related (only) psychologically . . . to the judgments they support" (Putnam 2012, 286). Whereas factual disagreements can be resolved inductively or deductively via rational argumentation, disagreements over values cannot; they rely upon our capacity to bring others to endorse our attitudes:

> *If* any ethical dispute *is* rooted in disagreement in belief, it may be settled by reasoning and inquiry to whatever extent the beliefs may be so settled. But if any ethical dispute is *not* rooted in disagreement in belief, then no *reasoned* solution of any sort is possible.
>
> (Stevenson 1944, 138)[10]

4.5 From Reference to Use

With Stevenson we see the most comprehensive emotivist attempt at reconciling a seemingly unbridgeable divide between the domain of factual, rationally assessable analysis, and the domain of persuasive and affectively conditioned evaluations. After Stevenson, from the early 1950s onwards, metaethics pursued a number of different paths, involving attempts at refining non-cognitivist views and defending it against the charge of irrationalism—see, e.g., the works of Richard M. Hare, John Mackie, Simon Blackburn, Allan Gibbard, David Copp, Mark Schroeder)—as well as plotting new alternative pathways for moral discourse and practice—most notably, new variations on naturalism (Aristotelean or Humean), the breakthrough of ordinary language philosophy (Wittgensteinian or Austinian), and the emergence of neo-pragmatism (of historicist or transcendental breed). While it is difficult, say impossible, to discern a single common thread amongst these often conflicting approaches, we can say, very generally, that they are all staunchly critical of the referential theory of meaning, which is discarded in favor of a use-theory of meaning, variously understood and put to work.[11]

Use-theory is a huge and sprawling topic, but what interests us about the shift away from referential theory are its implications for the way fact/dichotomy was perceived. To the degree it could continue to be sustained, the divide between facts and values became primarily a matter of linguistic *usage*, rather than denoting more serious semantic or metaphysical commitments—even when, as for example in Iris Murdoch's work, the distinction (or rather its very impossibility) is indeed garnished in a metaphysical language. Facts remain distinct from values (when they do) because of the way we treat them and talk about them, in everyday contexts as well as within specialist discourse. This novel approach was grounded in the conviction that our descriptions and evaluations alike are in the first place kinds of *activities*, and as such their justification should refer to the wider cluster of thoughts and deeds shaping up our human forms of life in which they take place. In assessing a certain situation and classifying a certain choice as concerning facts or rather values we do call in cause and employ distinctions that are linguistic in nature and pertain to the very texture of ordinary concepts. It is by using a certain word in certain ways that we confer the meaning it has and, conversely, its significance will be either granted or denied with reference to the way in which it is employed. It is thus the philosophical study of these transitions and their wider backgrounds that would hold the stage in the inquiry of the synergies, or lack thereof, between the descriptive and the evaluative.

Be that as it may, the dichotomy itself came under critical scrutiny, its validity gradually eroded from within: what were once seen as semantic or metaphysical convictions were now perceived as merely functional distinctions whose grounds should be traced back to the social dimension and texture of our ordinary and intellectual practices, and this process was only hastened by the contributions of a new generation of analysts, critical theorists,[12] and neo-pragmatists. However, it was classical pragmatism's frontal assault on fact/value dichotomy, at the very height of its popularity, which sowed the seeds of philosophical revolt, and which would later thrive by virtue of renewed attempts at critical cross-fertilization and transplantation—the fruits of which we witness in part II and III of the present volume.

5 The Entanglement of Facts and Values in Pragmatism

The most concerted effort to dissolve fact/value dichotomy and demonstrate their essential entanglement issues from the extended family of thinkers circling the pragmatist triumvirate of Peirce, James, and Dewey.[13] The debunking of the fact/value dichotomy is unquestionably one of the cornerstones of pragmatism from Peirce to Putnam—alongside anti-representationalism, holism, and fallibilism. What these four pillars have in common is their disregard for a picture of human understanding as a matter of faithful mirroring a reality which is brute and deaf to our requests (which are seen in fact as *imposition* on it), and replaced it with one in terms of an interactive coping with a world whose very shape is fashioned after this engaged activity (where our requests are seen as among the marks of our *responsiveness* to the world). In the light of this shift, pragmatism retooled the very epistemological inventory of modernity, revising our philosophical understanding of the very notions of nature, experience, and reason in the direction of their secular re-enchantment: the natural world is better accounted for as the realm of the statistically possible, experience depicted as the exhibition of our theoretical and practical abilities, and reason rendered as the responsive coping with one's environment. The normativity of our thoughts and talks would thus literally emerge from our socially-governed engagement with reality—which includes the world before us as well as our fellow inhabits. That being so, the very business of separating the descriptive from the evaluative component in knowledge becomes essentially a matter of paying attention to our ways of thinking and talking. In the same way in which we are responsible for keeping track of a certain aspect of reality or failing to, our responsibility extends to what we *take* to be a certain aspect of reality: factual, evaluative, *or* a blending of the two. The fact-value entanglement has been held as a ripest fruit of an anti-representationalist outlook, according to which our thoughts and words should not be assessed with reference to their ability to track down or not how things really are, but rather with reference to their capacity to guide us more or less successfully in the world. That being so, there would not be statements and judgments that by their own nature are qualified to ascertain truths about matters of facts, and statements and judgments that by their own nature are instead qualified to ascertain truths about matters of value.

The rejuvenation of interest in the debate over facts and values was in large part a result of pragmatism's questioning the very terms under which the debate was being conducted, and of its unmasking some of the more dubious pseudo-dualisms which lacked any connection to our ordinary and even scientific encounters with the world. Values permeate experience and are infused with the facts of science, with science itself understood as one among our human activities to which we granted a certain weight in selected contexts rather than as some kind of *über-activity* elevating us above contingency altogether and hence standing apart from the others. Schiller claims, for example, that our "valuations" pervade our whole experience, affecting whatever we consider a "fact", and can only be peeled off from the latter by means of artificial manoeuver (Schiller 1903, 10). Unlike logical empiricism and other varieties of scientific positivism, pragmatism accepts the need for "ground common to the conceptions and methods that are taken for granted in all physical matters and those that appear to be most important in respect to human activity" (Dewey 1939, 192). Even the most cursory look at the physical sciences reveals an absence of assertions expressing "value-facts or conceptions." Yet, as we grow more attentive, we realize how "all deliberate, all planned human conduct, personal or collective, seems to be influenced, if not controlled, by estimates of value or worth of ends to be attained" (Dewey 1939, 192). For Dewey, the rigid divide between facts and values will be dissolved once and for all when we realize that "not only is science itself *a* value (since it is the expression and the fulfillment of a special human desire and interest) but it is the supreme means of the valid determination of all valuations in all aspects of human and social life" (Dewey 1939, 250).

Pragmatism aims at renewing empiricism by way of a radicalization, or rather completion, of the Enlightenment project, which would lead it to the acknowledgment of deep intertwinement between the descriptive with the evaluative. Praising the many virtues and achievements of the Enlightenment, pragmatism calls for a "second Enlightenment" wherein morality would not only be freed from theology, but science would be freed from scientism—so that *both* knowledge and action would be made subject to collectively constituted normative assessment, rather than some supra-human totality lurking behind ordinary interactions and intellectual practices (cf. Rorty 1999 and Brandom 2009). It is only when science itself comes to realize how its own inquiries are value-laden that this most dogmatic dualism of modernity will be made to disappear. To this end, Dewey asserts that "there is nothing whatever that methodologically . . . marks off 'value judgments' from conclusions reached in astronomical, chemical or biological inquiries" (Dewey 1949, 357). Accordingly, the alleged opposition between facts and values will be dissolved when we "shall acknowledge the identity of Logical procedure" (Dewey 1903, 59) lying at the basis of each judgment: descriptive *and* evaluative.

Now we find this idea cashed out in a number of ways throughout the history of pragmatism, so much so that value-ladenness constitutes a *leitmotiv* for the tradition. Morris observes, for example, that "judgments of value are empirical in nature" as their scientific legitimacy and cognitive character remain "amenable in principle to control by scientific method" (Morris 1963, 94). Each objectively true statement has a normative aspect to it, and each experience is imbued with values and norms. For Schiller, facts, knowledge, and values are intertwined and interdependent: as values permeate experience as a whole, so knowledge itself operates as a kind of value. Facts without values are nothing more than sterile abstractions: "our 'Facts' when analyzed turn out to be 'Values,' and the conception of 'Value' therefore becomes more ultimate than that of 'Fact.' . . . Our valuations thus pervade our whole experience, and affect whatever 'fact' " (Schiller 1903, 10). To this end, Dewey adds that "*judgments about values are judgments about the conditions and the results of experienced objects; judgments about that which should regulate the formation of our desires, affections, and enjoyments*" (Dewey 1929, 212). Knowledge claims can be *more* or *less* evaluative in character, but no species of judgment—not even purely factual assessment, is value-free: "since selections of existences to serve as subject-data and of ideas to serve as predicate-possibilities (or ends in view) is necessarily involved in every judgment, the valuation operation is inherent in judgment as such" (Dewey 1938, 181).

Pragmatists, both classical and contemporary, acknowledge only too well how sometimes it might indeed be useful and prove congenial to trace more or less mobile lines of demarcation between different kinds of thoughts and talks, and yet they refuse to turn such demarcations into philosophical dualisms with their uncomfortable semantic and ontological baggage (Putnam 2002, 9–11, 19). Allowing the exigencies of social practice to lead us helps determine when a particular philosophical framing will be helpful—and when it might become an obstacle to our understanding. It may be useful at times to delineate evaluative considerations from other kinds of judgments and to prioritize the perspective of a theoretical physicist over that of a layperson. However, for the pragmatist "*nothing metaphysical follows from the existence of a fact/value distinction in this (modest) sense*" (Putnam 2002, 19). According to Putnam, the most unnerving aspect of fact/value dichotomy is how it has "corrupted our thinking about both ethical reasoning and description of the world," thereby preventing us from appreciating the deeper entanglement between the way the world naturally confronts us and the way we creatively act upon our surroundings. This divide, if anything, needs to be seen as another tool for helping clarify selected aspects of our variegated experiences, practices, and judgments.

Understood in this way, our deriving values *from* encounters with facts, and our reckoning with facts *as* embodiments of value can help us ascertain the significance and urgency of problematical situations in life. It is not the case that values offer "magical" implements of suasion for targeting our innermost, irrational selves, as this would suggest our falling back upon some version of emotivism (even of a different, milder kind). Rather, values and facts must be seen as cooperatively contributing to our mindedness and worldliness; they must be understood as equal partners in the fine-tuning of reality, and in our attunement to the significance and urgency of the problematic situation we find ourselves in. It is in this sense that distinguishing "facts" can be seen as part and parcel of our embeddedness within a particular evaluative horizon, and, conversely, that our endorsement of "values" can be seen as enabled by a universal responsiveness to facts.

6 Recent Attempts at Overcoming the Dichotomy

As we shift our focus to contemporary debates, we observe fact/value dichotomy under siege from all number of philosophical corners, with the old strategies being replaced or retooled to suit new discursive territories. The permafrost of metaphysical certitude could not survive the new philosophical climate, and there are few visitors willing to tour the old exhibition halls of fundamental ontology—although we still find a number of spirited revivals in analytic metaphysics (a telling event in itself whose bearing on the issue at hand is, however, only tangential[14]). We also note how the moral landscape has taken on a decidedly naturalistic tint, as witnessed by the proliferation of attempts to reconstruct morality from the vantage point of humanity as a biological species and as a cluster of *natural* forms of life. Naturalism has managed to unite friends and foes of the dichotomy, in lockstep with similar developments by researchers in "naturalized" metaphysics, epistemology, linguistics, and ethics—all of whom seem intent on claiming the mantle of "genuine" naturalism. It has become increasingly difficult to imagine what *else* one might be if not a "naturalist" of one kind or another, given the suspicion surrounding any "supra-" or "anti-naturalist" theories.

In light of the current ethical and meta-philosophical constellation, even attempts to peel-off normative and evaluative elements from the factual core of reality are now presented under the banner of naturalism—and as a crudely reductionist kind at that. This, largely internal, clash between reductionist and non-reductionist (or liberal)[15] forms of naturalism currently holds center stage in philosophical discussions about fact/value dichotomy. Nowadays, amongst both philosophers and the culture at large, we are less likely to hear the asserted impossibility of our drawing evaluative conclusions from factual descriptions, and we may note a greater willingness to grant core ethical beliefs genuine ontological legitimacy. Nevertheless, the dichotomical way of thinking remains with us, often disguised as an innocuous naturalism—whereby the intertwinement of values and facts allows some to conclude the former are completely reducible to the latter. Even if this more reductive naturalism still seems alien to our everyday, intuitive grasp of values, this has not prevented its resurgence amongst those we consider to be "specialists" in the field, and to whom we have often deferred judgment. Little wonder that reductionism is presented in heroic opposition to non-cognitivism (eliminativism, fictionalism and expressivism being among the usual suspects), despite their both sharing in the same representationalist semantics, according to which the purpose of statements is that of faithfully picturing reality (and where any divergence between the two factions applies mainly to criteria of inclusion).

One strategy for ending the civil war within naturalism amicably has been skillfully explored by Huw Price (2011), who, in the wake of Wittgenstein, Rorty, and Hume (here interpreted as an anti-representationalist thinker), questions the representationalist epistemology at the heart of both reductionist naturalism and its non-cognitivist opponents. Price draws a distinction between *object naturalism* (which defers to scientific considerations regarding the objects and methods of philosophical inquiry), and *subject naturalism* (which vindicates the legitimacy of philosophical inquiries as achievements of "natural" creatures). Whereas the former remains stuck in "placement problems" concerning the successful "matching" between our conceptions and language and the outside world, the latter merely replaces metaphysical speculation with philosophical investigations more akin to anthropological surveys. In contrast to object naturalism, subject naturalism does not ask *if* our concepts and words track objects in the world but rather *how* we arrive at thinking and talking in certain ways, and *why* we persist in this manner.

This radical reframing of naturalism has significant consequences for the way the dichotomy is conceived and challenged: by debunking the very possibility of an epistemological survey of the ontological underpinnings of thinking and speaking, we are freed from the old preoccupation with distinguishing between the descriptive and the evaluative, which can now be accounted for from the vantage point of their "social" role in human activity. The point is not that there are certain areas of discourse and practice in which our cognition takes hold, and others in which it is our evaluations that take center stage; rather, we are faced with a multitude of ways of thinking and speaking which differ both in their superficial grammatical appearance (some are in the indicative mode, some in the interrogative, while some in the prescriptive) and in their deeper pragmatic role (some are used to point our attention to an event, some invite us to reconsider a certain aspect of a situation, while others express a certain leaning)—all of which we try to make sense of in reference to the way this multitude is understood by practitioners themselves. According to this picture, *both* facts *and* values are a function of our activities of making sense of the world and caring for it.

Alternative strategies for overcoming dichotomy include the ever-expanding family of constructivists—which recently witnessed an upsurge in original contributions furthering the Kantian path opened by Rawls in the 1980s, and Christine Korsgaard in the 1990s. Moral constructivism reconfigures the relationship between facts and values by reconstructing the models of rationality and normativity to which they refer (cf. Bagnoli 2013). According to constructivists, it is by looking at the way we come to agree or disagree as *rational agents*—in the sense of our practical responsiveness to reasons (as opposed to a merely causal responsiveness)—that we arrive at a notion of objectivity robust enough to resolve our ethical and political problems (since it is assumed that we lack access to any transcendent or metaphysical criterion "beyond" social practice). It is thus by investigating what it means to be an agent who follows the rules of reason (variously specified in terms of consistency, responsibility, and autonomy), that we can begin to figure out which particular thought and actions qualify as intersubjectively valid and which do not.

Despite there being greater and lesser versions of "universalistic" constructivism (Aristotelean, Humean, and Kantian, each representing varying degrees of objectivity), the common mark of this approach is a firm rejection of the claim that practical normativity operates in a cognitive or theoretical register that is somehow detached from the domain of factual assessment. For constructivists, to count as a fully-fledged "rational" agent, we must be as well-acquainted with the empirical facts of the matter as we are to the normative expectations we incur by our engagements. From a practical, engaged point of view, there can be no discernment of "relevant" facts without acknowledgment of the values "at stake," nor any normative commitment wholly in isolation from the exigencies of particular situations. There can be no metaphysical predetermination of which particular considerations (factual or evaluative) will be relevant for thinking and acting morally *here and now*—what we require is an account of practical rationality that helps guide us in our evaluative grasping of facts and the concrete realization of values.

Part II of this collection offers the reader a wide selection of strategies for dissolving dichotomy, employing many different philosophical languages drawn from heterogeneous sources. Pragmatism as well as Aristotelianism are the key sources for Ruth Anna Putnam; Wittgenstein and Hilary Putnam, as well as gender and critical race theory, inform Scheman's chapter; Bagnoli blends Iris Murdoch with moral constructivism; Miller tackles analytic non-cognitivism, from Mackie to Terence Horgan and Mark Timmons; the two, allegedly opposed streams of sentimentalism and value realism find a congenial synthesis in Rossi and Tappolet's contribution; while inferentialism lurks behind Taylor's defense of relativism. Despite such a multitude of authors and positions, when read together, these chapters betray more than a hint of continuity and cross-reference, as all variously attempt to rehabilitate the domain of the evaluative by demonstrating its inbuilt objectivity. The chapters comprising part III of the collection explore some of the applications of the fact-value entanglement, engaging respectively with the conceptualization of science (Macarthur), with environmental ethics (Donatelli), and the philosophy of law (Lind) from broadly Wittgensteinian quarters. Each selected intervention calls upon the reader to urgently rethink the relations between science, nature, and the law in non-dichotomic terms, as cooperative fields of inquiry which are equally indispensable in our effective engagements with the world. The reading of the four more historical chapters on pragmatism (by Hilary Putnam and Robert Schwartz) and critical theory (by Maeve Cooke and John McGuire) comprising part I, which do offer an enthralling picture of the views on the dichotomy of some of the key figures in the two philosophical movements, shall offer a nice entry into these intricate matters and also prove quite useful for those who are relatively new to the debate and to these authors—and this is particularly true of the two chapters on critical theory, which deal with often overlooked philosophical synergies and legacies.

In closing, we have been arguing that the fate of the dichotomy hinges upon the viability of the meta-philosophical and representationalist conceits at its core, which, despite having endured a progressive erosion under pressure from anti-dualistic philosophies, remains surprisingly resilient within a number of prominent philosophical debates. Each of the thirteen original essays comprising this volume offers further evidence (historical and theoretical) of the untenability and undesirability of dichotomy, coupled with defenses of alternative conceptions that lend further credence to the ethical and metaphysical entanglement of facts and values. Having launched such a robust, multipronged attack on dichotomy, we await the equally forceful counterattack by its defenders—although we nurture a less-than-secret hope of having inflicted upon them a philosophical *coup de grâce*.[16]

Notes

1 To single out but one guide for thinking these topics through, the finest source is the work of John McDowell, and in particular his seminal *Mind and World* (1994). For a genealogical overview, see also Bilgrami (2010).

2 For a Wittgensteinan account of the fact/value entanglement, see Corradi Fiumara (2003).

3 If Hume's law "were simply one about the *form* of certain inferences, it would prohibit one from ever inferring 'you ought to do *x* in such-and-such circumstances' from 'for you to do *x* in such-and-such circumstances is good, and for you to refrain from doing *x* in those circumstances is bad." Many philosophers claimed in reply that this is indeed an instance of a derivation of an "ought" from an "ought," and hence not in conflict with Hume's Law. To them, Putnam replies observing that " 'for you to do *x* in a such-and such circumstances is good, and for you to refrain from doing *x* in those circumstances is bad' as a case of an 'ought' turns not on any feature of the form of the statement but rather on an understanding of its content" (Putnam 2002, 14).

4 Here Putnam follows Millgram's reading (Millgram 2005, ch. 6–7).

5 For an overview of emotivism, see Urmson (1968); for a rich and lively survey of the logical positivist's program in ethics, see Aa.Vv. (1948).

6 For the classical statement of this thesis, see Carnap (1932).

7 For an overall analysis of Russell's ethics, see Aiken (1963).

8 It is often said that Stevenson's emotivism is but a re-elaboration of Ayer's and logical positivist's (such as Schlick's or Carnap's). However, despite belonging to the same family of emotivists families, Ayer's and Stevenson's conceptions are so different that their purported affinities move to the background: Ayer's "radical and simplified" emotivism, in fact, does not even seem to belong to the same family of Stevenson's emotivism. Satris (1987, 20–1) holds in fact that Stevenson's emotivism does not depend nor draw from Ayer's or from the logical empiricists's, and rather points to their independence: "Stevenson never claims in his written work to be following Ayer or the logical positivists"). In his PhD dissertation, *The Emotive Meaning of Ethical Terms*, defended in 1935 at Harvard, Stevenson held that his subscription to emotivism owes to the long conversations with H. N. Kerns while an undergraduate at Yale in the late 1920s, that is, well before Ayer's *Language, Truth, and Logic*.

9 Those unconvinced by this specification shall likely object that it is indeed impossible to isolate such fact from any evaluative component, hence subscribing to a version of the fact-value entanglement already.

10 Putnam criticizes this assumption in Putnam (2004, 75–8).

11 For a careful reconstruction of this particular shift in the 1950s and 1960s, see Lecaldano (1970, ch. 5). For a concise yet lively picture of the wider metaethical changes in Anglophone twentieth-century moral philosophy, see Darwall, Gibbard, and Railton (1992). For an alternative reading of such reshaping in the light of the importance of the shift from reference to use, see Donatelli (2015, ch. 1–3).

12 For a lively presentation and discussion of the seminal contribution of some key figures of Critical Theory to the fact-value entanglement, see the chapters by Maeve Cooke and John McGuire in the present volume.

13 The chapters by Hilary Putnam and Robert Schwartz offer in-depth accounts of several aspects of Dewey's pragmatist critique of the dichotomy.

14 For a sympathetic account of this recent metaphysical trend in analytic philosophy written by perhaps its most representative practitioner, see

Williamson (2014). What is most interesting for our purposes is that this new (or renewed) fashion in metaphysics has interested only slightly moral philosophy, whose metaphysical grounding, when considered at all, makes reference not so much to the project of modal metaphysics launched by Lewis and developed by Williamson, but rather to the good old idea of conceptual analysis of which the latter are quite skeptical.

15 For this terminology, see De Caro and Macarthur 2004.

16 Despite this chapter has been in all respects a collaborative effort, Giancarlo Marchetti wrote §§2, 3, 4.2, 4.3, and 4.4, while Sarin Marchetti wrote §§1, 4.1, 4.5, 5, and 6. We would like to thank John McGuire for his precious suggestions and editorial assistance.

References

Aa. Vv. 1948. *Logical Positivism and Ethics*. Aristotelean Society Supplementary Volume XXII. London: Harrison and Sons.

Aiken, Lillian W. 1963. *Bertrand Russell's Philosophy of Morals*. New York: Humanities Press.

Ayer, Alfred. 1952. *Language Truth, and Logic*. New York: Dover Publications.

Ayer, Alfred. 1954. *Philosophical Essays*. London: Macmillan.

Bagnoli, Carla (ed.). 2013. *Constructivism in Ethics*. Cambridge: Cambridge University Press.

Bilgrami, Akeel. 2010. "The Wider Significance of Naturalism: A Genealogical Essay." In *Naturalism and Normativity*, edited by Mario De Caro and David Macarthur, 23–54. New York: Columbia University Press.

Brandom, Robert. 2009. "Classical American Pragmatism: The Pragmatist Enlightenment—and Its Problematic Semantics." In *Perspectives on Pragmatism: Classical, Recent, & Contemporary*, 35–55. Cambridge, MA: Harvard University Press.

Carnap, Rudolf. 1932. "Überwindung der Metaphysik durch logische Analyse der Sprache", *Erkenntnis*, 2: 219–41.

Corradi Fiumara, Gemma. 2001. *The Mind's Affective Life: A Psychoanalytic and Philosophical Inquiry*. London-Philadelphia: Brunner-Routledge.

Corradi Fiumara, Gemma. 2003. "Connessione tra fatti e valori in una prospettiva wittgensteiniana." In *Normatività, fatti, valori*, edited by R. Egidi, M. Dell'Utri, and M. De Caro, 23–30. Macerata: Quodlibet.

Darwall, Stephen, Gibbard, Allan, and Railton, Peter. 1992. "*Toward Fin de Siècle Ethics: Some Trends*", *The Philosophical Review*, 101: 115–89.

De Caro, Mario, and Macarthur, David (eds.). 2004. *Naturalism in Questions*. New York: Columbia University Press.

Dewey, John. 1903. "Logical Conditions of a Scientific Treatment of Morality." In *The Middle Works of John Dewey*, 1899–1924, vol. 3, edited by J. A. Boydston. Carbondale: Illinois University Press.

Dewey, John. 1929. "The Quest for Certainty: A Study of the Relation of Knowledge and Action." In *The Later Works of John Dewey* 1925–1953, vol. 4, edited by J. A. Boydston. Carbondale: Illinois University Press.

Dewey, John. 1938. "Logic: The Theory of Inquiry." In *The Later Works*, 1925–1953, vol. 12, edited by J. A. Boydston. Carbondale: Illinois University Press.

Dewey, John. 1939. "Theory of Valuation." In *The Later Works of John Dewey 1925–1953*, vol. 13, edited by J. A. Boydston. Carbondale: Illinois University Press.

Dewey, John. 1949. "The Field of Value." In *The Later Works of John Dewey. 1925–1953*, vol. 16, edited by J. A. Boydston. Carbondale: Illinois University Press.

Donatelli, Piergiorgio. 2015. *Manières d'être humain. Une autre philosophie morale*. Paris: Vrin.

Hudson, W. Donald (ed.). 1969. *The Is-Ought Question: A Collection of Papers on the Central Problems in Moral Philosophy*. London: Macmillan.

Hume, David, 1978. *A Treatise of Human Nature*, edited by L. A. Selby-Bigge and P. H. Nidditch. Oxford: Oxford University Press.

Kincaid, Harold, Dupré, John, and Wylie, Alison (eds.). 2007. *Value-Free Science? Ideals and Illusions*. Oxford-New York: Oxford University Press.

Lecaldano, Eugenio. 1970. *Le analisi del linguaggio morale. «Buono» e «dovere» nella filosofia inglese dal 1903 al 1965*. Roma: Edizioni dell'Ateneo.

McDowell, John. 1994. *Mind and World*. Cambridge, MA: Harvard University Press.

Millgram, Elijah. 2005. *Ethics Done Right: Practical Reasoning as a Foundation for a Moral Philosophy*. Cambridge, NY: Cambridge University Press.

Morris, Charles, 1963. "Pragmatism and Logical Empiricism." In *The Philosophy of Rudolf Carnap*, edited by P. A. Schilpp, 87–97. La Salle, IL: Open Court.

Potter, Michael K. 2006. *Bertrand Russell's Ethics*. London, NY: Continuum.

Price, Huw. 2011. *Naturalism without Mirrors*. Oxford: Oxford University Press.

Putnam, Hilary. 2002. *The Collapse of the Fact/Value Dichotomy and Other Essays*. Cambridge, MA: Harvard University Press.

Putnam, Hilary. 2004. *Ethics without Ontology*. Cambridge, MA: Harvard University Press.

Putnam, Hilary. 2012. *Philosophy in an Age of Science: Physics, Mathematics and Skepticism*. M. De Caro and D. Macarthur, eds. Cambridge, MA: Harvard University Press.

Rorty, Richard. 1999. "Pragmatism as Anti-Authoritarianism", *Revue Internationale de Philosophie*, 1: 7–20.

Russell, Bertrand. 1914. *On Scientific Method in Philosophy*. Oxford: Clarendon Press.

Russell, Bernard. 1997. *Religion and Science*. Oxford and New York: Oxford University Press.

Santayana, George. 1913. "Hypostatic Ethics." In *The Essential Santayana: Selected Writing*, 2009, edited by M.A. Coleman, 452–60. Bloomington: Indiana University Press.

Satris, Stephen. 1987. *Ethical Emotivism*. Dordrecht: Martinus Nijhoff Publishers.

Schiller, Ferdinand Canning Scott. 1903. *Humanism: Philosophical Essays*. London, New York: MacMillan.

Schwartz, Robert. 2007. "Fatti che riguardano fatti", *Iride*, 50(1): 67–82.

Stevenson, Charles L. 1937. "The Emotive Meaning of Ethical Terms", *Mind*, 46(181): 14–31 (now in Stevenson 1963).

Stevenson, Charles L. 1938a. "Ethical Judgements and Avoidability", *Mind*, 47(185): 45–57 (now in Stevenson 1963).

Stevenson, Charles L. 1938b. "Persuasive Definitions", *Mind*, 47(187): 331–50 (now in Stevenson 1963).

Stevenson, Charles L. 1944. *Ethics and Language*. New Haven: Yale University Press.

Stevenson, Charles L. 1963. *Facts and Values: Studies in Ethical Analysis*. New Haven: Yale University Press.

Urmson, James O. 1968. *The Emotive Theory of Ethics*. London: Hutchinson.

Williamson, Timothy. 2014. "How Did We Get Here from There? The Transformation of Analytic Philosophy", *Belgrade Philosophical Annual*, XXVII: 1–36.

Part I

A Counter-History of the Dichotomy

1 The Fact/Value Dichotomy and the Future of Philosophy[1]

Hilary Putnam

1 Introduction

The present essay is titled "The Fact/Value Dichotomy and the Future of Philosophy." I did not entitle it "The Fact/Value Dichotomy and the Future of Pragmatism" because I do not want to revive pragmatism as a movement. I don't believe in "movements" in philosophy, except as temporary expedients for bringing new ideas to the attention of the philosophical public. But there are ideas in classical pragmatism that I think deserve to be part of "the future of philosophy." The rejection of the claim that there is an absolute dichotomy between descriptions of facts and value judgments is perhaps the most important idea of Deweyan pragmatism that I hope philosophers will come to accept in the future.

This dichotomy has by now become something familiar to lay people and not just a matter for discussion by philosophers and social scientists. Every one of us has heard someone ask, "Is that supposed to be a fact or a value judgment?" The presupposition of the question is that if it's a "value judgment" it can't possibly be a "fact," and, all too often, a further presupposition is that value judgments are "subjective."

This dichotomy was already widely accepted by analytic philosophers when I was a graduate student more than a half-century ago. The logical positivists, including my teacher Hans Reichenbach, claimed to have shown that ethical propositions only appear to be bona fide assertions; in reality, they lack truth-value, they said, and, indeed, they are outside the sphere of rational argument altogether. Important social scientists accepted the dichotomy as well; Lionel Robbins, one of the most influential economists of the 1930s, gave this view one of its most aggressive formulations:

> If we disagree about ends it is a case of thy blood or mine—or live or let live according to the importance of the difference, or the relative strength of our opponents. But if we disagree about means, then scientific analysis can often help us resolve our differences. If we disagree about the morality of the taking of interest (and we understand what we are talking about), then there is no room for argument.
>
> (Robbins 1932, 132)

Yet this "self-evident" dichotomy is one that Dewey regards as baseless:

> Articles frequently appear that discuss the relation of fact and value. If the subject discussed under this caption were the relation of value-facts to other facts, there would not be the assumption of uniqueness just mentioned. But anyone reading articles devoted to discussion of this issue will note that it is an issue or problem just because it is held that propositions about values are somehow of a unique sort, being inherently marked off from propositions about facts. I can think of nothing more likely to be clarifying in the present confused state of the subject than an explicit statement of the grounds upon which it is assumed that propositions about values are not propositions about space-time facts, together with explicit discussion of the consequences of that position. If a question were raised about the relation of geological propositions to astronomical propositions, or of meteor- propositions to comet-propositions, it would not occur to anyone that the 'problem' was other than that of the connection between two sets of facts. It is my conviction that nothing would better clarify the present unsatisfactory state of discussion of value than definite and explicit statement of the reasons why the case is supposed to be otherwise in respect to value.
>
> (Dewey 1944, 106)

By "propositions about values" Dewey means here not just propositions about which human beings or communities have what values under what circumstances—sociological or psychological propositions—but propositions about what is valuable. Now that the fact/value dichotomy has become so widely accepted, it may appear naïve of Dewey to claim that there can be facts about values. How can there be facts about what is and what is not valuable? What is more, Dewey suggests that such facts are just "space-time facts," that is, they do not depend on anything supernatural or "non-natural." (And he wrote this in 1944, long after G. E. Moore argued in his famous *Principia Ethica*, published in 1903, that if there are such facts, they must involve a "non-natural property"!)

But Dewey was not naïve. He knew very well that his was a minority position. The first sentence of the essay from which I just quoted reads: "When I analyze the discouragement I have experienced lately in connection with discussion of value, I find that it proceeds from the feeling that little headway is being made in determining the questions or issues fundamentally involved rather than from the fact that the views I personally hold have not received general approval" (Dewey 1944, 101). Let us see what Dewey meant by "the questions or issues fundamentally involved."

The fundamental idea in all of Dewey's writing about the theory of value was to distinguish between what is valued in the sense of evoking a mere feeling of liking or enjoyment, and that which has been critically evaluated and studied. In Dewey's view, it is only when we have acquired knowledge of the relevant causes and effects and relations to our other values and our doings that what is valued become valuable or what is satisfying become satisfactory. Or, as he himself puts it:

> To say that something satisfies is to report an isolated finality. To say that it is satisfactory is to define it in its connections and interactions. The fact that it pleases or is immediately congenial poses a problem to judgment. How shall the satisfaction be rated? Is it a value or is it not? Is it something to be prized and cherished, to be enjoyed? Not stern moralists alone but everyday experience informs us that finding satisfaction in a thing may be a warning, a summons to be on the lookout for consequences. To declare something satisfactory is to assert that it meets specifiable conditions.
>
> (Dewey 1929, 208)

It is because he consistently rejected the idea that attitudes such as liking, preferring, approving, commending, and the like are sufficient to constitute value, and because he wished to emphasize the role of intelligent evaluation in value judgment that he offered what he called a "preliminary rough listing" of the questions he thought we should be discussing in the essay I mentioned earlier (Dewey 1944, 101–2). Here is Dewey's list:

I. What connection is there, if any, between an attitude that will be called prizing or holding dear and desiring, liking, interest, enjoying, etc.?

II. Irrespective of which of the above-named attitudes is taken to be primary, is it by itself a sufficient condition for the existence of values? Or, while it is a necessary condition, is a further condition, of the nature of valuation or appraisal, required?

III. Whatever the answer to the second question, is there anything in the nature of appraisal, evaluation, as judgment or/and proposition, that marks them off, with respect to their logical or their scientific status, from other propositions or judgments? Or are such distinctive properties as they possess wholly an affair of their subject-matter—as we might speak of astronomical and geological propositions without implying that there is any difference between them qua propositions?

IV. Is the scientific method of inquiry, in its broad sense applicable in determination of judgments and/or propositions in the way of valuations or appraisals? Or is there something inherent in the nature of values as subject-matter that precludes the application of such method?

—And, very importantly, in connection with the fourth of these questions, Dewey tells us that he inserted the phrase "in its broad sense" after "the scientific method of inquiry" to make it clear that the word "scientific" is "not assumed in advance to signify reduction to physical or biological terms, but, as is the case with scientific investigations of concrete matters generally, leaves the scope of the subject-matter to be determined in the course of inquiry" (Ivi, 102).

Dewey's responses, in his work as a whole, to the four questions he raised in this short article will be my topic. But first I want to look at certain well-known philosophical views that are diametrically opposed to Dewey's.

2 Dewey's Naturalist Opponents

One sort of opponent that Dewey recognizes but does not discuss in "Some Questions About Value" is the non-naturalist, the philosopher who holds that value propositions presuppose a non-natural or "transcendent" source of knowledge; in a footnote to that essay he writes that that view has been omitted "so what is said will not appeal to those who hold that view" (Ivi, 101). In a sense, the work of both Dewey and James (cf. James 1897) is intended to show us that (1) assuming that we need a transcendent justification for value propositions is only to make them seem occult, and they are anything but that—they are rooted in real natural facts about human nature and about real human environments; and that (2) those who take the "transcendent" route in ethics have historically been opponents of fallibilism in ethics, and fallibilism, for pragmatists, is inseparable from democratic ethics.

So this essay is addressed to naturalists in ethics. Dewey was well aware that among his fellow naturalists there were philosophers who regard ethical utterances as "pseudo-propositions."

In *The Unity of Science*, for example, after explaining that all nonscientific problems are "a confusion of... pseudoproblems" (Carnap 1934, 22), Rudolf Carnap wrote as follows:

> All statements belonging to Metaphysics, regulative Ethics, and (metaphysical) Epistemology . . . are in fact unverifiable and, therefore, unscientific. In the Viennese Circle, we are accustomed to describe such statements as nonsense . . . This terminology is to be understood as implying a logical, not to say a psychological distinction; its use is intended to assert only that the statements in question do not possess a certain logical characteristic common to all proper scientific statements [i.e. verifiability—HP]; we do not intend to assert the impossibility of associating any conceptions or images with these logically

> invalid statements. Conceptions can be associated with any arbitrarily compounded series of words; and metaphysical statements are richly evocative of associations and feelings both in authors and readers.
>
> (Carnap 1934, 22)

And Dewey discusses this view (although he does not mention Carnap by name) in "Some Questions About Value." How then, could Dewey write (in the very same article) the words I quoted earlier, namely, "It is my conviction that nothing would better clarify the present unsatisfactory state of discussion of value than definite and explicit statement of the reasons why the case is supposed to be otherwise in respect to value"? Didn't Carnap "state the reasons?"

Well, Carnap did give a supposed reason. He said that value propositions "are in fact unverifiable." This is certainly a reason that Dewey would have accepted, had Dewey agreed that it was the case. But the only reason Carnap had for saying that value propositions are unverifiable in 1934, the year that *The Unity of Science* was published, was the claim, accepted by the Logical Positivists and their "Vienna Circle" at that time, that the only verifiable propositions are observation reports such as "this chair is blue" (or, in an alternative version, "I have a blue sense-datum") and logical consequences of such propositions—a claim so extreme as to rule out all of the propositions of theoretical physics, as Carnap later came to realize. (cf. Putnam 2002, 22–3) From Dewey's perspective this is no reason at all.

Other defenders of the "emotivist" or "expressivist" account of ethical sentences simply claimed that it was the very "grammar" of ethical sentences to express emotions or "attitudes" (or, in Hans Reichenbach's version, to "command"), and not to state facts. But this is certainly not the surface grammar of such utterances: we do speak of some valuations as correct or true and others as incorrect or false, and we also discuss whether they are warranted or unwarranted. The emotivists' reply was that the surface grammar is "misleading," but again Dewey would want to hear a reason for this claim that a non-reductive naturalist should accept, and he claimed (and I think that he was right) that no such reason had been offered. (That one can be a naturalist in philosophy without being a reductionist is another idea of the classical pragmatists that certainly needs more advocates today.)

Perhaps just because the attempt to show that the "logical" or linguistic properties of value sentences support emotivism has collapsed (it was brilliantly criticized as early as 1960 by Paul Ziff in the final chapter of his *Semantic Analysis*), philosophers who deny that value propositions can genuinely state knowable facts more and more rely on purely metaphysical arguments. Thus, the late John Mackie argued in a book rather provocatively titled *Ethics: Inventing Right and Wrong* (1978) that although value sentences do indeed have the linguistic form of propositions, and are supposed to be capable of truth and falsity, warrant or lack of warrant, this is a metaphysical mistake. According to Mackie's famous "error theory" all ethical talk rests on an error. The supposed error is the belief that there could be such properties as good and evil, right and wrong. Although Mackie published this claim 13 years after Dewey's death, it is safe to assume that Dewey would demand a reason for this claim. And the only reason Mackie offered was that these ethical properties are too "queer" to exist (Mackie 1978, 41)!

Mackie, (who was a metaphysical materialist), tried to prove ethical properties exhibited this "queerness" by pointing to a property that he claimed ethical judgments possess: namely, that one cannot make an ethical judgment and mean it as a sincere ethical judgment, unless one thereby expresses an actual desire or preference. (The British philosopher Richard Hare earlier made the same assumption, and also came to the conclusion that ethical sentences do not state facts, although he did not speak of an "error.") Since descriptions of fact cannot, according to Mackie and Hare, be expressions of actual desires and preferences, it follows that ethical judgments are not descriptions of fact. *Quod erat demonstrandum.*

The origin of this supposed property of ethical judgments is clear: it comes from the older emotivism of the logical positivists! For the logical positivists and their emotivist followers expressing actual desires and preferences was the very "function" of ethical judgments. But, as Elizabeth Anderson (1993) has remarked, "Boredom, weakness, apathy, self-contempt, despair and other motivational states can make a person fail to desire what she judges to be good or desire what she judges to be bad. This prevents the identification of value judgments with expressions of actual desires and preferences, as Hare insists" (Anderson 1993, 102). And without this identification, Mackie has no reason at all for his "queerness" argument, and the famous "error theory" that he based upon it.

(Recall that the first of Dewey's four questions asks us what the connection is between valuing ["prizing"] and desiring. And it is clear from the essay as a whole that Dewey does not think the answer is simple.)

This is not to deny that believing an ethical proposition often has motivational force, nor is it to deny that our aim in uttering such propositions is often to motivate. It is simply to deny that they have the mysterious property of always and necessarily motivating whenever they are sincerely believed.

In his brilliant study, *What We Owe to Each Other* (1999), Thomas Scanlon discusses many of the questions I have just briefly discussed concerning the normative and motivational force of ethical assertions and the complexity of the connections between valuing and desiring. On Scanlon's "contractualist" theory, the moral motivation par excellence is the desire to avoid an action if the action is such that any principle allowing it would be one that other people could reasonably reject. And Scanlon (1999, 155) responds to the question, "Why accept this account of moral motivation?" by saying,[2]

> According to the version of contractualism that I am advancing here, our thinking about right and wrong is structured by . . . the aim of finding principles that others, insofar as they too have this aim, could not reasonably reject. This gives us a direct reason to be concerned with other people's point of view: not because we might, for all we know, actually be them, or because we might occupy their positions in some other possible world[3], but in order to find principles that they, as well as we, have reason to accept . . . there is on this view a strong continuity between the reasons that lead us to act in the way that the conclusions of arrive at those conclusions.
>
> (Scanlon 1999, 191)

I am not quoting Scanlon because I think that Dewey was a "contractualist" *avant la lettre*, but because Scanlon has well described how one sort of ethical claim can have motivating force in any community which shares one of the basic interests of morality. And the explanation he gives does not presuppose anything we ought to regard as metaphysically "queer."

"But the motive Scanlon describes won't motivate anyone who is indifferent to what others believe and desire!" someone will object. True, but the claim that ethical utterances have to motivate even those who are indifferent to the interests and beliefs of others to count as genuine propositions is just the claim that Mackie and Hare made, the claim for which I said no good reason has been offered. Ethical utterances do have various kinds of motivating force, and Scanlon has well illustrated how one kind of motivating force can be accounted for.

What Dewey would have disagreed with Scanlon about is (1) the idea that there is a unique motive for ethics, and (2) that idea that all ethical judgments depend on "principles."[4] But even more than Scanlon, perhaps, Dewey would have emphasized that ethics depends on a concern—perhaps "concern" is too weak a word—identification with the interests of others. Like Aristotle, Dewey believes that the reasons for being ethical are not apparent from a non-ethical or pre-ethical standpoint; one must be educated into the ethical life, and this means that one's interests must be transformed. In that process, Dewey tells us, one does not simply acquire an interest in helping other people alongside of and independent of one's various interests in art, in work, in recreation, etc.; rather all of those interests are likewise transformed. In Dewey's account, the natural impulse of sympathy itself is transformed by being fused with our other impulses, and our other impulses and interests are transformed by being fused with sympathy. As he writes,

> What is required is a blending, a fusing of the sympathetic tendencies with all the other impulses and habitual traits of the self. When interest in power is permeated with an affectionate impulse, it is protected from being a tendency to dominate and tyrannize; it becomes an interest in effectiveness of regard for common ends. When an interest in artistic or scientific objects is similarly fused, it loses the indifferent and coldly impersonal character which marks the specialist as such, and becomes an interest in the adequate aesthetic and intellectual development of the conditions of a common life. Sympathy does not merely associate one of these tendencies with another; still less does it make one a means to the other's ends. It so intimately permeates them as to transform both into a new and moral interest.
>
> (Dewey 1908, 272)

3 The Deweyan Alternative to Non-Cognitivism

One thing all three of the "classical pragmatists" agreed on was that *we have to learn in the course of inquiry* what can and cannot count as a verification, and that there is no way of delimiting in advance what we can and cannot learn from inquiry and from the application of the results of inquiry. (The latter—the application of the results of inquiry—is, for Dewey, a part of inquiry, and, indeed, the most essential part). For this reason, unlike Carnap, Dewey never appeals to a theory that purports to specify in advance what can and what cannot be "verifiable." And, unlike Mackie, Dewey is not a reductive materialist, and he does not pretend to have a principle which determines what is too "metaphysically queer" to be a real property of things. What he did instead was develop a naturalistic picture of the ways in which intelligence can be applied to ethical problems, and especially to social problems. For Dewey that there can be objective ethical truths follows from the fact that ethical problems are simply a subset of our practical problems, in the ancient sense of "practical"—problems of how to live—and it can be a fact that a certain course of action or a certain form of life solves, or better resolves, what Dewey called a problematic situation.[5]

Dewey explains and develops his view in detail in a number of works, including the *Ethics*, *The Quest for Certainty*, and *Human Nature and Conduct*, as well as many, many articles and responses to his critics. He did not try to present it in "Some Questions About Value," because his purpose there was only to say what he thought were the important questions to ask, but inevitably some features of his positive view can be perceived in his brief remarks about these questions.

In particular, when he discussed question I., Dewey drew an interesting distinction between two ways one can understand the notion of an attitude "such as prizing." To use a terminology that has come into use in recent decades, but which Dewey did not possess, we might call the two ways "internalist" and "externalist." An *internalist* philosopher in the philosophy of mind thinks of mental states as entirely "inside" the mind or brain. An *externalist* philosopher argues that the very identity of mental states depends on the actions and things to which they are directed. When applied to ethical theories, this distinction is relevant in more than one way. First of all, there are ethical theories according to which the only good (or bad) things are, in the last analysis states of mind. One example is G. E. Moore's theory in his 1903 *Principia Ethica*, a book whose tremendous influence on the members of the Bloomsbury Group is described in by J.M. Keynes (1949) in a wonderful essay titled "My Early Beliefs." Another example is Kant's ethical theory, in which only our acts of willing are ultimately good or bad, and a third is the Utilitarianism of Jeremy Bentham, according to which what is good is maximizing pleasure, conceived as a homogenous mental quantity capable of being added and even of being numerically measured (in some later versions of the theory (cf. Edgeworth 1881). These are all internalist theories. Also they are all "cognitivist" theories, in that genuine ethical knowledge is possible, according to all three of these very different theories. But the "non-cognitivists," such as the emotivists I mentioned earlier, who thought that ethical sentences do not really state anything that is rationally evaluable, but only express emotions or "attitudes,"[6] thought of these emotions and attitudes as mental states in an "internalist" sense, and one of the things Dewey wishes to do in "Some Questions" is to ask them *why* they think of them in this way. He asks this because he thinks that if we recognize that "prizing,"[7] for example, is bound up both with the things prized and with the actions we take in connection with those things, then we will be less likely to think that it must be one and the same "internal" mental state no matter what is prized and no matter what behavior the prizing leads to in a particular context. And this connects with his other questions, because if prizing is complex, external-world involving, and context-dependent, then it is plausible that it should be susceptible to rational evaluation, evaluation as to its appropriateness in a given context. On the other hand, if, as some "internalist" and non-cognitivist theories hold, prizing (or "commending," or "approving") is simply an irreducible and homogenous state in the private theater of the mind, and the function of ethical utterances is simply to express this state and to bring it about in one's hearers, then the social and cultural origins and functions of valuation become completely obscured. So conceived, in Dewey's view, ethical utterances become merely expressions of what he calls "bias," and he closes "Some Questions" by writing: "In this connection it seems worthy of note that those writers who hold to the completely a-rational character of valuing begin by accepting the "internal" mentalistic theory of value, and then proceed to endow this quasi-gaseous stuff with powers of resistance greater than are possessed by triple-plate steel" (Dewey 1944, 108).

Dewey's criticism of emotivism, that it has an untenable "internalist" view of our mental states, is not a very well-known one, even today. But it is continuous with the criticism of Bentham's Utilitarianism that he had made many years earlier, in the 1908 *Ethics*, a criticism stated much more vividly by Robert Nozick many years later. (Nozick was, as far as I know, unaware that Dewey had discussed the same issue.)

4 Dewey's and Nozick's Critique of Utilitarianism

The criticism I am referring to turns on the claim that happiness is not, contrary to Bentham, just a matter of subjective experiences. In the famous thought experiment Nozick employed, you have the option of spending your life lying in a tank connected to a super-duper machine that gives you whatever experiences you might like to have. Nozick asked, "Would you plug in? *What else can matter to us, other than how our lives feel from the inside*?"

And Nozick answered his own question:

> What does matter to us in addition to our experiences? First, we want to do certain things and not just have the experience of doing them. In the case of certain experiences, it is only because first we want to do the actions that we want the experiences of doing them or thinking we've done them . . . A second reason for not plugging in is that we want to be a certain way, to be a certain sort of person. Someone floating in a tank is an indeterminate blob. There is no answer to the question what a person is like who has been long in the tank. Is he courageous, kind, intelligent, witty, and loving? It's not merely that it's difficult to tell; there's no way he is. Plugging into the machine is a kind of suicide. It will seem to some, trapped by a picture, that nothing about what we are like can matter except as it gets reflected in our experiences. But should it be surprising that what we are is important to us? Why should we be concerned only with how our time is filled, but not with what we are?
>
> (Nozick 1974, 43)

And here is Dewey (1908, 256) making a similar point: "happiness consists in the *fulfillment in the appropriate objects* (or the anticipation of such fulfillment) of the powers of the self manifested in desires, purposes, efforts" (emphasis added). Dewey immediately proceeds to contrast this externalist conception of happiness "with the notion that it is a sum or collection of separate states of sensation or feeling." He describes precisely the "picture" that Nozick describes some of us as trapped by, according to which "nothing about what we are like can matter except as it gets reflected in our experiences" in the following words:

> [On the conception in question, according to which happiness is a sum or collection of separate states of sensation or feeling] it is the pleasure alone, when dissociated, which is the real end of conduct, an object being at best an external means of securing it. It is the pleasurable feeling which happens to be associated with food, with music, with a landscape, that makes it good; health, art, are not good in themselves. The other view [Dewey's own] holds that pleasure has no such existence by itself; that it is only a name for the pleasant object; that by pleasure is meant the agreement or congruity which exists between some capacity of the agent and some objective fact in which this capacity is realized.
>
> (Dewey 1908, 257).

The work that this conception of happiness as the satisfaction, actual or anticipated, of a capacity of an agent by an "objective fact" (as opposed to a mere subjective feeling) does for Dewey's critique of utilitarianism is enormous. In the utilitarian conception, as Dewey writes, "When happiness is conceived of as an aggregate of states of feeling, these are regarded as homogenous in quality, different from one another only in intensity and duration. Their qualitative differences are not intrinsic, but are due to the different objects with which they are associated (as pleasures of hearing, or vision). Hence they disappear when the pleasure is taken by itself as an end" (Dewey 1908, 257). This disappearance of the qualitative differences (as far as importance to the agent's "happiness" is concerned), is, of course, what makes it possible for the utilitarian to speak of "summing" pleasures, "maximizing" them, etc. But if Dewey is right, and if "agreeableness is precisely the agreeableness or congruence of some *objective condition* with some impulse, habit, or tendency of the agent" (Dewey 1908, 257)[8], then, "of course, pure pleasure is a myth. Any pleasure is qualitatively unique, being precisely the harmony of one set of conditions with its appropriate activity. The pleasure of eating is one thing; the pleasure of hearing music, another; the pleasure of an amiable act, another; the pleasure of drunkenness or of anger is still another" (Dewey 1908, 257).

And Dewey continues, "Hence the possibility of absolutely different moral values attaching to pleasures, according to the type or aspect of character which they express. But if the good is only a sum of pleasures, any pleasure, so far as it goes, is as good as any other—the pleasure of malignity as good as the pleasure of kindness, simply as pleasure" (Dewey 1908, 257–8).

Of course, Dewey was not a Nozickian libertarian. While Dewey and Nozick shared an important criticism of utilitarianism, Dewey defended social democracy all his life, as the best expression in our time of what he called "the democratic personal way of life" (Dewey 1939b, 228)[9] as well as the only way of truly applying cooperative intelligence and fallibilistic experimentation to the solution of our social problems,[10] while Nozick offered an a priori defense of free-markets and the minimal state. And when Dewey defends what he called "the democratic personal way of life" by claiming that "to cooperate by giving differences a chance to show themselves because of the belief that the expression of difference is not only a right of the other persons but is a means of enriching one's own life-experience" (Dewey 1939b, 289), he cannot convince the emotivists, who will regard Dewey's talk of "enriching" one's life experience as just more "exclamations, expressing only the dominant emotional state of the one from whom the ejaculation issues" (Dewey 1944, 107).[11] As Dewey says, the question which view is right, Dewey's or the emotivists, has great "practical import." "If valuing consists *wholly* and exclusively of something inherently recalcitrant to inquiry and adjudication, then it must be admitted that it cannot rise above the brute-animal level—save with respect to the *means* most likely to secure its victory over conflicting valuations and values. But, if in answer to the third question, it is decided that there is some element or aspect of valuation on 'objective' grounds in every case of prizing, desiring, etc., etc., then it is possible that this element or aspect may itself become so prized, desired, and enjoyed that it will gain in force at the expense of the brute and non-rational factor" (Dewey 1944, 455).

5 The Difference It Makes

How great a difference it makes if we treat ethical disagreements as Dewey thought we should, as disagreements that are not only rationally discussible, but as the disagreements which it is most important to discuss intelligently, is something I have discussed elsewhere.[12] But I am sure the question that will occur to many of my readers, is, "by what criteria can we tell who is right when we have an ethical disagreement?" And in response to that question, I will just make three remarks:

In the course of a genuine inquiry, one which is provoked by what Dewey would have called a "problematical situation" and not by what the pragmatists would have regarded as a mere "paper doubt," we always bring to bear an immense stock of both valuations and descriptions which are not in question in that problematic situation. The pragmatists—*all* the pragmatists—have taught us that any assumption can be questioned, but that that does not mean that all assumptions can be put in question in any one inquiry. We are never in the position, imagined by too many philosophers, of having a large stock of factual beliefs and no value judgments and having to decide whether our first value judgment is warranted. We are never in the position of having to infer our very first "ought" from a whole lot of "ises."

We neither have nor need one single "criterion" or "decision method" for judging warranted assertability in ethics any more than we do in history or in economics or in literature or, for that matter, in science. Rather, what we need is, first, an interest in what Dewey called *effectiveness of regard for common ends*, commitment to the ethical life itself, and, second, as Dewey also said, we need "the authority of intelligence, of criticism" (Dewey 1925, 305). (Dewey once defined philosophy itself as "criticism of criticisms.")

If Dewey did not believe that inquiry requires "criteria," in the sense of decision procedures, he did believe that there are some things that we have learned about inquiry from the conduct of inquiry. And what holds good for inquiry in general holds for value inquiry in particular.

Among the things that hold good for inquiry in general are the Peircean principles that inquiry which makes full use of human intelligence must not "block the paths of inquiry" by preventing the raising of questions and objections, or obstructing the formulation of hypotheses and the criticism of the hypotheses of others. Inquiry in any area should avoid relations of hierarchy and dependence, for these always lead back to the failed methods of inquiry that Peirce described in "The Fixation of Belief"—to the failed "method of tenacity," the failed "method of authority" and the failed "method of what is agreeable to reason." Rational inquiry in any area relies upon both the careful evaluation of experience and the invention of new descriptions of experience (and sometimes of new experiences), as Dewey explains in *The Quest for Certainty*. By appeal to these and similar standards we can often tell that views are irresponsibly defended in ethics as well as in science. The right approach to our ethical problems is neither to give up on the very possibility of intelligent discussion nor to seek a metaphysical foundation outside of (or "above") all problematic situations, but to investigate and discuss and try things out cooperatively, democratically, and above all fallibilistically. The terrible thing about the fact/value dichotomy is that by denying that there is such a thing as a responsible and rational ethical discussion, it "blocks the path of inquiry" from the very start.

Notes

1 An earlier version of this essay was delivered as one of my David Ross Boyd Lectures at the University of Oklahoma, October 28, 2005.

2 Scanlon 1999, 155.

3 Rawls's "veil of ignorance" and "initial position" are to what Scanlon is referring.

4 Why this is not the right way to think of ethics is something I discuss in the opening and closing chapters of Part I of *Ethics Without Ontology* (Putnam 2004).

5 This is how I understand Dewey's answer to the non-cognitivists in his 1939a *A Theory of Valuation*.
6 "Attitude" is a key notion in Stevenson's (1944) influential emotivist book.
7 "Prizing" resembles Hare's "commending."
8 Emphasis added.
9 This passage was called to my attention by Naoko Saito (2003, 289).
10 On this see "A Defense of Deweyan Democracy," chapter 9 (Putnam 1992).
11 This is Dewey's—accurate!—description of the emotivist position.
12 See my Spinoza Lectures, collected as Part II (Putnam 2004).

References

Anderson, Elizabeth. 1993. *Value in Ethics and Economics*. Cambridge, MA: Harvard University Press.

Carnap, Rudolf. 1934. *The Unity of Science*. London: Kegan Paul, Trench, Hubner & Co.

Dewey, John. 1908. *Ethics*. In *The Middle Works of John Dewey, 1899–1924*, vol. 5, edited by Jo A. Boydston. Carbondale: Southern Illinois University Press.

Dewey, John. 1929. *The Quest for Certainty*. In *The Middle Works of John Dewey 1899–1924*, vol. 4, edited by Jo A. Boydston. Carbondale: Southern Illinois University Press.

Dewey, John. 1939a. *Theory of Valuation*. In *The Later Works of John Dewey, 1925–1953*, vol. 13, edited by Jo A. Boydston. Carbondale: Illinois University Press.

Dewey, John. 1939b. "Creative Democracy—the Task before Us." In *The Later Works of John Dewey, 1925–1953*, vol. 14, edited by Jo A. Boydston. Carbondale: Illinois University Press.

Dewey, John. 1944. "Some Questions about Value." In *The Later Works of John Dewey 1925–1953*, vol. 15, edited by Jo A. Boydston. Carbondale: Southern Illinois University Press.

Edgeworth, Francis Y. 1881. *Mathematical Psychics: An Essay on the Application of Mathematics to the Moral Sciences*. London: Kegan Paul.

James, William. 1897. "The Moral Philosopher and the Moral Life." In *The Will to Believe and Other Essays in Popular Philosophy*, 184–215. New York: Longmans Green and Co.

Keynes, John M. 1949. *Two Memoirs: Dr. Melchior a Defeated Enemy, and My Early Beliefs*. New York: A.M. Kelley.

Mackie, John L. 1978. *Ethics: Inventing Right and Wrong*. Harmondsworth: Penguin Books.

Moore, George E. 1903. *Principia Ethica*. Cambridge: Cambridge University Press.

Nozick, Robert. 1974. *Anarchy State and Utopia*. New York: Harper and Row.

Putnam, Hilary. 1992. *Renewing Philosophy*. Cambridge, MA: Harvard University Press.

Putnam, Hilary. 2002. *The Collapse of the Fact/Value Dichotomy and Other Essays*. Cambridge, MA: Harvard University Press.

Putnam, Hilary. 2004. *Ethics without Ontology*. Cambridge, MA: Harvard University Press.

Robbins, Lionel. 1932. *On the Nature and Significance of Economic Science*. London: Macmillan.

Saito, Naoko. 2003. "Transcending the Tragic with Dewey and Emerson: Beyond the Morse-Boisvert Debate", *Transactions of the Charles S. Peirce Society*, XXXIX: 275–92.

Scanlon, Thomas. 1999. *What We Owe to Each Other*. Cambridge, MA: Harvard University Press.

Stevenson, Charles L. 1944. *Ethics and Language*. New Haven: Yale University Press.

2 Pragmatic Constructivism

Values, Norms, and Obligations*

Robert Schwartz

1 Introduction

The *Stanford Encyclopedia* entry "Constructivism in Metaethics" (Bagnoli 2011) begins thus, "The term 'constructivism' entered recent debates in moral theory with John Rawls' seminal article "Kantian Constructivism in Moral Theory." I wish to examine an earlier twentieth-century constructivist theory of ethics, Pragmatic Constructivism, as associated especially with the writings of John Dewey.[1] However, I will not enter directly into current debates swirling around constructivist theories. For one, 'constructivism' is a broad label for a number of approaches to ethics that vary in significant ways. Second, the issues now under debate differ depending on whether the constructivist procedures apply to political institutions or individuals, and on whether the constructivist theory is taken to be an ethical theory or a metaethical theory. It is possible to accept constructivism as an ethical theory, yet deny that it provides an acceptable metaethical account of morality.

Perhaps the main obstacle to relating pragmatic constructivist ideas to contemporary controversies in moral theory is that pragmatists, like Dewey, reject many of the core assumptions that underlie the entire problematic. Pragmatic constructivists argue that it is not possible to give a substantive account of the supposed dualisms of: reason and desire, thought and action, is and ought, means and ends, the individual and society, and more, that provide the pivots around which current battles are fought. Dewey, for example, is quite explicit that he is not a Realist or an anti-Realist. Nor is he a Kantian or Utilitarian.[2]

Dewey does not deny that the concepts these dualisms contrast play important roles in ethical discourse. He argues instead that the concepts and contrasts they are used to make are neither well defined nor appropriately deployed. Moreover, it is not possible to characterize one pole of these dualisms without incorporating critical features of the other. When properly understood, the contrasts to be drawn are context relative, matters of degree or emphasis, and lack the epistemic and metaphysical significance usually accorded them. Dewey agreed with his contemporary critics, and would agree with critics now, that his theory often fails to speak to, let alone answer, the questions that concern them. For Dewey challenges the presuppositions that give content both to the questions raised and the answers proffered.

Pragmatic constructivists, like other constructivist theorists, insist that the source and justification of ethical claims lies within. As Dewey says, they are not the "idle mirroring of preexisting facts" (Dewey 2008, 45). They evolve in response to the ever changing "*interaction* between elements of human nature and the environment, natural and social" (Dewey 2008, xxv). Ethical inquiry that does not reflect and speak to these empirical facts about the nature of human life as lived is pointless.

Human conduct, though, is not reducible to either behavioral dispositions or attitudes toward propositions. Pragmatic constructivists are functionalists. The content of mental states and the intentions that are constitutive of action can only be understood in terms of these worldly interactions. That moral judgments are evaluative and inherently relative to changes in both the human condition and our knowledge of it, is no indication that such judgments are subjective. It is misleading to think that ethical propositions are non-factive or non-cognitive in ways expressivists or emotivists are wont to do (see Dewey 1939, sections II and III).[3]

Where Pragmatic Constructivism tends to diverge from other constructivist approaches to moral theory is in its account of *inquiry*. Although pragmatists acknowledge there are differences in concern and subject matter, ethical inquiry *per se* is not significantly different from theoretical deliberation. Pragmatic constructivists argue that many of their fellow constructivists fail to appreciate the parallels, because they too presuppose a number of the untenable dualisms their opponents assume. As a result, these constructivists believe they must take a stance and defend positions that their pragmatic allies argue are non-substantive and result in unending debate. According to Dewey, misconceptions about the goals and practices of theoretical inquiry are as much to blame for this unhealthy state of affairs as misconceptions about ethical inquiry. "The question fundamentally at issue is nothing more or less than whether moral values, regulations, principles and objects form a separate and independent domain or whether they are part and parcel of a normal development of a life process" (Dewey 2008, 122).[4]

2 Pragmatic Constructivism

Given the centrality of this question, a brief, selective summary of the pragmatic account of theoretical inquiry is in order.[5] It goes something like this. As far as we know, the scientific experimental method is the best way to find out about the world. The method is not a special kind of thinking occurring only in scientific inquiry rather it is a subspecies of intelligent deliberation across the board. Substantive claims cannot be justified either by *a priori* reasoning or by intuition. Nor is it possible to begin inquiry with a clean slate or from a God's-eye perspective. All inquiry starts in the middle of things, from where we are at the time. The concepts constructed in the articulation of theories do not pick out natural kinds, for there are none. Nor are the meanings and reference of these concepts fixed. Conceptual content evolves in step with the theories in which they are embedded. As a result, there is no empirically significant meaning/fact or analytic/synthetic distinction.[6] Judgments yielded by inquiry that employs the experimental method are *ipso facto* objective.

Pragmatists, too, are anti-foundationalists. There is no unique base upon which to construct theories. Just as all arguments start with premises, and all definitions in formal systems assume some terms as primitive, all construction starts from something unconstructed. Construction, then, is always a matter of *re*construction. What is taken though as unconstructed, a premise, or primitive in one system, may be a constructed item, conclusion, or defined term in another.[7] While all construction must start somewhere, no point of origin has epistemic priority. None are metaphysically basic or privileged, and there may be alternative theories that can do the job.

As fallibilists, pragmatists allow that even our best theories are likely to be challenged, and in turn improved or dismissed. In *practice*, the goal of inquiry is to settle specific problems at hand, not to find hypotheses that will survive forever or be on the books at the end of inquiry. "The scientist finds no help in determining the probable truth of some proposed theory by comparing it with a standard of absolute truth and immutable being" (Dewey 1960, 264). Pragmatists agree that it is possible to think of the goal of inquiry in terms of such idealizations, but it can be misleading to do so.

Useful scientific idealizations, such as ideal gasses, pure metals, and frictionless surfaces present clear targets. We know what we are aiming at and what it would be like to approach them, even if we know the goals are unachievable. By contrast, the goal of final or absolute truth offers no such directive or guidance to on-going deliberations. It cannot serve as a target to aim at or get closer to. For we do not know, in advance of inquiry, what such ultimate theories would look like or where, if anywhere, they might be found. At any given time, the best we can do is the most we can do, and in inquiry, the most we can do is make progress. In practice, progress is what we settle for and evaluate. It is possible to have fact-based reasons for believing that new hypotheses are better than the ones on hand, without assuming there must be best or eternal solutions. This is not a pessimistic analysis of inquiry; it is a realistic explanation of how inquiry is conducted.

3 Value Inquiry as Empirical

For a person to rely on something does not mean it is reliable. That conclusion depends on the facts. Similarly, a student is often satisfied with the term paper turned in, but someone more knowledgeable on the topic may correctly decide the work is unsatisfactory. And what a person desires need not be in either her best interest or those of society. 'Reliable,' 'satisfactory,' and 'desirable' have both descriptive and evaluative functions. The former, Dewey labels *de facto*, the latter *de jure*. Both, though, are truth-apt. *De facto* statements are simply reports of peoples' present attitudes. *De jure* statements are evaluative judgments of such attitudes. Their justification requires inquiry that is based on specifiable empirical evidence, i.e., the established facts that support the positive *de jure* assessments. Evaluative statements are, therefore, factive. They make claims that are either true or false. Widespread agreement as to the truth of such statements, however, does not mean they are true, any more than everyone's belief that the earth is flat makes it so.

"Valuations are empirically observable patterns of behavior and may be studied as such. The propositions that result are *about* value but are not themselves value-propositions in any sense marking them off from other matter-of-fact propositions" (Dewey 1939, 50). Although judgments of the reliable, satisfactory, and desirable are evaluative, they do not assert that items so judged possess any abstract property "value," over and above the empirical properties cited to justify the evaluations. In fact, it is questionable whether it is useful to posit a general concept of value or a corresponding mental state of "valuing." We do value the reliable, the satisfactory, and the desirable, but it is for the properties they have that make them so. Moral inquiry will make more progress by focusing attention on the nature of the actual processes of evaluative deliberation and the evaluations rendered, than by pondering the nature and essence of value, *tout court*.

While *de jure* judgments can be objectively grounded, the facts relevant to determining their correctness will shift with the context and over time. The better car for short trips may not be the better for long hauls, and the better tomato for eating, may not be the better for cooking. In addition, gains in knowledge and the discovery of hitherto unknown options are likely to alter today's evaluations. Permanence of evaluative judgments and the standards used for assessment are not to be expected. Nor is such permanence a realistic aim for evaluative inquiry.

The pragmatic analysis of the relationship between *de facto* valuing and *de jure* judgments differs from various dispositional accounts of value.[8] For example, many claim that value attributions are like the attribution of color to an object. Something is red, if and only if, it is disposed to appear red to normal perceivers under standard conditions. 'Reliable,' 'desirable,' and 'satisfactory', though, are not helpfully analyzed in this way. A normal person may not be able to determine correctly whether a car is reliable, a paper satisfactory, or a desire desirable. Perhaps more significantly, the idea of normal perceivers and the standard conditions for such evaluations are not readily stated.[9]

Similarly, the claim that we can determine what is truly valuable by imagining what someone would or should value with "complete" information or from the standpoint of an "ideal" observer, is just hand waving. As with imaginations of eternal truth or the end of inquiry, such imaginings of the ideal provide no real guidance to current deliberation. They do not provide a usable target at which to aim. Experiences and assessments of values are always subject to change. There is no single preordained correct conclusion or even a natural termination to evaluative inquiry.

Our ability to see into the future, especially with respect to human affairs is quite limited. An imagined final agreement, as imagined absolute truth, cannot serve as a goal of value inquiry. In practice, we settle for progress in dealing with the problems at hand. We judge the worth of new evaluative claims in terms of progress made, not perfection found. This is a welcome, not a pessimistic conclusion. For it is possible to have objective grounds indicating a proposed solution *improves* on a situation found unsatisfactory. While it is not obvious what more could be offered to convince someone that a favored position is the one that will be accepted given complete information and faultless reason, or from a God's-eye point of view. It is not necessary to assume there even is a best, in order to determine that it is better to prefer some things or hypotheses to others. We may hope to uncover the best, but in practice, we must settle for the better. As Dewey continually reminded us, we must not make the best the enemy of the better.

Intelligent inquiry is required to turn *de facto* valuings into objective *de jure* evaluations. This is not to suggest that present attitudes occur in a vacuum, for no rhyme and reason. They evolve from prior evaluations. The process is one of *re*construction. To determine what is objectively reliable, satisfactory and desirable we start in the middle of things. We accept what is so taken as data and seek to refine these convictions when difficulties arise. Although evaluative experiences temporally precede their reflective evaluation, neither factor has explanatory priority. Without *de facto* valuings there would be no subject matter to evaluate, and without attempts to determine what is objectively reliable, satisfactory, and desirable such experiences lose their normative point.

For pragmatic constructivists, moral evaluations are of a piece. We must start with the facts of human moral valuing. If people did not appreciate some behaviors and attempt to distance themselves from others, there would be no subject matter for ethical deliberation to refine. As reflective beings, we evaluate what is morally valuable, by looking at the consequences of experiencing something as being of value. Judgments of what is valuable, though, are not fixed, for like all judgments they are subject to challenges that may lead to improvement or rejection. The stimuli to engage in moral inquiry are problems that arise with our current commitments. We find they are inconsistent, do not cover new cases, or conflict with insistent, often newly developed, moral sensitivities. Neither past convictions nor present urgencies have priority. To be applicable to life as lived, we must seek accommodations. Indeed, progress arises out of conflict and accommodation. When or where conflicts will arise, and how they will be resolved, is not predictable. Dewey frequently remarked that perhaps the only thing he was sure of was that things would change.

Pragmatists have another reason for refusing to draw a sharp distinction between moral evaluation and other evaluative judgments. Humans are social beings by birth, if not by choice. The goal of ethical inquiry is simply to figure out how *better* to live with one another. It is not to determine either what is best for me or to point me in the direction of some imagined self-perfection. Pragmatists believe that any action that has effects on the welfare of others has a moral dimension. Morality concerns, "[A]ll subjects of distinctively human import, all of the social disciplines as far as they are intimately connected with the life of man and as they bear upon the interests of humanity" [Dewey 1930, Forward,1]. Beyond such concerns, there is no distinctive domain of ethical inquiry.[10]

4 Norms

A major criticism of constructivist ethical theories is that although they may accurately describe how ethical consensus is best pursued and achieved, they fail to account for the prescriptive nature of moral judgments. To be normative or regulatory, consensus is not sufficient. Thus, critics maintain, "[T]o derive moral standards from social custom is to evacuate the latter of all authority. Morals, it is said, imply subordination of fact to ideal considerations, while the [pragmatic] view presented makes morals secondary to bare fact, which is equal to depriving them of dignity and jurisdiction" (Dewey 2008, 45). Agreement of minds cannot confer objective normative status.

To appreciate the pragmatists' response to this criticism, it is helpful to start by looking at some prescriptive claims that do not appear to have any obvious moral dimension. Consider the rules of grammar (Dewey 2008, 46). People spoke in ways that indicated they found certain sentences grammatically acceptable and others not, and they did so before anyone reflected on the difference. Over the years, implicit or explicit standards emerged capturing the intuitions in play. As standards or norms, the rules of grammar do not purport to describe actual human behavior, and to ask whether grammatical rules "correspond to reality" makes no sense. The rules that determine which sentences are grammatical are human constructions and in that sense "mind-dependent." Nevertheless, they are normative. Claims about what the rules are and which sentences meet them are empirical. It is an objective fact that the sentence, 'The cat is on the mat' is grammatical, and 'The is the cat mat on' is not.

Opponents of Pragmatic Constructivism are apt to be unimpressed by this example. No one doubts that when it comes to social practices like language, the norms and facts are socially constructed. Also, the rules of grammar are not the product of deliberation; they have evolved unconsciously without reflection. A different set of rules may just as well have evolved and then judgments of grammaticality would have been different.

Consider, though, the rules of deductive logic, inductive reasoning, and statistical analysis. These rules of inference are products of serious deliberation and few would claim they are not objectively justified. Yet, these rules have developed and changed over time. As in the case of language, people experienced some arguments as good and others bad long before anyone attempted to articulate criteria for distinguishing valid from invalid deductive, inductive, and statistical arguments. Without such initial intuitions, there would be no subject matter to codify. Early attempts to reflect on and codify practice captured some of these intuitions and not others. Inevitably, inconsistencies were discovered or new patterns of argumentation seemed needed that did not fall within the scope of the formulated rules. As the result of pressure from discovered inconsistencies, recalcitrant intuitions, and theoretical developments, new conceptions of deductive, inductive, and statistical validity evolved, along with the rules for evaluating such arguments.[11]

From Aristotle's syllogisms, to scholastic logics, to the propositional and predicate calculi with identity, things changed and changed largely for the better. The same can be said for advances in inductive and statistical inference. Progress has been step-wise, and the deliberations involved were not in any sense guided by substantive conceptions of ideal reasoning or ideal reasoners. Although these ideals may have been aspirational and inspirational, in practice, the aim of research was to solve specific problems. Clearly, such changes in the rules of deduction were not random, a matter of mere preference, immune to criticism. The same holds for the development of inductive and statistical inference. In response to discovered failures or expanded needs, the very concepts of validity have all been reconstructed, along with the rules that articulate them. There is no assurance they will not change again or that they will reach prefixed points of termination.

When serving as norms, the rules of inference are neither true nor false. They set the standards of best argumentation as currently conceived and objectively justified. They remain in play because experience shows they are useful tools for inquiry and better than others around. Claims that an argument violates or satisfies the rules are objectively true or false, and arguments are evaluated accordingly. Explanations of the norms or judgments of validity in terms of their "correspondence to reality" are of questionable content. The genesis and justification of norms is a process of "reflective equilibrium," pragmatically reconstructed as needed, and in this sense the norms are mind-dependent.[12]

If you wish to engage with others in discussion, deliberation, and action, you are obliged to follow the rules of language and argumentation that govern the practices. This is what fellow participants demand. Unsatisfactory performance in either may result in failing a course, not getting a job, or no longer being welcome as a participant. And even for actions having no consequences for others, deliberation and action will be better when these norms are recognized and given their due.

According to Dewey, ethical deliberation and the normativity of standards revealed by moral inquiry are not different, in principle, from the sorts of rules just canvassed. When it comes to matters of morals, the starting point is the fact that people welcome some behaviors and shun others. Based on these intuitions, we construct the standards for what we take to be appropriate ways to live together. These standards represent the demands society places on its members. Just as the norms of deductive, inductive, and statistical validity define acceptable inferences and place demands on argumentation, so do moral rules place demands on how we should interact with others. When serving as norms, moral principles are prescriptive and do not purport to state the facts. It is, nevertheless, a fact whether a person's actions accord with the rules, and it is objective and proper to judge them accordingly.

Moral principles will change in response to newly gained knowledge of nature, human nature, and actual flesh and blood changes in living conditions. Sometimes they will occur because conflicts arise among the principles accepted. Quite frequently, the stimuli for change are evolving intuitions and convictions that are incompatible with the dictates of the principles embedded in current practice. One should expect that principles for correct human behavior are not very stable. How could they be? "Conflict and uncertainty are ultimate traits" (Dewey 2008, xxvi). We cannot hope to prepare now for all unknowable future contingencies and ever evolving moral sensitivities. It is a hindrance to successful inquiry to think that we can. "[P]erhaps the most pervasive of all fallacies in philosophy . . . [is] . . . the supposition that whatever is true under certain conditions can forthwith be asserted universally or without limits and conditions" (Dewey 2008, 115).

Norms are a pervasive and significant factor in the conduct of human affairs. For pragmatic constructivists, they have a moral dimension to the extent that the decisions made and actions taken have consequences for the lives of our self and others. Norms having such import cannot be justified by appeals to a special faculty of moral perception, *a priori* principles, or idealized conceptions of the human mind, will, and rationality not founded on empirical inquiry. Morals are "ineradicably empirical" (Dewey 2008, 198). Regulative standards evolve from reflection on the consequences of practice. There is no other way to establish what the norms should be, and no realistically alternative route to justify them.

5 Motivation: Reason and Desire

Critics see another problem for pragmatic constructivist theories. They say constructivists cannot account for the motivational force of moral principles. Dewey, though, wishes to treat the issue of moral motivation as he treats our motivation to meet evaluative standards in general. Empirical inquiry concerning the nature of human deliberation and action in real life situations is a major focus of his analysis.

For those who see the point of determining what is objectively better and better to do, the answer to questions of motivation seem relatively unproblematic. It would appear contrary to commonsense to ignore the conclusions of reflective inquiry. Indeed, what would motivate someone to not rely on the reliable, write papers that are satisfactory, pursue the desirable, speak grammatically, and make valid inferences? It would be foolish/irrational to rely on something you know to be unreliable, be satisfied with what you take to be unsatisfactory, or not alter an argument being convinced that it is invalid.

Of course, this response does not address what many feel to be a more serious philosophical problem of motivation, the one brought prominently to the forefront by Hume. Is it desire that motivates action, or is it reason? Pragmatic constructivists like Dewey, see no need to take sides in this debate, for they contend that the question presupposes a dualism they reject. Desire and reason are inextricably linked in determining the content, cause, and rationalization of voluntary actions. Consider, for example, the joint role desire and reason play in actions influenced by the emotions. Emotions are states of mind that motivate, and they have a cognitive component that is essential to their meaning and individuation. The claim that Sam is jealous of George's new job presupposes that Sam has certain relevant beliefs. For instance, Sam must believe that George got the job, that it is better than his, et cetera. Without such evaluative and non-evaluative beliefs, Sam would have no reason to be jealous and could not be in a state of jealousy. In addition, Sam would not be jealous, if he did not care about the perceived difference between his and George's situation, and did not have desires relevant to that care. It is not surprising, then, that emotions such as jealousy are subject to criticism. We can find fault with Sam's jealousy, if either his judgments concerning George's job are unjustified or the strength of his caring is disproportionate to the affects the difference in positions will actually have on his own well-being.

Analytically, of course, it is possible to separate emotions into independent components, desires, and reasons. The point is that an action does not count as a case of jealousy without both. Neither factor alone can be properly understood as the cause or rationale for jealous action. The meaning of a jealous act is an amalgam of desire and reason, and the reasonableness of the motivational pair can be critically evaluated.

Experience shows that by and large the consequences of actions motivated primarily by jealousy are not good.

For Dewey, deliberation and reason independent of any desire has no direction, and desire independent of all reason is more an appetite or instinct than a cause or rationale for a voluntary act. "There is no desire and no interest which, in its distinction from raw impulse and strictly organic appetite, is not what it is because of transformation effected in the latter by their interaction with the cultural environment" (Dewey 1939, 63–4). Aimless reason and unreasoned desire have no place in serious deliberation. In this context, little of importance is at stake in splitting them apart, since there are no grounds for assigning causal or justificatory primacy to one factor rather than the other.

Action is co-determined by desires along with evaluative and non-evaluative judgments of reason. What makes an act the type of act it is depends on its motivation, and the motivation of an act is important in evaluating its worthiness. "The first quality which is the object of judgment primarily resides in intention" (Dewey and Tufts 1906, 261). But it is the motivational package as a whole, not one of its components, that determines the content of the act's intention. '*Neither the inner* [desire and deliberation] *apart from the outer* [action] *nor the outer apart from the inner has any voluntary or moral quality at all*" (Dewey and Tuft 1906, 288). According to Dewey's pragmatic functionalist model of mind, the meaning of mental states and actions are set holistically in terms of the relations they bear to each other. Thus, Dewey asks, "Is there any intrinsic moral connection between the *mental* and the *overt* in activity?" He answers, while" capable of intellectual discrimination [they] are incapable of real separation . . . neither being real or significant without the other" (Dewey and Tuft 1906, 228).

Good conduct is behavior that is motivated by "reasonable desires." "[A] truly moral (or right) act is one which is intelligent in an emphatic and peculiar sense: it is a *reasonable* act" (Dewey and Tuft 1906, 307). Our interactions with others go better when they are motivated by what reason judges desirable, (satisfactory and valuable), when, that is, they are motivated by rationally endorsed desires (satisfactions and values). This requires reflective thought and evaluation of the consequences of our actions. The reasonableness of one's motivation is something that can be judged objectively.

This analysis of the relation between desire and reason has major implications for Dewey's account of practical reasoning, the details of which cannot be gone into here.[13] In sum, he argues that once we give up on a reason/desire dualism, along with the dualisms of fact and value, thought and action, means and ends, practice and theory, it can be seen that both Kantian and Utilitarian pictures of practical reasoning are not true to the facts about how humans actually deliberate and determine what to do. Hence, debates about which of these two pictures of motivation is *the* correct one cannot be resolved. A realistic theory of morals cannot be constructed on an *un*realistic view of deliberation. The common error of both views "lies in trying to split a voluntary act which is single and entire into two unrelated parts . . . *We may begin the analysis of a voluntary act at whichever end we please, but we are always carried to the other end in order to complete the analysis*" (Dewey and Tuft 1906, 239).

That a person's motivation reflects his reasoned desires is no indication that the motivation is for self-gain. It is a fallacy to transform "the (truistic) fact of acting *as* a self into the fiction of acting always *for* self" (Dewey 2008, 85). The honest man returns the money because he is an honest self, and he gives to charity because he has concerns for the welfare of others. That is just the sort of person he is. To act in accord with motivations *of* the self, does not entail the person is acting *for* self-benefit. Moral behavior does not require the absence of motives that contribute to self-satisfaction. Nor is deliberation dependent on such purification empirically possible. It is not a useful idealization, and when thought of as a target at which to aim, distorts the nature of deliberation actually involved in deciding on a course of action.

6 Obligation

It is often assumed that the grounding and force of obligation must ride above empirical facts. Facts describe the way the world is. Claims of obligation purport to tell how things should be and assert that we have a responsibility to bring them about. There is an unbridgeable gap between the two; it is not possible to derive is from ought. The only way to secure the binding force of obligations is to step outside experience, typically by appealing to Reason, personal identity, godly commands, natural law, and other abstract factors that transcend the factual.

The pragmatic constructivist rejects the need for such appeals in order to account for the emergence and force of obligations. Obligations are norms of conduct that are constructed and shaped by human practice. We start with the fact that people do feel they have responsibilities to others in light of their relationships to them. Parents, for example, have a tendency, perhaps genetically hard-wired, to take care of their children. They see this as something they not only wish to do, but have a responsibility to do. It is not imposed on them. Such is the starting point for the emergence of a concept of obligation. Inquiry begins in the middle of things. If humans had no such impulses or dispositions, there would be no phenomena upon which to construct a notion of the obligatory.

It is also a fact that humans are inescapably social beings. There is no living a human life in isolation. As a result, we find ourselves interacting with others who are not members of our immediate family. They, too, have needs, fears, and desires that cannot be accommodated without our cooperation. Furthermore, they are in the position to articulate their demands on our conduct and will not let their demands be ignored. "Every relationship in life, is, as it were *a tacit or expressed contract with others*, committing one, by the simple fact that he occupies that relationship, to a corresponding mode of action" (Dewey and Tufts 1906, 345). Social standards of obligation emerge from reflection on these interactions.

To appreciate that we have responsibilities to others does not mean we know what all our obligations are and to whom we have such responsibilities. This needs to be worked out. There are no ready-made answers for us to discover. They must be constructed, and the results of our construction will not be permanently relevant. The nature, details, and range of our objective obligations will evolve in light of reflection on the consequences of adopting or failing to adopt them.

To feel obliged, of course, is not to be obliged, to have an obligation. We can be mistaken here as we are when relying on the unreliable, being satisfied with the unsatisfactory, or desiring the undesirable. Empirical inquiry is required to flesh out the nature of obligations, their extent, and the appropriate responses to obligations that are not met. It should not be assumed that solutions that best fit one society will fit them all. There are no *a priori* or transcendental principles available to provide the answers needed to solve these problems. In addition, any substantive principle will have to be recalibrated in light of newly acquired knowledge of nature, human nature, and societal relations, as well as changes in living conditions and the range of people with whom we interact.

There is no point then looking for forces outside or over and above the empirical to resolve these issues, for there is nothing to see. Nor is there any point to looking for solutions that are eternal or will be on the books at the end of moral inquiry. The concept of obligation itself evolves with inquiry. The goal of such inquiry is not to determine final principles, but to construct and settle on obligations we can agree are objectively better to adopt than those to which we now adhere. As our obligations are articulated and new rules established to encompass them, the concept of obligation itself is reconstructed.

The fact that an act is obligatory and thus binding does not mean that everyone or every rational being will feel obliged to take on the responsibility. Nonetheless, it remains objectively binding. The responsibility to meet our obligations is not optional since other members of the community demand that we do. When failures to meet obligations have serious effects on the community we live in, laws may be passed to enforce conformity. "Morals *are* social. The question of ought . . . is a question of better or worse *in* social affairs." Our obligations and duties grow "*out of a course of action he is committed to by a regular, socially established connection between himself and others*" (Dewey 2008, 213).

The moral person is not moved to meet her responsibilities by demands of reason, personhood, or any external inducements or commands. There is, as Kant insisted, a difference between acting according to duty and acting for duty. "It is a sound principle to do our duty for duties sake. Duty for duty's sake means, in truth *an act for the acts own sake* . . . [however] . . . "*no moral act is a means to anything beyond itself,—not even to morality.* But upon Kant's theory, duty for duty's sake means a special act not for its own sake, but for the sake of an abstract principle" (Dewey and Tufts 1906, 352). Dewey argues that in the abstract, Kant's principle cannot gain real traction in determining how to act.

The moral person acts as a self, not for herself, or for anything else. She is a self who is disposed to act according to moral obligations and is impelled to do so. Given the self she is, it is reasonable desire that motivates her. A disposition to act in accord with one's objective obligations is an important "habit" that is constitutive of a person's character and identity. It is one, among other dispositions, that make someone a good person. If as pragmatists insist, there is no distinctive set of decisions or actions that are peculiarly moral, the centrality of dispositions that are typically labeled "moral dispositions," are less secure. Any decision or act that has an affect on peoples' well-being has a moral dimension. Norms of inference typically are not assumed to have moral dimensions, but in certain circumstances they will be so understood. The doctor who ignores them in reaching conclusions concerning diagnoses and appropriate treatments is morally remiss.

7 Constructivism and Decision Procedures

As mentioned at the start, current discussions of constructivism take their lead from Rawls. Put briefly, Rawls' central concern is to construct principles for a society to use in setting up its public institutions. The underlying intuition that guides his proposal is that the institutions must be fair. Rawls attempts to explicate his notion of fairness with the help of a thought experiment. He asks what sort of institutions people would to agree to, if they had to choose from behind a "veil of ignorance" (Rawls 1971). The principles implicit in their choices can then serve as a decision procedure, which when followed warrant the institutions so constructed. Rawls likens his project to Kant's use of the Categorical Imperative as a decision procedure to determine and warrant morally correct conduct.[14]

Pragmatists, of course, agree with Rawls that moral facts and the justification of norms are not found ready-made. They are empirically motivated and constructed. Like Rawls, pragmatists reject Kant's idea that such regulative principles are *a priori*, eternally fixed, universally applicable, and independent of experience. Dewey says, nevertheless, that he has no qualms accepting the Categorical Imperative as a constraint on ethical deliberation and conduct. He has reservations, though, of its usefulness in constructing answers to substantive moral questions.[15] As he conceives it, the Categorical Imperative, like imperatives to treat everyone as equal, or to do onto others as you would like them to do unto you, have an important role to play. The goal of ethical inquiry, after all, is to figure out what we can do to improve the way we live together. It is not to determine what is best for me or for my loved ones, but what is fair for everybody. Imperatives of fairness, then, are a constant of ethical inquiry, and the principles constructed to attain it may be understood as regulative. Hence, moral principles that turn out not to succeed in treating everyone impartially present a problem that calls for further inquiry to resolve.

Dewey, like many others, thinks Kant's abstract notion of a Categorical Imperative is of limited use. It provides no substantive guidance in determining where changes are needed, how to make such improvements, or what constitutes the relevant community. Indeed, the Categorical Imperative cannot actually serve to define or determine what constitutes fairness. The concept fairness, like the concepts of validity and the rules used to evaluate arguments, evolves over time. It must be constructed and reconstructed. Standards of fairness will not necessarily be the same for all times and places. Furthermore, the Categorical Imperative procedure cannot have the authority Kant supposes it has. Moral inquiry must start in the middle of things, and its focus is on the continuing need to adjust principles to accepted practices. When in the course of everyday affairs no problems arise, practice should yield to principle for this is what objective deliberation has indicated is the better way to proceed. Nevertheless, principles are not fixed; they evolve. Changes may be due to conflicts among the principles. Or, they may change because they are in conflict with new emerging moral knowledge and sensitivities that resist being ignored. In many such cases, principles will give way to practice. There are no principles that can resist forever the forces of human living that demand and propel change.

Rawls' project does take this into account. In writings subsequent to *A Theory of Justice*, Rawls makes clear that the procedure he recommends is meant to capture and articulate the commitments and intuitions of those living in a liberal democratic society. The concept of fairness does not stand outside of experience. It is instilled by, and dependent on social conditions and influences. Rawls' conception of fairness, as characterized in terms of decisions made in the original position, is not universal, eternal, or discoverable by *a priori* principles of reason.

For Dewey, the idea of a pure Cartesian soul and a nonsocial self or mind makes no sense. All thought is conditional. Kant's claim that "every rational being must wish to be wholly free from inclinations" (Dewey and Tufts 1906, 364) is not realistic. The idea of a completely autonomous decision maker is not a useful idealization. It provides no substantive guidance, and in the end distorts rather than clarifies our understanding of human nature and deliberation. The same can be said for a quite different picture of autonomous decision-making, namely that of the "rugged individual." It, too, is a myth. Without the help of others, we cannot live a human life. Indeed, without the help of others we cannot survive for long after birth. This is an empirical fact, which may not hold for other possible rational creatures that may be able to make it on their own. Thus, Dewey argues given the human condition, there cannot be a sharp individual and society dualism. Infants do not enter the world good or evil. The physical and social environment they inhabit shapes their character. All human thought, desire, and decision are infused with social ideas and influences.

Since society plays a significant role in determining the character and identity of its members, society cannot escape a portion of the blame that accompanies immoral behavior. The development of morally good persons depends crucially on their being raised in morally good societies. Education, broadly construed, is the key. Moral progress requires that members of a community work together to find ways to meet the evolving demands of their situation and living conditions. Inquiry, whether in ethics or science, is not the work of an individual. It is a shared enterprise with a socially constructed past and future. It is not possible to construct substantive concepts of objectivity or standards of inquiry individualistically.

8 The Good and the Right

This account of the emergence and justification of norms has an impact on how pragmatic constructivists see the relationship between the good and the right. Kantian and Rawlsian constructivists give priority to the right, while Humean constructivists give priority to the good. Dewey thinks it is misleading to privilege one over the other. They each depend on the other. There is no right way to act that is not an empirically good way to behave. And there is no way to determine what is objectively a good or better way to act that is independent of considerations of what it is right to do.[16]

As society evolves, our judgments of the good will frequently come in conflict with the principles of what is allowable or right to do. An accommodation must be reached. Sometimes considerations of good consequences win out, leading to accepted principles being overturned. Most of the time things go the other way. As long as social interactions run smoothly, principles have the upper hand. For they make explicit the demands society has objectively determined is the right way for its members to behave. There are no abstract algorithms or procedures, however, for settling such conflicts outside of their specific contexts. "Experimental empiricism in the field of ideas of good and bad is demanded to meet the conditions of the present situation" (Dewey 1960, 258).

For the pragmatic constructivist, the good and the right cannot be separated. Experiences of the good are normatively empty, independent of considerations of what it is right to do. While judgments of what it is right to do are blind, independent of considerations of experiences of the good. The assumption that either the good or the right must have epistemic, metaphysical, causal, or rational priority is due to the acceptance of untenable dualisms. "For Right is only an abstract name for the Multitude of concrete demands in action others impress on us and of which we are obliged, if we would live, to take some account. Its authority is the exigency of their demands, the efficacy of their insistencies" (Dewey 2008, 218). "A right, individual in residence, is social in origin and intent" (Dewey and Tufts 1906, 114).

9 Conclusion

As usually understood, Pragmatic Constructivism does not fit comfortably into the camp of either so-called "objectivists" or "subjectivists." Likewise, there is difficulty finding a place for its proponents on either side of the realist/anti-realist, cognitivist/non-cognitivist, and mind-dependent/mind-independent divides.[17] The inability to pigeonhole their position is not due to their desire to be quietists or to their inability to decide on which side of these divides to land. The reason, instead, is they believe these distinctions themselves are of questionable clarity and use. When taken to mark off substantive epistemic and metaphysical differences, they fail to capture the realities of lived experience that moral inquiry is intended to examine and critique. Pragmatic constructivists believe the situation is the same when these dualisms are employed to characterize theoretical inquiry. It leads to nonrealistic accounts of the goals, subject matter, and practices of science. Theoretical inquiry, too, is constructive, not the "idle mirroring of preexisting facts." Failure to appreciate the similarities between moral and theoretical inquiry distorts our understanding of both.

Notes

* I wish to thank Nataliya Platanik, Stan Husi, and Joshua Spencer for comments and objections.

1 Prominent statements of Dewey's position can be found in *Ethics* (coauthored with Tufts), *Human Nature and Conduct,Theory of Valuation*, and *Quest for Certainty* (especially in Chapter X, "The Construction of the Good"). I can offer here only a glimpse and cursory summary of his ideas. Although sympathetic to Dewey's pragmatic constructivist position, my goal in this paper is to explain, not defend it.

2 I wish to stress, that I use the names 'Kant', 'Hume' and 'Utilitarian' throughout to label broad positions often associated with these writers. I am not claiming that such attributions are always in keeping with their actual positions, which themselves were a matter of much controversy in Dewey's time and remain so now.

3 Nowadays, many expressivists and emotivists accept that 'P' is true if and only if P, whether 'P' is a statement of morals or a statement of science. Still,

they believe there is something less than fully factual or objective in the truth-conditions of ethical statements.

4 Dewey reads Hume as a constructivist, not a skeptic, but one who also falls into this trap (Dewey 1930). See Kitcher (2011) for an important, Dewey-inspired, ethical theory. From my perspective, in places Kitcher, too, seems wedded to assumptions about inquiry a pragmatic constructivist should reject.

5 For a fuller account of the pragmatic constructivist's account of theoretical inquiry, see Schwartz (2012).

6 See Dewey's comments on the evolution of the concepts "water" and "stars," and the anti-essentialist implications he draws (1960, 105 and 207). See Schwartz (2012) for an elaboration and defense of this analysis.

7 See Schwartz (2000).

8 For an influential discussion of this approach, see the symposium papers of M. Smith, D. Lewis, and M. Johnston (1989) in "Dispositional Theories of Value."

9 Critics of Dewey, such as White (2005), assume his analysis of desirable is along the lines of the one given for red, and find it faulty. I do not think their interpretation of Dewey's account of the relationship between the desired and desirable is correct.

10 See Schwartz (2007).

11 With some alterations, similar stories could be told about the current standards of medical and legal practice or the evolution of double entry bookkeeping as a norm in business accounting.

12 See Goodman's (1955, 65–8) account of the development of both deductive and inductive logics, and his argument for their normative status. See Rawls's note in *A Theory of Justice* (p. 18) where he endorses Goodman's reflective equilibrium approach to the justification and warrant of norms. Scanlon (2003, 2014) provides an in depth examination of the proper interpretation and use of the method, especially with regard to Rawls's theory.

13 See especially Dewey (2008, Part III).

14 See Rawls's Dewey Lectures (1980), where he traces the Kantian roots of his theory of justice, but points out that he, like Dewey, deviates from Kant in not accepting the dualisms that underpin Kant's analyses and claims. See Rorty's (2007) challenge to the very idea of Kantian projects for this reason.

15 I think this is an accurate account of Dewey's reading of Kant. More recently, Kant scholars have argued against this austere understanding of Kant's approach and its supposed detachment from the realities of life as lived. If these readings are correct, the distance between Kant and Dewey's positions considerably narrows. Dewey was aware of the possibility. He says in various places that Kant and the Utilitarians must confront and account for the same empirical facts. He argues that to the extent they may succeed, the differences between their positions will converge and the epistemic and metaphysical dualisms that separate them break down. Each theory "tends inevitably to make concessions and to borrow factors from its competitor, and thus insensibly to bridge the gap between them" (Dewey and Tufts 1906, 238). See Onora O'Neill's (2003) on the differences between Kant and Dewey.

16 It is essential to keep in mind that for Dewey the good is not to be equated with maximizing happiness and the right is not transcendent.

17 I think "irrealist," in the sense Goodman uses it, is a useful way to describe their position, but not if understood in the way many quasi-realists' do when attempting to accommodate realist intuitions that moral discourse is somewhat less than fully objective or truly factive.

References

Bagnoli, Carla. 2011. "Constructivism in Metaethics." In *Stanford Encyclopedia of Philosophy*. http://plato.stanford.edu/entries/constructivism-metaethics/

Dewey, John. 1930. "Forward." In *Human Nature and Conduct*. New York: Random House.

Dewey, John. 1939. *Theory of Valuation*. Chicago: University of Chicago Press.

Dewey, John. 1960. *Quest for Certainty*. New York: Putnam/Capricorn Books.

Dewey, John. 2008. *Human Nature and Conduct*. New York: Barnes & Noble.

Dewey, John, and Tufts, John. 1906. *Ethics*. New York: Henry Holt.

Goodman, Nelson. 1955. *Fact, Fiction and Forecast*. Cambridge: Harvard University Press.

Kitcher, Philip. 2011. *The Ethical Project*. Cambridge: Harvard University Press.

O'Neill, Onora. 2003. "Constructivism in Rawls and Kant." In *Cambridge Companion to Rawls*, edited by Samuel Freeman, 347–67. Cambridge: Cambridge University Press.

Rawls, John. 1971. *A Theory of Justice*. Cambridge: Harvard University Press.

Rawls, John. 1980. "Constructivism in Moral Theory." *Journal of Philosophy*, LXXVII(9): 515–72.

Rorty, Richard. 2007. "Kant vs. Dewey." In *Philosophy as Cultural Politics*. Cambridge: Cambridge University Press.

Scanlon, Thomas M. 2003. "Rawls on Justification." In *Cambridge Companion to Rawls*, edited by Samuel Freeman, 139–67. Cambridge: Cambridge University Press.

Scanlon, Thomas M. 2014. *Being Realistic about Reasons*. Oxford: Oxford University Press.

Schwartz, Robert. 2000. "Starting from Scratch: Making Worlds", *Erkenntnis*, 52: 151–9.

Schwartz, Robert. 2007. "Fatti che riguardano fatti." *Iride*, 50.1: 67–82

Schwartz, Robert. 2012. *Rethinking Pragmatism*. Oxford: Wiley/Blackwell.

Smith, Michael, Lewis, David, and Johnson, Mark. 1989. Symposium "Dispositional Theories of Value", *Proceedings of the Aristotelian Society: Supplement*, 63: 89–174.

White, Morton. 2005. "Desire and Desirability." In *From a Philosophical Point of View*, 167–77. Princeton: Princeton University Press.

3 Contingency and Objectivity in Critical Social Theory

Horkheimer and Habermas

Maeve Cooke

The contingency of facts and objectivity of values is built into the very premises of critical social theory in the Frankfurt School tradition (henceforth "Critical Theory"). The contingency of facts on human activity enables Critical Theory to reveal possibilities for social transformation within existing social structures and social relations. The objectivity of values enables it to attribute rationality or truth to its claims that the advocated social transformation is genuinely progressive—that it is social change for the better. But the situation is more complicated. Critical Theory takes a materialist view of reality. This means that it sees the objectivity of values not as a matter of metaphysical speculation but as dependent on the objectivity of facts. In its case, the relevant kinds of facts relate to the natural condition of humans, the development of human history, and the social conditions that the theory criticizes. One of the theory's tasks, accordingly, is to explain why the contingency of facts on human activity nonetheless permits it to treat facts as objective and, on this basis, also the values that guide it.

Importantly, in this tradition, the objectivity of values has a meaning over and beyond their rationality or truth: in line with the Hegelian idea of objectivity as actualization (*Verwirklichung*), it also means the realization in human history of the basic values guiding the theory, above all the value of individual human freedom. In contrast to Hegel, however, Critical Theory does not subscribe to a philosophy of history that allows it to see the realization of freedom in human history as inevitable;[1] instead it insists that its realization is contingent on the historical struggles of human agents who seek to overcome socially produced obstacles to achieving the realization of freedom in the world, aided by the diagnostic and enlightening powers of the theory itself.

Thus, Critical Theory understands its own activity as a complex interplay of contingency and objectivity. Furthermore, it does not regard its self-understanding as a matter of abstract interest, for the theory itself is an agent of social transformation—part of the historical struggle for a social order in which freedom would be actualized.

However, this self-understanding presents Critical Theory with a number of challenges. On the one hand, given the practical intentions of the theory, certain basic facts must be taken as effectively indisputable; otherwise the struggle to realize freedom in the world would be unable to move beyond epistemic debates as to the rationality of the theory's intentions and the truth of its analysis of the prevailing social conditions. On the other hand, even the basic facts are not indisputable; indeed they came to be disputed among theorists writing in the Frankfurt School tradition. This raises the question of how to assess the claims the theory makes for the truth of its understanding of the human species and its view of the development of human history. An added complexity is that for the early Critical Theorists the social conditions of the time have rendered people's actual perceptions of their needs and interests epistemically unreliable; in consequence, the theory claims to possess an illuminating power, an ability to disclose to its addressees the rationality or truth of its guiding values and diagnostic claims. This raises the practical question of how to do so successfully and, in addition, opens the theory to the objection that it treats its addressees not as subjects but as objects, denying their capacity for freedom. It also complicates assessment of the theory's claims to rationality or truth, for cognitions gained through disclosure may be difficult to discuss argumentatively. These are challenges facing any attempt to account for the general structure of the world, materialist or idealist, secular or religious. But they arise with particular urgency for a social theory that aims to change the world. For, if the theory fails to speak to those who suffer under the prevailing social conditions, its practical role as part of the struggle for a better social order is undermined. But if it seeks to uphold this role by revealing the truth about existing social conditions to those who suffer most from these conditions, it seems to claim an epistemically privileged position that opens it to the accusation of what I call "epistemological authoritarianism."[2] Furthermore, there is a problem of validation: the theory's practical role as part of the struggle for freedom depends not just on its motivating power; it also depends on the validity of its guiding conception of freedom and of the facts it asserts about the human species and human history. A flawed idea of freedom and/or erroneous factual basis would make the theory an agent not of liberation but of enslavement. Given the materialist view of reality on which the theory rests, validity cannot decided by way of theoretical speculation; rather it must be justified by arguments with a factual basis. Not surprisingly, therefore, questions of contingency and objectivity have troubled Critical Theory from the outset, and continue to do so today.

In the first section, starting with the early writings of Max Horkheimer, I show both how the contingency of facts and objectivity of values is an essential part of the self-understanding of Critical Theory and generates the challenges outlined above. I argue that these challenges become serious problems when the relationship between social reality and critical theorizing becomes a perniciously circular one. By this I mean a circular relationship in which the facts of social life determine the theory's evaluative perspective and the theory's evaluative perspective determines what counts as facts. The relationship is perniciously circular since, on the one side, certain general facts are taken as indisputable by the theory and, on the other, the theory cannot be challenged by any facts. This leads to the problems of authoritarianism and validation mentioned above. What is missing is an opening in the loop that allows for a third term in the relationship: an "outside" to the circular relationship, the objectivity of which is not itself determined by way of the relationship, yet is epistemically significant for the two terms within it. In the second section I argue that the writings of Jürgen Habermas, a second-generation Critical Theorist, offer resources that are helpful in this regard; ultimately, however, I find his position unsatisfactory and in need of further development. In the final section I sketch the direction in which Critical Theory should move if it is to avoid a perniciously circular relationship between contingency and objectivity, with the problems of authoritarianism and validation to which this gives rise.

1

Originally based at the independently funded Institute of Social Research, established in Frankfurt am Main in 1923, theorists such as Max Horkheimer, Theodor Adorno and Herbert Marcuse sought to rescue Marxism from the rigid and authoritarian form it had acquired in the Communist parties of the Soviet Union.[3] Their aim was to develop a critical Marxist theory that would contribute to genuinely progressive revolutionary change. Taking their inspiration from the dialectical version of materialism propounded by Marx in the 1840s, the first generation of Critical Theorists understood the facticity of the social world as the product of human activity over time, driven not just by interests in physical survival but also in freedom and happiness. While human knowledge of the world, including all factual knowledge, is held to be dependent on human activity, some facts are regarded as primordial, having their origins in the human species as such. The human drive for happiness, understood as the satisfaction of fundamental human needs, is one such natural fact. In Marx's account of human "species-being," happiness is entwined with freedom. Like Marx, the early Critical Theorists conceived of freedom as the conscious development of human powers to produce and create: this is the distinctive mark of human life-activity. In Marx's words: "The animal's product is directly part of its physical body, while man steps out freely to confront his product . . . man creates . . . according to the laws of beauty."[4] In this conception, humans can relate consciously to their life-activity only through their relationships with other humans. This is because each human being is able to see its own life-activity as consciously creative only in relationships with other beings, who recognize it and their own productive activity as such. In

Marx's words once more: "man's relationship to himself only becomes objective and actual for him through his relationship to the other man."[5] This means, in turn, that the freedom of each human being is the condition for the free development of all.[6] The early Critical Theorists regard these elements of Marx's anthropology, too, as natural facts.

In Horkheimer's programmatic writings from the 1930s, it is clear that he attributes a kind of objectivity to natural facts that non-natural facts must first acquire. He writes, for example, that "it has always meant to materialists that man's striving for happiness is to be recognized as a natural fact requiring no justification." [7] By contrast, non-natural facts acquire objectivity by proving themselves empirically to be true. Horkheimer appears to allow for two ways in which they may do so.[8] First, the truth of facts may be established by the methods of inquiry appropriate in a particular scientific field (for example, physics or psychology); here, truth is understood in the narrow sense of logical consistency. Second, the truth of facts may be established in historical struggles for a better society; such struggles prove the truth of Critical Theory's account of the structure of social reality, understood as "the development of men acting in history."[9] In other words, historical struggles for a better social order, motivated by real experiences of material deprivation and injustice, confirm the objectivity of the general facts of Marxist political economy. It is important to notice here that Horkheimer acknowledges the valuable contributions to Critical Theory of the sciences of his time and, more generally, of what he calls "traditional" theories.[10] Such theories operate as closed systems of propositions without attention to the connections between their theoretical endeavors and the social life-process.[11] He criticizes them primarily for having lost sight of the concern for a better society that animated their development in the European Enlightenment, and the subsequent self-limitation of scientific activity to the description, classification, and generalization of phenomena, with no care to distinguish the unimportant from the essential. In consequence, the sciences of his time have become incapable of grasping the structure of social reality. This has led to an abdication of scientific responsibility to "cast light on the social life-process in its totality."[12] Horkheimer nonetheless acknowledges the contribution to Critical Theory of traditional theories. Despite their avoidance of any theoretical consideration of society as a whole, traditional scientific theories, including the applied social sciences, establish a store of information on nature and the human world. Without this there could be no Critical Theory. Horkheimer writes: "every step [of Critical Theory] rests on knowledge of man and nature, which is stored up in the sciences and in historical experience."[13] These constitute the basic tenets of Marxist theory. They include the general fact that no human society so far has managed satisfactorily to establish conditions that would universally satisfy the natural human interests in happiness and freedom as well as physical survival. A further general fact is that modern industrialized societies are the first form of social order with the capacity to satisfy these needs, but have not fulfilled their promise due to the capitalist economic system and accompanying social institutions, in particular bourgeois private property.

Thus, Critical Theory starts from certain general facts, comprising natural facts and proven facts of political economy. This gives it its basic structure. Horkheimer acknowledges the need for on-going revision of various elements of Marxist theory due to changes in historical circumstances and under the pressure of historical struggles.[14] He suggests, however, that the basic structure is invariant.[15] But since all facts are contingent on human activity in the world, we must assume that even these general facts are open to revision in response to human experiences, illuminated by theoretical reflection, as manifested in historical struggles. Horkheimer seems to acknowledge this implicitly when he speaks of Critical Theory as the "unfolding of a single existential judgment,"[16] as constructing "a developing picture of society as a whole, an existential judgment with a historical dimension."[17] On the one hand, therefore, he attributes stability to the theory; this is because amidst all change in society, and further specification and correction of the theory, its decisive substantive elements remain invariant (these are the basic economic structure, the class relationship in its simplest form and the contradiction between the two). On the other hand, in describing the theory as constructing a "picture" and as the unfolding of an "existential judgment," he appears to recognize that the objectivity of the basic structure is itself a social construction and, hence, open to theoretical reconstruction if new historical conditions or the pressure of social struggles make this necessary.

What would count as rational grounds for theoretical reconstruction of Critical Theory's picture of society as a whole? In other words, what counts as a rational challenge to the theory's basic structure? Horkheimer does not address this question directly. Nonetheless, it looms on the horizon for two reasons. First, we have seen that Horkheimer is alert to the deficiencies of "traditional" theories of his time; despite their valuable contributions to the store of human knowledge, which is required for any genuinely progressive social theory, "traditional" theories share a failure to "grasp the totality"; in other words, they fail to understand the contradiction between the capitalist economy's promise of a rational society and the wretchedness of actual social conditions. Thus, Horkheimer claims for Critical Theory a superior understanding of society that cannot *in principle* be challenged by traditional theories. The second reason has to do with the development of capitalism itself. He holds that one of the elements of Marxist theory that needs to be revised is its view that capitalism will inevitably be overcome through revolution by the proletariat.[18] In the 1930s Horkheimer still shared the early Marx's belief that the proletariat formed a social class with the potential to bring about universal human emancipation. In the *Communist Manifesto* Marx and Engels declare: "The proletariat "heralds the dissolution of the order that has existed hitherto . . . it is the factual dissolution of this order".[19] However, benefiting from the hindsight afforded by the Russian Revolution and its aftermath, he did not share Marx and Engel's view that the dissolution of the bourgeois capitalist order (or the monopoly capitalist order that he saw as its successor) would inevitably result from the contradictions of capitalism: unlike them he did not hold that bourgeois capitalism produces its own gravediggers.[20] Nor did he share their view that the situation of the proletariat was a guarantee of correct knowledge.[21] Indeed, he took the view that Critical Theory should be "a critical, promotive factor in the development of the masses."[22] For, while its concerns were those of most men, they were not recognized by them as such.[23] One important consequence is an enhanced role for Critical Theory as an agent of social transformation. Whereas Marx and Engels describe communist theory as merely *expressing* in general terms actual relations springing from an *existing* class struggle,[24] Horkheimer sees Critical Theory as itself a force within the real historical situation for emancipatory social change.[25] In other words, the transformative agency Marx and Engels attribute to the proletariat is now exercised by the theory; the theorist's own thinking becomes part of the historical struggle.[26]

Taken together, these two factors increase the epistemic burdens assumed by Critical Theory vis-à-vis early versions of dialectical materialism. Indeed, they seem to lead Critical Theory into the troubling situation that it now becomes the ultimate arbiter of its own claims to truth and rationality. This situation is worrying since it casts doubt on the validity of its claims to promote universal emancipation and also gives rise to the danger of epistemological authoritarianism. Both the problem of validation and the risk of authoritarianism are evident in the following passage from his 1936 essay "Traditional and Critical Theory."

Horkheimer writes that the transmission of critical theory:

> will not take place via solidly established practice and fixed ways of acting but via concern for social transformation. Such a concern will necessarily be aroused ever anew by prevailing injustice, but it must be shaped and guided by the theory itself and in turn react upon the theory . . . A version of [the theory] that has the propaganda apparatus and a majority on its side is not therefore the better one. In the general upheaval the truth may reside with numerically small groups of men. History teaches us that such groups, hardly noticed even by those opposed to the status quo . . . may at the decisive moment become the leaders because of their deeper insight."[27]

In the model presented here, Critical Theory is driven by a concern for social transformation (its value component), which is determined by the prevailing injustice (its factual component); at the same time its value component determines its factual component. This is because the factual component has to be constructed by a theory that is capable of grasping the life-process in its totality. Only Critical Theory meets this requirement, for given the social conditions of the time, neither "traditional" theories nor the proletariat, the class that suffers most from the prevailing injustice, is able to grasp the totality. This explains the "deeper insight" he grants to "small groups of men"—presumably "men who adopt the critical attitude."[28]

In ascribing deeper insight to small groups of men, Horkheimer builds an epistemologically authoritarian moment into the very activity of critical social theorizing. The problem is not simply that the theory grants epistemic privilege to one social group vis-à-vis another; Marx granted to the proletariat knowledge of the unsustainability of bourgeois capitalism—a knowledge that was unavailable to the bourgeoisie due to their interest in perpetuation of that social order. But he did not grant Communist *theory* superior knowledge of the social order's unsustainability; Communist theory, as we saw, merely *gives voice* to the contradictions inherent in bourgeois capitalism. In contrast, Horkheimer seems to confer on the theory itself an unquestionable epistemic authority. This epistemologically authoritarian view of theory is out of tune with the democratic spirit of socially engaged critique in general and, indeed, is at variance with Critical Theory's own commitment to individual freedom.[29]

On the other hand, Horkheimer is right to expect Critical Theory to have a "critical, promotive" power, if we understand this power to be a disclosive, enlightening one.[30] Let us assume that he is correct to hold that the social conditions of his time prevent the realization of individual human freedom, also that they are capable of being changed through human activity. Let us then assume that he is correct to hold that the proletariat no longer occupies a position of epistemic privilege enabling it to see this; let us assume further, that no other group—or theory—can step into the proletariat's shoes. Let us grant, in addition, that he is right to view "traditional" theories as lacking the practical intention that drives Critical Theory. In this case, disclosive thinking, which would open our eyes to a better way of imagining freedom and would make us aware of real possibilities for achieving the social conditions corresponding to it, would be of vital importance. Indeed, it would be a crucial part of any theory engaged in the task of social transformation.

A theory that opens people's eyes to values and facts they had been unable to see is not in itself epistemologically authoritarian. It becomes authoritarian only when it cannot be challenged by arguments that unsettle its basic structure: when it sets itself up as the final authority for truth. In his account of Critical Theory Horkheimer seems to succumb to this danger. How might Critical Theory maintain its disclosive moment, thereby upholding its practical intention, while avoiding epistemological authoritarianism?

The cited passage also draws attention to the related problem of validation. This problem arises from the circular relationship Horkheimer draws between the fact of an unjust social reality, which motivates the concern for social transformation, and the theory that criticizes the prevailing injustice. On the one side, the theory guides and shapes the concern for social transformation; from the essay as a whole we know that it does so by virtue its grasp of the life-process in its totality; this allows it to construct a valid picture—and to make an existential judgment—on the basis of the available facts. On the other side, the concern for social transformation, which is based on the facts of actual social life, reacts back upon the theory. But since only the theory can make sense of the facts, it is hard to see how the facts could give rise to arguments that would challenge its existential judgment. The relationship between facts and values now looks perniciously circular, for the theory's values are validated by facts, the validity of which it itself determines.

In order to avoid this perniciously circular relationship, and the related problem of epistemological authoritarianism, Critical Theory must allow for something outside the relationship between theory and actual social reality. It must posit a third term, something the epistemic status of which is at once relatively independent of the relationship yet relevant to the assessment of claims to truth or rationality made on behalf of the facts or values to which the theory appeals. For this we will have to move beyond the early Horkheimer.

2

The later Horkheimer,[31] Horkheimer and Adorno in their collaborative book *Dialectic of Enlightenment*,[32] and Marcuse[33] struggle unsuccessfully to avoid these problems. However, in the 1970s Jürgen Habermas introduces a paradigm shift into Critical Theory, which significantly changes the nature of the debate about the relationship between facts and values. The shift is from a subject-object model of cognition and action to an intersubjective, communicative model. In my view, the path opened for Critical Theory by Habermas's move is the best one to take if we wish to avoid epistemological authoritarianism and the related problem of the theory's validation. As things stand however, Habermas's intersubjective model of validation, although a move in the right direction, is unsatisfactory and in need of further development.

Habermas published a number of important studies in the 1960s, emphasizing Critical Theory's practical intention and its role as itself an agent of social transformation.[34] In particular, his books *Knowledge and Human Interests* and *Theory and Practice*, both originally published in 1968, sought to clarify the relationship between Critical Theory and practice. "Critique," understood as a practice of enlightening self-reflection modeled on Freudian psychoanalysis, is distinguished from "struggle," understood as the non-communicative, instrumental application of the theory in response to the incapacity of its opponents to recognize the validity of its message due to ideological deception and self-deception.[35] Although both critique and struggle are motivated by a concern for emancipatory social transformation, and although both may be necessary for this purpose, Habermas is alert to a danger inherent in struggle that is akin to what I described as epistemological authoritarianism. He cautions that the "objectivating" attitude adopted by proponents of the struggle for social transformation vis-à-vis their opponents should be a temporary one; moreover, it must be coupled with the expectation that a dialogue would someday be possible, in which opponents and proponents would together decide the truth of the theory.[36] Habermas writes: "the vindicating superiority of those who do the enlightening over those who are to be enlightened is theoretically unavoidable, but at the same time it is fictive and requires self-correction: in a process of enlightenment there can only be participants."[37]

At that point, however, Habermas had not yet spelled out what it would mean for the theory to be vindicated discursively by all concerned in their capacity as "participants." Moreover, his notion of critique as enlightenment, modeled on the psychoanalyst-patient relationship, seems vulnerable to the objection of pernicious circularity that I raised earlier against Horkheimer: the posited relationship between the theory's values, which determine what counts as "enlightenment," and the facts that are supposed to correct these values looks like a *closed* feedback loop. In other words, it is unclear how a successful process of self-reflection initiated by the theory could lead to a process of discursive validation in which the theory's truth would be an open question—one to be decided by all participants. Put differently again, it is not clear what rational grounds there could be for challenging the theory's basic ethical picture.

In the early 1970s Habermas saw the need for a fundamental switch in paradigm. In his view, most of the theoretical problems of earlier Critical Theory were the result of a subject-object model of cognition and action. In this model, knowledge and action are conceived of instrumentally as the imposition of will by a self-contained human subject (or group of subjects) on some object distinct from it (or them). This allows for no non-repressive relationship between the knowing or acting subject and the object of its thought and action. Habermas replaces this model with an intersubjective one. He proposes a model of cognition in terms of intersubjective processes of reaching understanding, and a corresponding model of communicative action as a freedom-enabling mode of social integration.[38]

What implications does this paradigm shift have for Critical Theory's self-understanding as an agent of social transformation and for assessment of the validity of its emancipatory claims? These are not questions to which Habermas gives a clear answer. In response to critics around 1980, he acknowledged the force of the objection that the new communication-theoretical basis he develops for Critical Theory threatens to open the gap between theory and practice that he had tried to close in his earlier work.[39] Since then, however, he has explicitly addressed neither the question of theory as an agent of social transformation[40] nor the related problem of validation of his theory's claims. As stated earlier, I regard his communication-theoretical approach as opening a fruitful new path for Critical Theory. However, if it is to provide satisfactory answers to the questions raised by our discussion of Horkheimer, it requires some further development.

Drawing on formal-pragmatic studies of everyday linguistic behaviour, Habermas claims that communicative action is the primary mode of language use.[41] Even in its everyday forms communicative action is a rudimentary practice of argumentation for it is interaction between human subjects that involves making claims to validity and responding to them.[42] Such interaction is guided by a host of idealizing suppositions concerning both the conduct of the interaction and its outcome. As regards its conduct Habermas maintains, for example, that participants in the discussion are guided by the assumption that all relevant arguments will be considered and that the only permissible force is that of the better argument. As regards its outcome he maintains, for example, that participants are guided by the assumption that a valid outcome—a true proposition or just norm of action—will command universal agreement as to its validity.

Habermas's thesis is that the idealizing suppositions guiding communicative action serve to establish an internal connection between everyday behavior and validity in a context-transcending sense. Thus, while humans can arrive at truth and justice only by way of practices of intersubjective action that are subject to the influences of history and context, truth and justice transcend every actual historical context.

In the 1970s Habermas defined truth and moral rightness (justice) as argumentatively reached agreement under ideal justificatory conditions.[43] Drawing on his formal-pragmatic analysis of everyday language use, he argued that an intersubjectively achieved agreement is a criterion of valid knowledge on condition that the agreement is reached through a process of argumentation that satisfies certain demanding procedural requirements (such as inclusion of all relevant arguments and granting equal voice to all participants in the discussion). In this conception procedure and outcome are intimately entwined. The validity of propositions or norms cannot be established purely on the basis of satisfaction of procedural conditions such as inclusivity and fairness; universal agreement is also necessary. Conversely, even universal agreement is not sufficient for establishing the validity of propositions or norms; the agreement has to be reached in processes of deliberation that satisfy the procedural conditions.

In the meantime, Habermas has modified his view of truth, while retaining his original constructivist conception of justice.[44] In the modified conception, truth is not *defined* as the outcome of an idealized intersubjective argumentation; rather it is described as a Janus-faced concept, with two aspects: a discursive one, which connects it to the argumentative exchange of reasons, and a pragmatic one, which ties it to everyday contexts of action. On the one side, truth is the concern of participants in argumentation who are guided by the idea that a proposition, if true, would withstand all attempts to refute it under ideal justificatory conditions.[45] On the other side, truth is a pragmatic fiction, an assumption made by participants in everyday communicative practices, who are guided by the need for behavioral certainty.[46] Truth's "Janus face" refers to the dynamic interplay between everyday behavioral certainties and processes of intersubjective deliberation about these certainties once they fail to prove reliable as a basis for everyday action. This happens for contingent empirical reasons. When everyday behavioral certainties no longer work, they become the subject of specialized argumentations. The results of these deliberations are then fed back as truths into everyday contexts of action; so long as they prove reliable, they can maintain their status as truths.

While this conception continues to uphold a conceptual connection between truth and argumentation, the newly introduced gap between justification and truth significantly changes the nature of the link. In the earlier version, truth was conceived of epistemically, as knowledge that is *constructed* by way of the exchange of reasons in idealized forms of intersubjective deliberation. Now, it is deemed to transcend the results of any such deliberation. In other words, his modified theory of propositional truth posits a gap between a discursively achieved rational consensus, which is properly described as justification, and truth, which transcends justification, no matter how ideal the justificatory conditions.

By contrast, Habermas' concept of moral validity remains an epistemic-constructivist one. In this case, he allows for no gap in principle between moral justification and moral rightness. Thus, while both propositional truth and moral validity are deemed to have a context-transcending force, and while both are still tied conceptually to rational agreement reached through argumentative justification, propositional truth is assumed to transcend justification, while moral rightness is *defined by* an argumentatively reached agreement as to its universal validity.

It should be noted that Habermas distinguishes sharply between moral rightness (justice) and ethical validity. By "moral rightness" or "justice" he means universally binding norms and principles. By "ethical validity" he means evaluative questions of "the good life" or (though he does not usually use this term) "the "good society." As we have seen, he defines moral rightness as the outcome of an idealized procedure of deliberation as to the universal validity of the matter under discussion. This general formulation holds equally for propositional truth. But moral rightness is distinguished from propositional truth by an additional moment of universality: moral norms and principles claim not only to be valid for everyone, they also claim to be equally in everyone's interests;[47] this moral principle of universalizability, too, is justified by way of formal-pragmatic analysis.[48] In the case of moral rightness, therefore, all participants in deliberation agree that a given norm or principle is equally in everyone's interests; moreover they do so for the same reasons. For Habermas, the modern plurality of worldviews means that this demanding requirement can be satisfied only if participants abstract from their particular conceptions of the good, focusing instead on those interests that are universalizable. Ethical questions—questions relating to the good—cannot expect to command universal agreement, either the less demanding form of agreement required for propositional truth or the more demanding form required for moral rightness. Accordingly, Habermas generally describes ethical validity claims as context-specific as opposed to universal. They are context-specific, not just in the sense that they emerge as questions only in specific historical and social contexts, but also in the sense that they admit of answers that hold only for particular individuals and groups and are thus not universally binding.[49]

I have raised the question of how to assess Critical Theory's validity claims: its claims as to what is wrong with a given set of social conditions, which are tied to its claims about human freedom and happiness. Habermas gives no clear answer as to how such claims may be validated. When asked a similar question about philosophical truth claims in an interview published in 1985, he confessed that this was a question that had occupied him for a long time and continued to perplex him.[50] His subsequent writings throw no light either on the validation of philosophical truth claims in general, or on the claims of Critical Theory in particular. Evidently, Critical Theory's claims include claims to various kinds of propositional truth and appeals to moral norms; however, they also seem to involve ethical validity claims—claims about how best to think of individual freedom and happiness ("the good life") and about the social conditions that would facilitate this ("the good society"). Indeed, ethical claims seem to take precedence, for they provide the general framework (Horkheimer's "picture") within which claims to truth and moral rightness are integrated into a "single existential judgment." As we have seen, Habermas offers theories that explicate the meaning of truth and moral rightness and provide guidelines for assessing validity claims of these kinds. But how should we assess claims to ethical validity?

In its current form Habermas' theory of discourse is unhelpful. By "discourse" Habermas means forms of argumentation that come sufficiently close to satisfying the specified idealized justificatory conditions. In *Theory of Communicative Action* he restricted use of the term "discourse" to forms of argumentation that, in addition to approximating these procedural conditions, also rest on the supposition that a rationally motivated consensus on the universally binding validity of the outcome is possible in principle.[51] In his view *only* claims to propositional truth and moral rightness meet this requirement. He has since expanded the category of discourse to include forms of argumentation that meet the procedural conditions, but in which there is no expectation of reaching a rational consensus.[52] In consequence, ethical deliberations, pragmatic deliberations, and legal-political deliberations (and presumably other kinds of deliberation such as religious or aesthetic ones) now qualify as discourses, when conducted intersubjectively and guided by the specified idealizing suppositions. Importantly, however, he does not attribute to *all* discourses an orientation towards reaching a universally binding consensus. His position in this regard remains unchanged. As before, he holds that only propositional truth and moral validity are *conceptually* tied to the expectation of a discursively achieved, rational consensus as to the validity of their outcome for everyone. Most importantly in light of our present concerns, he does not—as a rule—connect ethical validity claims with such an expectation. Rather, ethical discourses are usually described as intersubjective deliberations concerned with the hermeneutic explication of individual and collective value orientations and self-understandings, as opposed to seeking an outcome that is universally valid.

3

Clearly, this account of ethical deliberation is not well-suited as a model for assessing Critical Theory's claims to validity. Critical Theory offers a diagnosis of the social obstacles to freedom and happiness as part of an "ethical picture" that is held to be valid for all human beings. Thus, the claims it raises for the validity of its ethical picture are context-transcending, indeed universal. While hermeneutic explication of its value orientations is certainly important, intersubjective deliberation on the claims it raises could not be confined to this; it would have to extend to reflection on its claims to universal validity. Lacking an account of ethical validity in a context-transcending sense, Habermas's discourse theory is unable to deal with assessment of Critical Theory's ethical claims. However, his account of moral deliberation does not help us to in the attempt to fill this gap.

As we have seen, moral validity is constructed by the participants in a process of intersubjective deliberation. Under certain idealized justificatory conditions, the outcome of the deliberation is morally valid: the validity of moral norms and principles is *defined* by a discursively achieved consensus as to their universality (and universalizability). In this conception the moral principle of universalizability constitutes a standard external to the discourse, providing a formal basis for assessment of the arguments exchanged. At first glance this enables Habermas to avoid the accusation of pernicious circularity that threatens the early Horkheimer's view of the theory's vindication (and also the early Habermas's idea of critique as emancipatory self-reflection). On closer inspection it does not do so. The principle of universalizability is a product of Habermas's theoretical reflections, extracted from a rational reconstruction of the presuppositions of argumentation. Rational reconstruction is a theoretical endeavour based on empirical findings. As the name suggests it is a (re-)construction of certain such findings. Rational reconstructions thus inevitably involve something akin to what Horkheimer calls "existential judgment". The moment of judgment means that the validity of rational reconstructions is not a purely empirical matter, raising the question of how they may be challenged rationally. I find no clear answer in Habermas's writings. The problem of rational contestability affects his moral theory and, in particular, the principle of universalizability on which it relies. It is unclear whether empirical counter-evidence could challenge this element of his theory and, if not, what other kinds of arguments could call it into question. Indeed, Habermas's principle of universalizability seems immune to discursive contestation and vindication.[53]

A further difficulty is that the principle of universalizability is a *purely procedural* normative standard. It leaves open the content of moral norms and principles, stipulating only that they be genuinely universalizable. Thus, the transformative component Habermas attributes to moral discourses should not be confused with the enlightening moment he earlier attributed to critique in the sense of emancipatory self-reflection. To be sure, transformation is an important element in Habermas's account of moral deliberation. Moral discourses serve not just to test whether norms and principles are equally in everyone's interests; under conditions of value pluralism they are necessary in order to *establish* them. Under such conditions establishing genuinely universalizable norms and principles requires a transformative process in which participants in discourse come to see that satisfaction of their particular interests would negatively affect the interests of others and that they must reassess the validity of their interests accordingly. However, in this process of transformation only formal universalizability counts; the substance of participants' interests is irrelevant from the point of view of moral validity. For this reason, perfect argumentative conditions, in which there were no time-constraints, in which no-one's voice was excluded, everyone's voice was given an equal hearing, all participants were optimally informed, there was no self-deception and the guiding concern was the rightness of the outcome, would ensure universalizability. If the validity of the principle of universalizability is taken as established, therefore, and given its purely procedural character, Habermas can consistently maintain that a consensus reached under ideal argumentative conditions constitutes moral validity. But quite apart from the threat of pernicious circularity raised by appeal to the principle of universalizability, such a purely procedural external reference point is unsuitable for assessing the ethical claims raised by Critical Theory. In contrast to moral validity claims, the transformative power of ethical validity claims is not generated primarily within processes of argumentation, for as a rule transformation has to occur before argumentation is possible. The transformative power of ethical validity claims is often what I have called disclosive. I have suggested that Critical Theory seeks to be ethically disclosive: to open people's eyes to aspects of their social situation and to possibilities for the future that they had been unable to grasp hitherto. When it succeeds in doing so, its ethical picture engages the imagination, invoking ideas of freedom and happiness that are not purely abstract. Indeed, it not only sets in train imaginative processes of reflection; it impacts comprehensively on the subjectivities of those to whom it speaks, rendering them fluid and contributing to profound re-shapings of identity and to changed self-understandings. This may in turn lead to the transformations of accepted meanings and the generation of new ones, for the existing semantic reservoir is likely to prove inadequate in the attempt to articulate linguistically these changes in perceptions of self-identity and social life. In sum, Critical Theory's disclosive power is profound, far-reaching, and complex, and impacts on subjective identities as well as on the language use of the subjectivities affected. Furthermore, it typically brings about shifts in perception that are not attributable to the exchange of arguments, but are rather a prerequisite for any such exchange.[54]

Subjective experiences, no matter how profound, far-reaching, and complex, are not epistemically reliable. Subjective experience alone does not amount to reliable *knowledge* of the validity of what is disclosed. Habermas is in my view correct to insist on the linguistic mediation of all human experience. Furthermore, I endorse his view that intersubjective deliberation, along the lines of what he describes as discourse, is necessary for establishing the validity of assertions. However, as will become clear, I do not think that intersubjective deliberation is sufficient.

My argument for the necessity of intersubjective deliberation is also different to Habermas's. Unlike Habermas, my argument is not based on a formal-pragmatic reconstruction of the idealizing suppositions of everyday communication. My position, rather, is that a constellation of ideas connecting truth conceptually to public justification is an important element in the formation of (Western) modern identities as autonomous agents. I hold that the conceptual connection between validity and argumentation, though historically contingent, has a stable foundation.[55] It is based on a constellation of ideas relating to the absence of absolute, ahistorical, and decontextualized standards for adjudicating validity, and to the equal moral worth of human beings, that is constitutive of the evaluative horizon of Western modernity. To be congruent with this constellation of ideas, the concept of truth in its various domains must be tied to open-ended, inclusive, and fair practices of evaluating reasons in public spaces. Since the ideas in question are formative of individual and collective identities within the evaluative horizon of Western modernity, modern subjects cannot set them aside at will. This does not mean that these ideas cannot change over time; the foundation they provide, though stable, is always in motion. This thought is well-captured by Wittgenstein's metaphor of a "riverbed of thoughts," which like the bank of a river "consists partly of hard rock, subject to no alteration or only to an imperceptible one, partly of sand, which now in one place, now in another gets washed away, or deposited."[56] Nor does it mean that they cannot be challenged directly—Nietzsche's entire oevre could be read as challenging precisely these ideas. My point is rather that they change over time for contingent reasons and cannot be challenged without destabilizing the entire horizon of meanings within which humans develop as subjects in Western modernity.

If ethical validity is understood in a context-transcending sense then it, too, must be conceptually tied to public practices of intersubjective evaluation. The experiential dimension of ethically disclosive claims, together with the difficulties of linguistic articulation to which this gives rise, complicates these processes of evaluation.[57] While it by no means makes them impossible, it casts doubt on the suitability of the idea of a rational consensus as a yardstick for measuring their outcomes. For, given the experiential dimension, the reasoning powers of participants in even a perfect procedure of argumentation would not suffice for finding the right answer. In other words, if the rational acceptability of Critical Theory's ethical validity claims depend in significant measure on shifts in perception that are not attributable to the exchange of arguments, there can be no expectation that participants in deliberation would agree on their validity, *even under ideal justificatory conditions*.[58]

Importantly, however, this does not deny the idea of consensus a cognitive status.[59] It merely calls for a different interpretation of it. Whereas the epistemic-constructivist conception sees human beings as the makers of unconditional validity in the domain of practical reason, my alternative conception sees them as capable of *recognizing* truth in the practical sense. The relationship between consensus and practical truth is inverted: whereas in the epistemic-constructivist conception, a discursively achieved, rational consensus *produces* moral validity, in the alternative conception, validity *commands* a rational consensus.[60]

In sum, I propose the need to take seriously both the disclosive moment of Critical Theory and the risk of pernicious circularity to which it gives rise. Doing so calls for a model of ethical validation that is different from Habermas's model of moral validity in key respects. unlike his model, which relies on an apparently incontrovertible principle of universalizability, it allows for an epistemically relevant "outside" to deliberation that is open to rational contestation. A further difference is that in the alternative model validity is not *defined* by a discursively achieved rational consensus; while we may think of it as commanding consensus in some ideal future situation, consensus does not generate validity. My outlined model of ethical validation may seem closer to Habermas's revised conception of propositional truth than to his (unchanged) conception of moral rightness. Certainly, both his revised conception of truth and my proposed account of ethical validity posit a gap between justification and validity in a context-transcending sense. However, in my proposal only the justificatory face corresponds to an aspect of the Janus-faced conception of truth now offered by Habermas. As we have seen, Habermas describes the other face—its reference to a justification-transcendent reality—as a pragmatic fiction. In the account I propose, the second face is not a pragmatic fiction but an actual subjective experience of something rational or true that is disclosed to us by Critical Theory. As such, though epistemically unreliable, it is epistemically significant.

Notes

1 As we shall see, this also distinguishes Critical Theory from orthodox Marxist versions of dialectical materialism.
2 In general, authoritarian theories make assertions regarding the validity of a particular perspective or position that claim privileged access to truth (often they do so tacitly). Their assertions of validity block certain kinds of counter-arguments from the outset. More precisely, they deny a priori the relevance of certain counter-arguments for reasons that have to do with social class, gender, race, and the like or, indeed, the pervasiveness of an all-encompassing ideology (Cooke 2005 and 2006: 20–1).
3 The Institute was forced out of Germany in 1933, maintained its official headquarters in New York from 1934 onwards, before being re-established in Frankfurt in the early 1950s, where it continues its work today.
4 Marx (1983a, 139–40).
5 Marx (1983a, 142).
6 Marx and Engels (1983, 105).
7 Horkheimer (1972a, 44).
8 In Horkheimer's writings I do not find a clear account of the validation of non-natural facts. However, various passages seem to me to make sense only if we make a distinction between modes of validation along the lines I propose.
9 Horkheimer (1972b, 5).
10 Horkheimer (1972c).
11 Horkheimer (1972, 188–243).
12 Horkheimer (1935, 347); Horkheimer(1972b, 8).
13 Horkheimer (1972c, 226). See also Horkheimer (1972b).
14 Horkheimer (1935, 349–52); Horkheimer (1972c, 213); Horkheimer (1972a, 43).
15 Horkheimer (1972c, 225, 234, 238); Cf. Horkheimer (1935, 346).
16 Horkheimer (1972c, 227).
17 Horkheimer (1972c, 239).
18 Marx describes the proletariat as: "a class in civil society which is not a class of civil society, a class which is the dissolution of all classes, a sphere of society which has a universal character because its sufferings are universal" (Marx 1983b, 122–3). Cf. Horkheimer, who writes that "objectively considered," the proletariat embodies the power to change society (Horkheimer 1972c, 214–5).
19 Marx (1983b, 123).
20 Marx and Engels (1983, 94).
21 Horkheimer (1972c, 213).
22 Horkheimer (1972c, 214).

23 Horkheimer (1972c, 218).
24 Marx and Engels (1983, 95–6).
25 Contrast Horkheimer (72c, 215); but cf. also Horkheimer (1935, 344), where he attributes his view to Marxist theory.
26 Horkheimer (1972c, 216)
27 Horkheimer (1972c, 241).
28 Horkheimer (1972c, 207, 242).
29 Cooke (2005).
30 In Horkheimer (1935, 341), in criticism of pragmatism, he calls for an idea of truth as explosive and dangerous. This suggests a reading of Critical Theory's "promotive" power (1972c, 214) as disclosive. Admittedly, it could also be read as an instrumental one. Understood in this way, theory promotes the struggle for social transformation, but does so non-communicatively. Instead it imposes its views temporarily on those who are unable to hear its message because they are trapped within a system of ideological delusion. In his early writings Habermas seems to cautiously acknowledge the necessity on occasion of this kind of promotive power. I would argue that it in every case violates the core value of individual freedom to which Critical Theory is committed. See Habermas (1973a). Cf. Cooke (2005).
31 Horkheimer (1974).
32 Horkheimer and Adorno (1972).
33 Marcuse (1964).
34 Especially, Habermas (1970a, 1970b, 1973a, 1989).
35 The concern for emancipatory self-reflection was held to be one of three anthropologically deep-seated human interests, which were constitutive of knowledge in three domains. Habermas identified an "emancipatory" interest of reason in overcoming dogmatism, compulsion, and domination, alongside a "technical" interest in the mastery of nature and a "practical" interest in securing and expanding possibilities of mutual and self-understanding in the conduct of life.
36 Habermas (1973a, 39).
37 Horkheimer (1973a, 39)
38 The culmination of Habermas' work in the 1970s is his two volume *Theory of Communicative Action* (Habermas 1984 and 1987), originally published in 1981.
39 Habermas (1982, 233). The specific objection had been articulated forcefully by T. McCarthy. It concerns Habermas' endeavor in the 1970s to use a strategy of what he initially called "transcendental reflection," and later "rational reconstruction," as the objective basis for his critical theory, which would in turn justify its guiding values. McCarthy asks how this justificatory strategy relates to critique as an emancipatory practice of self-reflection (McCarthy 1978, 101–2).
40 The concluding essay in Habermas (2003) is entitled "The Relationship between Theory and Practice Revisited." But Habermas does not pick up the question of theory itself as an agent of social transformation, which was central for the early Critical Theorists and, indeed, as indicated earlier, for Habermas in his writings in the 1960s.
41 Habermas (1998b). See also Cooke (1994a).
42 The following paragraphs draw on my discussion in Cooke (2006, 106–13).
43 Habermas (1973b).
44 J. Habermas, "Rightness versus Truth," in Habermas (2003). See also Habermas (1998a).
45 Habermas (1998a, 367).

46 Habermas (1998a, 364, 370, 372).
47 Habermas (1990).
48 Habermas (1990, 86).
49 There is some ambiguity in Habermas' account of ethical validity. He rarely addresses the question of what it means to hold, develop, and pursue a conception of the good and, when he does so, is vague (sometimes even contradictory) as to whether the validity of conceptions of the good is to be determined through appeal to values specific to local contexts or through appeal to an idea of validity with context-transcending power. See Cooke (1994b).
50 Habermas (1985, 86).
51 Habermas (1984, 42).
52 Habermas (1993). See also Habermas (1996, 157–68).
53 See Cooke (2005, 2006, 184–5).
54 See Cooke (2013).
55 Cooke (2006, 132–3).
56 Wittgenstein (1969, §97, §99).
57 See Cooke (2014).
58 Cooke (2013, 271 and 2014).
59 Nor does it rule out the possibility of consensus.
60 The intuition here is that if something is true, every rational person who had undergone the requisite shifts in perception would have to agree to it. I leave open the question of whether a discursively achieved consensus indicates truth, in the sense of pointing towards it. This seems to be Habermas' position with regard to empirical truth. As he puts it: "[. . .] rational acceptability [. . .] indicates the truth of a proposition" (Habermas 1998a, 381).

References

Cooke, Maeve. 1994a. *Language and Reason*. Cambridge, MA: MIT Press.

Cooke, Maeve. 1994b. "Realizing the Post-Conventional Self", *Philosophy and Social Criticism*, 20(1–2): 87–101.

Cooke, Maeve. 2005. "Avoiding Authoritarianism: On the Problem of Justification in Contemporary Critical Social Theory", *International Journal of Philosophical Studies*, 13(3): 379–404.

Cooke, Maeve. 2006. *Re-Presenting the Good Society*. Cambridge, MA: MIT Press.

Cooke, Maeve. 2013. "Violating Neutrality? Religious Validity Claims and Democratic Legitimacy?" In *Habermas and Religion*, edited by C. Calhoun, E. Mendieta, and J. VanAntwerpen (249–275). Cambridge: Polity Press.

Cooke, Maeve. 2014. "The Limits of Learning: Habermas' Social Theory and Religion", *European Journal of Philosophy*. DOI: 10:1111/ejop.12099.

Habermas, Jürgen. 1970a. *Towards a Rational Society*. Boston: Beacon Press.

Habermas, Jürgen. 1970b. *Knowledge and Human Interests*. Translated by Jeremy J. Shapiro, Boston: Beacon Press.

Habermas, Jürgen. 1973a. *Theory and Practice*. Translated by John Viertel. Boston: Beacon Press.

Habermas, Jürgen. 1973b. "Wahrheitstheorien." In *Wirklichkeit und Reflexion*, edited by Helmut Fahrenbach (211–265). Pfullingen: Neske.

Habermas, Jürgen. 1982. "A Reply to my Critics." In *Habermas: Critical Debates*, edited by D. Held and J. B. Thompson (219–282). London: Macmillan.

Habermas, Jürgen. 1984 and 1987. *Theory of Communicative Action*. Translated by Thomas McCarthy. Boston: Beacon Press.

Habermas, Jürgen. 1985. "A Philosophico-Political Profile." *New Left Review* 151, May–June.

Habermas, Jürgen. 1989. *The Structural Transformation of the Public Sphere*. Translated by Thomas Burger. Cambridge, MA: MIT Press.

Habermas, Jürgen. 1990. "Notes on a Program of Discourse Ethics." In *Moral Consciousness and Communicative Action*. Translated by C. Lenhardt and S. Weber Nicholsen, 43–115. Cambridge, MA: MIT Press.

Habermas, Jürgen. 1993. "On the Pragmatic, the Ethical, and the Moral Employments of Practical Reason." In *Justification and Application*, translated by Ciaran Cronin, 1–19. Cambridge: Polity Press.

Habermas, Jürgen. 1996. *Between Facts and Norms*. Translated by William Rehg. Cambridge, MA: MIT Press.

Habermas, Jürgen. 1998a. "Richard Rorty's Pragmatic Turn." In Habermas 1998b.

Habermas, Jürgen. 1998b. *On the Pragmatics of Communication*, edited by Maeve Cooke. Cambridge, MA: MIT Press, 1998.

Habermas, Jürgen. 2003. *Truth and Justification*. Translated by Barbara Fultner. Cambridge, MA: MIT Press.

Horkheimer, Max. 1935. "Zum Problem der Wahrheit", *Zeitschrift für Sozialforschung*, 4: 321–63.

Horkheimer, Max. 1972a. "Materialism and Metaphysics." In Horkheimer 1972d.

Horkheimer, Max. 1972b. "Notes on Science and the Crisis." In Horkheimer 1972d.

Horkheimer, Max. 1972c. "Traditional and Critical Theory." In Horkheimer 1972d.

Horkheimer, Max. 1972d. *Critical Theory: Selected Essays*. Translated by Matthew J. O'Connell and others. New York: Continuum.

Horkheimer, Max. 1974. *Critique of Instrumental Reason: Lectures and Essays since the End of* WWII. Translated by Matthew J. O'Connell and others. New York: Seabury Press.

Horkheimer, Max, and Adorno, Theodore W. 1972. *Dialectic of Enlightenment*. Translated by John Cumming. New York: Continuum.

Marcuse, Herbert. 1964. *One Dimensional Man*. Boston: Beacon Press.

Marx, Karl. 1983a. "From the First Manuscript: 'Alienated Labour'." In Marx 1983c.

Marx, Karl. 1983b. "Contribution to the Critique of Hegel's *Philosophy of Right*." In Marx 1983c.

Marx, Karl. 1983c. *The Portable Karl Marx*, edited by Eugene Kamenka. Harmondsworth: Penguin.

Marx, Karl, and Engels, Friedrich. 1983. *The Communist Manifesto*. Harmondsworth: Penguin.

McCarthy, Thomas. 1978. *The Critical Theory of Jürgen Habermas*. Cambridge, MA: MIT Press.

Wittgenstein, Ludwig. 1969. *Philosophical Investigations*, edited by G.E.M. Anscombe and Rush Rhees. Translated by G.E.M. Anscombe. Oxford: Blackwell.

4 From the *Positivismusstreit* to Putnam

Facts and Values in the Shadow of Dichotomy

John McGuire

Hilary Putnam begins his contribution to this volume with the suggestion that we have all at one time or another been confronted by the question: "Is that supposed to be a fact or a value judgment?" This allegedly common refrain poses a twofold threat in his view: first, it imposes a dubious standard of objectivity that corrals ethical statements into a semantic category shared by irrational and purely subjective beliefs—the implication being that our ethical convictions are not germane to rational argumentation (Putnam 2016a, 2). Second, the rigidity of the fact/value dichotomy itself threatens to override subtler distinctions within natural language—between what we "value" as a defining principle versus what we simply prefer (Putnam 2016a, 4). Yet, despite these reasonable concerns, Putnam's framing smacks of proof surrogacy. However heated the debate may be regarding the role of non-epistemic values in scientific explanation, the real world consequences of fact/value dichotomy are rather difficult to pin down. To a skeptical reader, the danger posed by a quasi-metaphysical partitioning of observable descriptions and ethical convictions loses all urgency when considered in relation to the state of public debate in places like the United States—a country currently leading the world in climate change denialism, anti-vaccination hysteria, and where the current presidential election cycle witnessed the resistible rise of a demagogic billionaire whose popularity appears based on eschewing comprehensible policies in favor of doom-laden screeds against non-white races.[1]

Of course, Putnam never presented himself as a social theorist or political commentator. Nor did he seem interested in connecting his vision of philosophy's offering "meaningful orientation in life" to more targeted diagnoses of social pathologies (Putnam 1999, 52). Still, when it comes to addressing the contaminating effects of bad metaphysical reasoning, Putnam builds his case upon a rather opaque notion that *certain* philosophical ideas are *especially* prone to trickling-down into ordinary moral vocabularies. The aim of this essay is to retrace the toxic legacy of positivism in order to arrive at a clearer picture of how the products of philosophical reasoning (both good and bad) escape the cloistered discourse of expertise. Alongside Putnam, we will consider a parallel critique of positivism advanced by the Frankfurt School social theorist Jürgen Habermas, who (at least initially) portrays the threat as a hollowing out of interpretive understanding by the singular rubric of technical-instrumental control. My concern here is not whether Habermas's diagnosis is more reliable than Putnam's, it is rather to show how their respective critical strategies reveal a deeper dispute concerning the social role of practically-oriented philosophies, both pragmatic and critical.

1 Pathogenesis

Discussions of Putnam's philosophical legacy tend to neglect the abiding influence of his historiographic narratives. A case in point is his account of the evolution of fact/value dichotomy, in which David Hume is identified as the original proponent, based on his partitioning "relations of ideas" (pertaining to "ought" judgments) from matters of fact (observations of what "is"). According to Putnam, Humean cognition entails that all conceptual contents boil down to pictorial representations or aroused sentiments; and, because they are assumed to be a reactive processing of original sense impressions, ethical evaluations are definitionally disconnected from factual descriptions (Putnam 2002, 14–16). But here the aim is not to present an authoritative interpretation of Hume, nor to demonstrate his influence upon Kant's delineation of maxims for acting. The point, rather, is to frame the logical positivists' denigration of ethical statements (as non-ampliative tautologies or cognitively meaningless commands) as marking a regression to an eighteenth-century phenomenology of sense impressions wholly out of step with contemporaneous developments in biology and theoretical physics (Putnam 2002, 17). Having made a principle of disavowing metaphysical extravagances like the synthetic a priori, the positivist's verifiability theory of meaning fell back on an insupportably crude model of cognition and a surprisingly vague notion of what constitutes a discoverable "fact" (Putnam 2002, 21). And it was not long before the doctrinal core of positivism was itself consigned to obsolescence by way of W.V. Quine's unraveling of the notion that precisely rendered statements clearly demarcate evidential observation from socially-mediated convention.[2]

At least within philosophical circles then, positivism's influence was brief and largely contained—which might lead us to conclude that fact/value dichotomy has had precious little impact upon ordinary ethical reasoning. After all, Putnam's defense of "entanglement" appeals directly to everyday descriptive-normative blends like *generous*, *elegant*, *skillful*, *strong*, *gauche*, *weak*, or *vulgar*.[3] What then is the danger of dichotomy if everyday understanding remains one step ahead of philosophical dogma? Interestingly, the decisive moment in Putnam's historical reconstruction arrives after the controlled demolition of the positivist criterion of cognitive significance. Towards the end of the 1930s, against a backdrop of calamitous global economic uncertainty and political instability in Europe, a number of prominent social scientists (most notably the British economist Lionel Robbins) overturned the reigning orthodoxy of Benthamite utilitarianism with a grand positivistic dismissal of interpersonal comparisons, which they declared to be cognitively empty and socioeconomically irrelevant:

> With one stroke, the idea that the economist could and should be concerned with the welfare of the society in an evaluative sense was rejected, and in its place was inserted the positivist idea that such a concern was "meaningless," at least from a scientific point of view. [. . .] After they had been persuaded to accept these views . . . economists did not simply conclude that there was no such field as "welfare economics." [. . .] What the positivist views that came to dominate economic thinking did was to proscribe the very idea of a "reasoned" consensus on any value question.
>
> (Putnam 2002, 54–6)

On the one hand, by confronting the bogeyman of allocative efficiency, the prescriptive role of the ethicist enjoys a certain clarity of purpose: "Matters of—literally—life and death may well be at stake" (Putnam 2002, 2). On the other hand, as Putnam elsewhere concedes, unmasking the murkier presumptions of policy science does not conclusively prove any link between "bad metaphysics" and the "instrumentalisation of human beings."[4] Given that the value-neutral criterion of optimal economic functioning (initially ascribed to Pareto optimality) is also dismissed by Putnam as a transparently absurd metric for assessing socioeconomic states of affairs, how do we account for positivism's paradoxical insurgence at the precise moment of its philosophical bankruptcy? Putnam's apparent conclusion is that it was all a case of tragic *irony*—the "wrong" idea taking hold at the "right" time within the "right" group of policymakers (Putnam 2003, 398). While it may have originated, innocently enough, as a misguided attempt to make public policy a more rigorous science, the language of dichotomy has now attained the solidity of a cultural institution, stubbornly resistant to challenge or debate. Putnam despairs at the prospect that, were he to succeed in mounting an unassailable refutation of the fact-lessness of values, the dualistic spirit of positivism would persist within the grammar of everyday political and social life (Putnam 1981, 128). As far as specifying the negative consequences of positivism, the reasoning seems to be as follows: by narrowing interpretation to the verification of general laws, positivist-inspired theorists sever the essential edifying link between theoretical "knowledge for its own sake" and practical "problems of use and enjoyment."[5]

How does this narrative framing help dissolve dichotomy? Lord Robbins's positivist commitments are said to consist in his banishing ethical talk about distributional unfairness from policy debates—a pretense to objectivity Putnam shows us is spurious, since assessments of preference satisfaction inevitably smuggle ethically "thick" notions about agent rationality and the optimality of states of affairs (Putnam 2002, 50–4). Putnam's narratives should not be seen as neutral reconstructions but as active interventions through which the history of philosophy is divided into two clear categories: theories still offering edifying "expedients" for addressing contemporary problems; and cultural monuments best left to sentimental or aesthetic "appreciation." But I believe he overlooks another, equally salient aspect of positivism's allure, particularly from a public policy perspective. I refer here to Robbins's hopeful boast that the methodological scope of economic science will no longer be restricted to the "procuring of material welfare":

> The conception we have adopted may be described as *analytical*. It does not attempt to pick out certain *kinds* of behaviour, but focuses attention on a particular *aspect* of behaviour, the form imposed by the influence of scarcity. It follows from this, therefore, that in so far as it offers this aspect, any kind of human behaviour falls within the scope of Economic Generalisations. We do not say that the production of potatoes is economic activity and the production of philosophy is not. We say rather that, in so far as either kind of activity involves the relinquishment of other desired alternatives, it has its economic aspect. There are no limitations on the subject-matter of Economic Science save this.[6]
>
> (Robbins 1932, 16)

"Preference ordering under conditions of scarcity" is presented as the new *Punctum Archimedis* from which to adjudge all domains of human activity. This also opens up the prospect of less politically innocent co-optations, where, for example, econometric projections of expected future earnings are used to undermine the "value" of a humanities degree, or where the rhetoric of fiscal "responsibility" is used to block renewed investment in public infrastructure and social services. This amounts to much more than *ironic* happenstance; it portends the deformation of social life by an ideologically charged criterion ("value for money" on educational "investments") whose rhetorical currency is all but immune to the objections Putnam raises against its descriptive accuracy. We will see in the next section how Habermas's early critique of positivism follows this mode of ideology critique.

For Putnam, by contrast, the danger of fact/value dichotomy appears to consist mainly in its *extended latency*: whereas most other metaphysical or epistemological debates have remained benignly indecipherable to non-specialists, the question of whether something is "really" factual or "merely" valued is still a forced choice imposed on all reflective persons—despite no longer having any defensible justifications to support it (Putnam

1981, 127). Here we need to get a clearer idea of what Putnam understands to be the enabling conditions of reflexive flourishing. In a lecture on ethical objectivity, he suggests that the critical faculty needed for dissolving dogma like fact/value dichotomy remains reassuringly ineradicable:

> Philosophy is an enterprise that aims at what I called "reflective transcendence"—the act of standing back from conventional beliefs, and even received practices, and asking a penetrating "Why should we accept this as right?" At certain crucial moments in history, the result is a profound revaluation of our ways of thinking, which we may call "enlightenment."
>
> (Putnam 2004, 148)

In a footnote, Putnam claims this transformational reflexive capacity is *anthropologically rooted*, and draws support from the early work of Jürgen Habermas: "The impulse to reflective transcendence is closely related to what Habermas calls 'the emancipatory interest' in *Erkenntnis und Interesse*" (Habermas 2004, 148, note 1). "Reflective transcendence" denotes an innate ability, but also a transhistorical social status by which philosophers (although it remains unclear how this vocational identity emerges) contribute directly to historical learning processes. This puts a new gloss on Putnam's various collaborations with philosophically-minded economists and socially-conscious ethicists, because it suggests an underlying pedagogical intent through which tragically ironic confluences of bad reasoning and policymaking will be avoided.[7] Thus, with Martha Nussbaum, Putnam has sought to develop universally defensible indicators of flourishing, starting from a threshold of basic entitlements and building towards more substantive expectations concerning property ownership, inclusion in political decision-making, and supportive institutions for cultivating self-reflection, ethical respect, and aesthetic taste.[8] Although this list of central capabilities is not presented as a civilizational blueprint, Putnam does make the strong claim that denying or obstructing their attainment constitutes an unjust deprivation (Putnam 2003, 403). Even the critical assault on fact/value dichotomy is infused with heuristic potential. Here, Putnam avoids the temptation to indulge in philosophical "obituary" when it comes to discussing the work of his own former teachers, Hans Reichenbach and Rudolph Carnap, who are most remembered for their role in promoting logical empiricism:

> What overcame my "philosophy is over" mood, what made the field come alive for me, made it more exciting and more challenging than I had been able to imagine, was Reichenbach's seminar, and his lecture course, on the philosophy of space and time. And it was not the technical details of Reichenbach's philosophy that did this, although I pondered those details with great excitement, but the sense of a

> powerful philosophical vision behind those details [. . .] Even if today one cannot accept either Carnap's or Reichenbach's solutions to the problems of philosophy, their attempts represent contributions of enormous value, and there is much to be learned from their study . . . [O]ne does not understand logical empiricism itself if one simply identifies it with any of Carnap's successive programs of rational reconstruction. The fact that the movement had room for metaphysical visions as different as those of Carnap and Reichenbach is itself a fact that has to be appreciated if simplistic pictures of 'positivism' are to be overcome.
>
> (Putnam 1994, 99–101)

Putnam's painstaking reconstructions of relatively obscure corners of analytic philosophy could be seen as a deliberate countermove against positivist reductionism, fleshing out more nuanced portraits of an easily caricatured, *dépassé* philosophy; retracing Carnap's attempts to redeem his epistemology of experience against charges of solipsism; sympathetically unpacking Reichenbach's flawed (but "magnificent") defense of induction as an approach to probability logic (Putnam 1994, 109).

Nevertheless, it remains curious that, despite the positivists' failure to secure a viable methodology, despite the sterility of their semantical universe (in which language was nothing more than a tool for transmitting the results of scientific experiments), Putnam does not countenance severing the link to his philosophical past and to those figures who awakened him to philosophy's vitality as a discipline. This curatorial gesture seems to belie his claim that philosophers offer only "suggestions" for resolving ethical dilemmas and should not be seen as proffering "expert testimony"—which implies suggestions are *dispensable* "advice" (Putnam 2016b, 129). Nor does this square with the apparent indispensability of knowledgeable curators of defunct philosophical schools. Although we can be certain that Putnam would object to this formulation, it is as though philosophical "suggestion" does not aim to provide answers to ethical questions so much as it induces a kind of *suggestibility*, through which "curators of sapience" determine, on behalf of non-philosophers, who is still worth reading and how they should be read. As we will see, Putnam's critical exchange with Jürgen Habermas revolves not just around the question of the status of ethical principles, but also the question of how these insights are best preserved and transmitted.

2 Eine Trügerische Tagung in Tübingen

The *Positivismusstreit*, as it came to be known, refers to an acrimonious, decade-long dispute between members of the Frankfurt School of critical theory (including T.W. Adorno, Jürgen Habermas, and Albrecht Wellmer) and proponents of critical rationalism (Karl Popper and

Hans Albert).[9] The initial trigger was Karl Popper's twenty-seven theses on the methodology and scope of social science, presented to the 1961 meeting of the German Sociological Association in Tübingen. This paper was in no way intended as a defense of positivism, which in Popper's view promoted a naïve empiricism that collapsed into a metaphysics of induction.[10] However, in remonstrating with his Frankfurt School accusers, Popper's consignment of sociology to piecemeal troubleshooting (subserving the truth-disclosing role of physics) clearly evinces the language of rational preference ordering which had already overtaken neoclassical economics:

> The logical investigation of economics culminates in a result which can be applied to all social sciences. This result shows that there exists *a purely objective method* in the social sciences which may well be called the method of *objective* understanding, or situational logic. [. . .] Objective understanding consists in realising that the action was objectively *appropriate to the situation.* In other words, the situation is analysed far enough for the elements which initially appeared to be psychological (such as wishes, motives, memories, and associations) to be transformed into elements of the situation.
>
> (Popper 1976, 102)

In this, Popper retains positivism's defining claim that truths of scientific theory are established through frictional encounters with the world rather than *conceptual* clashes with competing theories.[11] "Progress" is defined by the gradient of successive historical disappointments, encompassing everything from failed experiments to civilizational collapse. To those who insist on an independent interpretive logic for the social sciences, and who argue that "society" expresses an idea that supervenes upon the desires and actions of individuals, Popper offers high-handed dismissals of trivialities wrapped in "impressive talk" (Popper 1976, 297).

Both in his initial response and in his introduction to the collected volume for the conference, Theodor Adorno defends an interpretive approach that considers society as a conceptual whole, and questions whether the mapping of individual behavior patterns is really the rigorous science it claims to be. Focusing solely on the latter ignored the obvious behavior-conditioning effects of political ideologies and socializing institutions—just as adopting a purely holistic societal image overlooked potential conflicts between the collective "needs" of society and the needs of individual citizens:

> Society is a total process in which human beings surrounded, guided and formed by objectivity do, in turn, act back upon society; psychology, for its part, can no more be absorbed into sociology than can the individual being be absorbed into its biological species and

> its natural history. Certainly, fascism cannot be explained in social-psychological terms . . . [b]ut if the authoritarian character type had not been so widespread for reasons which, in their turn, are sociologically intelligible, then fascism, at any rate, would not have found its mass basis, without which it would not have achieved power in a society like that of the Weimar democracy.
>
> (Adorno 1976, 119–20)

By supplanting the utopian perspective of philosophical critique with an allegedly value-free metric of preference-ranking, the new technocratic model of sociology risked becoming an unthinking appendage of *any* governmental regime, democratic or fascist: “Even in the rationality of a scientific management of the whole society which had apparently thrown off its shackles, domination would survive” (Adorno 1976, 34). Interestingly, Adorno’s anxieties about the alienating—and potentially authoritarian—character of this societal self-image overlap with Putnam’s own philosophical hero, John Dewey, who in his “Theory of Valuation” expressly links the splitting of affective and cognitive interests with the rise of dictatorship.[12] Whereas applied science might have once been seen as the seedbed for breakthroughs in theoretical science, the two modes of inquiry now increasingly resembled the figures of Echo and Narcissus, with critical rationalism offering only the shadowy repetition of a “true” science increasingly enamored of its own magnificent purity. Far from streamlining scientific research, the collapse of interpretative understanding into a single mode of falsificationism evidences a deformed self-understanding on the part of its practitioners; and Popper’s dismissal of “dialectical” thinking only betrays a benighted faith in the progressive character of Western liberalism.

A more concise diagnosis of positivism, and of Popper’s affinities with its conceits, emerges in Jürgen Habermas’s contribution to the debate. Like Adorno, Habermas initially identifies the problem as the inability (or unwillingness) of theorists to situate their interpretive praxis within the society under investigation. Value-freedom offers no normative purchase upon which the interpreter can assess societal developments as progressive or regressive (Habermas 1976, 147). Presenting his critique as an unmasking of positivist ideology, Habermas sets out to defend the dialectical mode of analysis through which conceptual weaknesses in an existing framework are illuminated, and the potential for transforming our social practices is revealed:

> In the evolution of consciousness, the problem of identity presents itself as a problem of survival and, at the same time, of reflection. From here dialectical philosophy once developed. In the shirt-sleeved world picture of many a positivist, dialectics plays the part of a bogeyman. For others, who occasionally become aware of the

> fact that they lapse into dialectical trains of thought, dialectics only expresses the fact that we think and are able to think when, according to the traditional rules of logical inference, we really should not be able to do so . . . The self-reflection of the strict empirical sciences . . . strikes a cautionary note as far as positivistic expectations are concerned. It includes the realisation that our theories do not simply describe reality.
>
> (Habermas 1976, 223–4)

What really distinguishes Habermas's critique is his appeal to the tradition of American pragmatism as a corrective for the "scientistic" inflation of empirical research. Here, Charles Peirce's consensus-driven model of scientific investigation allows the practical continuation (rather than theoretical *correction*) of everyday problem-solving (Habermas 1976, 150). The adequacy of any methodology is then judged by socially-mediated ends, linking theoretical knowledge to pre-scientific experience (which, as we saw earlier, is important for Putnam as well). Peircean problem-solving (at least in Habermas's reading) does not need to establish free-standing truths *about* the world, but instead offers coping strategies for addressing challenges and uncertainties encountered in the world.[13] Truth connotes the fixation of beliefs that have a successful track record of predicting, directing, and enhancing natural and biological processes.

In response, one of his critical rationalist interlocutors, Hans Albert, accuses Habermas of compounding Hegelian obscurantism with a reductive image of scientific research as instrumental manipulation (Albert 1976, 170). But this misses the point of Habermas's appeal to Peirce's social epistemology, which simply claims that the situatedness of scientific inquiry is what ensures technical progress unfolds in an intelligible way:

> In opposition to positivism, I should like to justify the view that the research process, which is carried out by human subjects, belongs to the objective context which itself constitutes the object of cognition, by virtue of cognitive acts. The dimension in which this combination of the research process with the social life-process is formed belongs neither to the sphere of facts nor to that of theories. It stands apart from this dualism, which only has meaning for empirical scientific theories. Rather, in the comprehensive communicative context of scientific criticism one moment links itself to another [. . .]. I am concerned with knowledge-guiding interests which, in each case, form the basis for a whole research system. In contrast to positivistic self-understanding, I should like to point out the connection of the empirical-analytical sciences with a technical cognitive interest. But this has nothing to do with 'denunciation'.
>
> (Habermas 1976, 220–1)

Habermas does not dispute the need for law-like hypotheses; his point is that this technical-cognitive interest is only *one* aspect of a larger sociocultural learning process which inevitably extends beyond the narrow scope of logical inference and empirical disconfirmation. Habermas's concern here is to recuperate a manifest image of human agency and experience that is threatened by a scientific perspective evincing high-minded indifference towards socially and culturally-mediated goals. In Habermas's first systematic work, *Knowledge and Human Interests*, he expands upon his initial sketch of deep-seated cognitive interests governing practical engagements, and offers a "weak" philosophical anthropology of technical mastery, symbolic interpretation, and emancipation, which are depicted as the driving impulses behind developments in empirical-analytic, historical-hermeneutic, and critically-oriented sciences.[14]

I should emphasize here that I am not trying to present Habermas's early research program as an alternative to Putnam's critique of positivist dichotomy. There is no smooth line of continuity that can be drawn between Habermas's early model of quasi-transcendental interests and his later reconstructions of the pragmatics of social interaction, speech acts, and the normative foundations of democratic society. Nor am I interested in weighing in on whether Habermas's constructivism should be rescinded in favor of a stronger realist or anti-realist stance. There is already a well-developed line of commentary concerning Habermas's eventual abandonment of the Peircean/epistemic understanding of "truth," and of the implications this has for his cognitivist conception of normative rightness—and for his *non-cognitivist* disavowal of "rightness" being equivalent (rather than merely analogous) to truth.[15] I am following a much humbler historiographic strategy here, which takes Habermas's initial response to positivism as presaging a significant development in the self-understanding of his own philosophical practice. Like Putnam, Habermas initially assumes a strong link between theoretical innovations and social practices. But just a few years later, he comes to see the supposedly universal character of "knowledge-constitutive interests"—in particular the *emancipatory* interest—as a deeply problematic assumption. Whereas Putnam is still happy to appeal to a universal cognitive foundation for reflective transcendence, Habermas has all but severed the connection between philosophical edification and reflective transformation:

> Till now I have not adequately distinguished posterior reconstruction [*Nachkonstruktion*] from reflection on oneself. [. . .] In the philosophical tradition these two legitimate forms of self-knowledge have generally remained undifferentiated and have both been included under the term of reflection. However, a reliable criterion of distinction is available. Self-reflection leads to insight due to the fact that

> what has previously been unconscious is made conscious in a manner rich in practical consequences: analytic insights intervene in life, if I may borrow this dramatic phrase from Wittgenstein. A successful reconstruction also raises an "unconsciously" functioning rule system to consciousness in a certain manner; it renders explicit the intuitive knowledge that is given with competence with respect to the rules in the form of "know-how." *But this theoretical knowledge has no practical consequences.* By learning logic or linguistics I acquire theoretical knowledge, but in general I do not thereby change my previous practice of reasoning or speaking.
>
> (Habermas 1973, 22)

Using the model of psychoanalysis for comparison, Habermas points out a key distinction between "ordinary" philosophical discourse and the "therapeutic" discourse held between the analyst and patient. Although philosophical reconstruction can still be said to arise out of everyday life (at least to the extent that within modern sociocultural lifeworlds a reflexive attitude towards inherited beliefs and conventions is increasingly commonplace) this does not constitute the same system of action and experience we find operating between analyst and patient (Habermas 1973, 23). Participants in philosophical discourse are assumed to initiate their discussions as equals; the goal is not for one participant to enlighten the other or for one participant to seek out their interlocutor's help in identifying the roots of their self-deception. By contrast, psychoanalysis establishes asymmetry (and even conflict) from the outset, although the analyst's clinical diagnosis of the patient's self-deception, and the patient's own worked out self-interpretation, may eventually overlap. In an essay written long after he had ceased to invoke the emancipatory interest or the exemplarity of psychoanalytic critique, Habermas reaffirms his rejection of the notion that the theorist—or even the scientist—can claim the status of "usher of reason":

> Ultimately, there is only one criterion by which beliefs can be judged valid, and that is that they are based on agreement reached by argumentation. This means that everything whose validity is at all disputable rests on shaky foundations. It matters little if the ground underfoot shakes a bit less for those who debate problems of physics than for those who debate problems of morals and aesthetics. The difference is a matter of degree only [. . .] [U]nless I am completely mistaken, it makes sense to suggest that philosophy, instead of just dropping the usher role and being left with nothing, ought to exchange it for the part of stand-in [*Platzhalter*]. Whose seat would philosophy be keeping; what would it be standing in for? Empirical theories with strong universalistic claims. As I have indicated, there have surfaced and will continue to surface in non-philosophical

> disciplines fertile minds who will give such theories a try . . . Marked down in price, the venerable transcendental and dialectical modes of justification may still come in handy. All they can fairly be expected to furnish, however, is reconstructive hypotheses for use in empirical settings.
>
> (Habermas 1990, 15–16)

Whereas during the *Positivismusstreit*, Habermas seemed willing to defend a therapeutic image of philosophers rekindling emancipatory energies, his critical strategy now takes the form of a rearguard action against all pretenders to an as-yet unrealized "empirical theory with strong universalistic claims" (an ever-expanding group which includes positivistic interpreters of cognitive science, as well as French post-structuralists, Heideggerians, Nietzscheans, neo-conservatives, Left-Schmittians, and free market fundamentalists).[16] If any impetus for social progress is to be found in the realm of "theory," it will not appear in the traditional guise of philosophical or poetical world-disclosure. Philosophical critique can no longer lay claim to unique criteria of validity that somehow exist independently of science morality and art (Habermas 1992, 17–20).And Habermas is no longer willing to countenance the possibility that there is some universal cognitive capacity that can *only* be serviced by dialectics; philosophy is now a relentlessly (self)-critical, defensive gesture which awaits its replacement by a future hybridizing reconstructive science (possibly in the genre-bending spirit of psychoanalysis or historical materialism). In this regard, Habermas has remained quite consistent—he does not offer his dialogical principle of universalization as a heuristic tool to be bestowed upon real world actors to expedite their problem-solving.[17] The examples of discursive "repair work" found in *The Theory of Communicative Action* occur entirely beyond the purview of linguistics and speech-act theory.[18] There is no suggestion that the naïveté of the lifeworld must first be therapeutically unlocked in order for problematizing and problem-solving to commence—philosophical "mediation" is now mostly a matter of defending the lifeworld "against extreme alienation at the hands of the objectivating, the moralizing, and the aestheticizing interventions of expert cultures" (Habermas 1992, 18). This does not mean that philosophers retreat from public debate, it means that, when they intervene, they do so in the self-effacing guise of concerned citizens, who only take advantage of their elevated profile (as public intellectuals) to draw attention to issues they feel merit greater public scrutiny (Habermas 2003, 289).

Both Putnam and Habermas engage in historiographic neutralizations of philosophical legacies they believe threaten preferred forms of moral agency and sociocultural understanding. The difference is that, for Habermas, the unraveling of metaphysical conceits means there are no exemplars left to whom we can still appeal directly for moral, cultural, or political guidance. Habermas exhibits none of the reverence towards his own supposed lodestars that Putnam exhibits towards Dewey and Aristotle. If Habermas identifies with Kant, it is only in the sense that he feels we can sufficiently "detranscendentalise" Kant's moral theory, so that it may serve to reinforce the pragmatic presuppositions already found within everyday language use and social interaction.[19] In all other respects philosophers are treated by Habermas as cautionary footnotes in an aporia-riddled history of Western metaphysics and the "philosophy of consciousness" (Habermas 1990, 138). Habermass strongly deflationary reading of philosophical praxis makes dialogic exchange (including intercultural and intergenerational "dialogues") the *only* route through which novel insights can emerge to unsettle take-for-granted certainties. In the concluding section we will see how this final postmetaphysical abandonment of philosophy's curatorial role informs what I take to be the unspoken disagreement between Putnam and Habermas concerning the continued "vitality" of philosophy.

3 Putnam, Habermas, and the Elusive Pedagogy of Enlightenment

Habermas' critical exchange with Putnam exhibits none of the hostility and contempt that colored proceedings during the *Positivismusstreit*. Nevertheless, there is a sense that both thinkers are talking past one another. Ostensibly, their disagreement centers on whether ethical convictions and moral principles are fundamentally distinct—an argument Putnam rejects on the grounds that it exacerbates a "creeping" metaphysical dualism already found in the work of Neo-Kantians like Christine Korsgaard, and even "minimalist" Kantians like Habermas.[20]

For Habermas, norms act as a constraint on collective decisions and individual plans. In order to be recognized as legitimate, the reasons behind our various pursuits must be responsive to potential objections from those who might be "affected":

> [Universalisation Principle] A norm is valid when the foreseeable consequences and side-effects of its general observance for the interests and value-orientations of *each individual* could be *jointly* accepted by *all* concerned without coercion.
>
> (Habermas 1998, 42)

For Putnam, this demarcation of universally binding "norms" from ethical "values" (as though the former were a special subset of the latter) is both undesirable and untenable: *undesirable*, because it wrongly suggests that we can avoid controversial ethical claims by treating their non-universal content as an expression of merely subjective preferences.

Untenable because it fails to acknowledge the ways in which even our purest procedural norms remain value-laden. The operability of normative assessments and procedural guidelines inevitably relies upon some minimal criterion of agent rationality and exemplarity (Habermas 2003, 410). Even the criterion of "affectedness" seems open to the charge that it incorporates a particular interpretation of what kinds of effects and what degree of affectedness should be considered morally salient. Thus, Habermas' defense of the unconditional character of moral obligations does not, in Putnam's view, escape the needful endorsement of a particular set of cognitive and cultural capacities drawn from the sociocultural inheritances of European enlightenment. To illustrate his point, Putnam offers the following thought experiment, in which a procedurally ideal consensus fails to overcome the moral obtuseness of the participants:

> A father engages in psychological cruelty by teasing his child, while denying (either because he is obtuse or because of a streak of sadism) that the child's tears are really "serious." "He has to learn to take it," the father says [. . .]. Suppose most of the members of the community or even a significant minority share the father's obtuseness in the case described. They are not, we suppose, bad people in most other respects. They genuinely want to do what is right, and they love rational argument. Indeed, they regard the question as to whether the case is a case of "cruelty" as a fascinating one, and they discuss it endlessly. No one tries to manipulate anyone else, and everyone listens patiently to everyone else's arguments. But we can perfectly well imagine that the father and others like him never "get it" [. . .]. What is wrong with the discussants in the above situation isn't that they aren't obeying the norms of discourse ethics. What is wrong is statable using the thick ethical vocabulary appropriate to the particular ethical problem. They are "obtuse" (the opposite of having *Menschenkenntnis*), they have a "trace of sadism," and so on. To describe what an "ideal discussion situation" would be in this case, one would have to use (and presuppose one's audience had mastered) thick ethical concepts. In *The Theory of Communicative Action*, it seems at times as if Habermas, at least, wants to build in the additional requirement that, in effect, the participants in an ideal discussion have achieved the equivalent of an ideal psychoanalysis (whether through actual psychoanalysis or otherwise).
>
> (Putnam 2002, 127–8)

Putnam claims that Habermas fails to resolve a fundamental ambiguity, whereby norms governing discourse are said to be discovered *within* the process of deliberation, but are also somehow established as criteria for valid ethical judgments *prior* to any discourse taking place (Putnam 2003, 410). This leaves his model open to the familiar charge that it is emptily formal and unable to confront the authoritarian viewpoints of the father and his sympathizers.

Without pretending to speak for Habermas, we can point to a number of problems with Putnam's construal. It is one thing to suggest that generality and unconditionality must first attain a certain sociocultural salience before they can serve as standards for public discussion. This does not however, establish a pattern for civilizational development through which the "value" of equitable understanding is established and disseminated. "Democracy" and "capitalism" are phenomena that can each be retraced along particular historical trajectories. But this does not mean their constituent elements (the demand for more equitable access to decision-making or the functional exchange of privately maintained wealth) were "discovered" or "invented" by Pericles or Adam Smith. Putnam's thought experiment lacks any temporal or intergenerational dimension, whereby we might imagine the son one day confronting his father over the oppressive conditions of his upbringing. The counterfactual considerations of the universalization principle are not restricted to a particular time or place: as far as the normative rightness of their claims is concerned, the participants remain beholden to *potential* objections by anyone *foreseeably* affected by the general observance of a deliberative outcome. Even for the father and his authoritarian sympathizers, the future self-identity and moral character of the child are a recognizable, ethically salient concern.

The intergenerational and intercultural transmissibility of moral principles, through which our convictions face the risk of being exposed as parochial (and possibly injurious) conceits, are a significant theoretical concern for Habermas.[21] As he explains elsewhere in vivid detail, Habermas attributes his own philosophical awakening to having been born into a generation of Germans who witnessed first-hand the dramatic unmasking of their society as a murderous and criminally authoritarian regime.[22] These considerations inform Habermas's response to Putnam—which also raises the issue of intellectual "expertise":

> The morality of equal respect and joint responsibility for everyone can no more be justified from the ethical perspective of a single community concerned about its common good than can human rights or liberal democracy. Putnam, who would like to deal with both under the heading of cognitive realism, likes Dewey's epistemological justification of democracy: "Democracy is not just one form of social life among other workable forms of social life; it is the precondition for the full application of intelligence to the solution of social problems." This argument is a double-edged sword if we consider that Dewey explains intelligent behaviour in terms of the model of scientific method. Contrary to Dewey's own intention, it also allows for a reading according to which a scientific expertocracy is the superior

> form of social organisation. If it is implemented with sufficient discursive structures, the democratic process has epistemic functions also and especially with regard to questions of justice that cannot be resolved either by merely compromising on interests or by appealing to a common ethos.
>
> (Putnam 2003, 235)

What is described in the above passage is a tendency (by no means unique to Putnam) in which particular sociocultural artifacts are claimed to embody the fruits of "intelligent behavior," and thereby become a global standard of sociocultural flourishing and ripeness for improvement. The problem here is not just the familiar worry about ethnocentrism (which Putnam has already addressed), but the problem of an equally overweening *paternalism*.

Consider Putnam's return within the present volume to the work of John Dewey, from whom he borrows the notion of "prizing," as an evaluative term that encapsulates our direct engagement with the objective world (Putnam 2016a, 14–15). The context-dependence of anticipation and satisfaction-in-attainment avoids any positivistic retreat from ethical objectivity (Putnam 2016a, 19). But prizing also relies on our having a certain skill set (devising feasible plans of action, managing our expectations) as well as a particular attunement to the object of our prizing, for which we must first become acculturated.

> Like Aristotle, Dewey believes that the reasons for being ethical are not apparent from a non-ethical or pre-ethical standpoint; one must be educated into the ethical life, and this means that one's interests must be transformed. In that process, Dewey tells us, one does not simply acquire an interest in helping other people alongside of and independent of one's various interests in art, in work, in recreation, etc.; rather all of those interests are likewise transformed. In Dewey's account, the natural impulse of sympathy itself is transformed by being fused with our other impulses, and our other impulses and interests are transformed by being fused with sympathy. As he writes, "What is required is a blending, a fusing of the sympathetic tendencies with all the other impulses and habitual traits of the self. When interest in power is permeated with an affectionate impulse, it is protected from being a tendency to dominate and tyrannise; it becomes an interest in effectiveness of regard for common ends."
>
> (Putnam 2016a, 12)

It is strange to see Dewey's conception of power presented so unproblematically, as though there were nothing controversial about suggesting we can inoculate against authoritarian tendencies by blending its controlling impulse with *affection*—as though a healthily hortatory encouragement of common ends is the most natural result. Be that as it may, what we see here is the route by which philosophical "suggestions" garner a status that goes well beyond a mere "linguistic division of labour."[23] Ethical guidance in early socialization is not administered as a temporary expedient, but rather becomes part of the architecture of selfhood—all of which depends upon a level of normative and epistemic expertise that would make Putnam uncomfortable in any other context. It is not simply a matter of "suggesting" the scientific image and the manifest image "interpenetrate" one another. When philosophers make themselves curators of culutral and intellectual inheritances they create the facts on the ground that decide the trajectory of subsequent "learning processes." The problematically paternalistic nature of this "education into the ethical life" stems not from any quasi-objective "fact" of developmental psychology—it stems from the presumptive authority by which the more "enlightened" disposition is advanced:

> More enlightened values always depend on more enlightened view of the facts as to how things are and as to how things have been in the past, and particularly facts about injustices and their consequences.
> (Putnam 2016b, 129)

Both of the above passages treat "success" in ethical learning as an affirmation of prevailing sociocultural exemplars, with little consideration for the kind of moral progress that requires the critical unraveling of "false" or "oppressive" ideals. My point here is simply that the way ethicists detect and confront pathologies like positivist dichotomy is significantly shaped by their own self-conception as practitioners, and by their willingness or unwillingness to drop the traditional philosophical privilege of "ushering" reason. Leaving philosophical analysis and emancipatory self-reflection undifferentiated (as I am suggesting Putnam tends to do) has the effect of inflating any perceived contamination from "bad metaphysics." That is, if we assume from the outset that philosophical insights make lasting contributions to cultural self-understanding, it will matter a great deal whether the "right" ethical and empirical understandings holds sway. By contrast, for the Habermasian *Platzhalter*, there is no straightforward way of funneling insights from specialist discourses (articulated from a "third-person perspective") into the self-understanding of agents (operating primarily from a "first-person perspective").

My claim in this essay is that the philosophical worry about the persistence of dualistic thinking in ethics is, at bottom, a concern about the way this dualism is being preserved and disseminated. Although I have presented Habermas' postmetaphysical deflation of praxis as a much needed break with traditional, self-aggrandizing conceptions of emancipatory critique, what ultimately needs to be overturned is not any one particular intellectual legacy (whether positivist or Kantian or Neo-Aristotelian) but rather the underlying passive model of acculturation itself. What we need is not another Sellarsian account of "how things in the broadest possible sense of the term hang together in the broadest possible sense of the term"—but rather a new way of defining (and defending) the "edifying" authority of philosophy itself. A new model of "self-cultivation" would require abandoning the historiography of philosophical heroes and villains and the curation of their intellectual reputations; it would also require finding a new pathway for philosophical teaching beyond the cul-de-sac of Habermas's ethical-political abstemiousness. I believe there is still room for a "theoretical" defense of pedagogy made in in the voice of the philosopher, rather than the voice of the concerned citizen or public intellectual. Confronted by a world in which both scientific dogmatism and irrational demagoguery hold sway, it is tempting to invoke the ideal (and ever diminishing capacity) of the university classroom as a place where students are encouraged to discover their own critical voice, rather than parroting that of their professors—where they are invited to experiment with a wider array of critical and rhetorical strategies than are available within other institutional spaces. This is obviously not the place to articulate a new pedagogy that actually manages to avoid terminating in stale platitudes of healthy socialization or cultural enrichment.[24] However, I do think we discover an important starting point in Putnam's earlier description of his seminars with Reichenbach, and of the passionate exploratory spirit that prompted his younger, seemingly more jaded self to overcome his "philosophy is over mood." My contention, that the "spirit" of Putnam's unwearying curiosity is more philosophically valuable than the "technical details" of his realism, is therefore intended as the highest possible compliment.

Notes

1 Regarding the rise of anti-establishment (and often virulently xenophobic) populism in the United States and in Europe, it could just as easily be argued that what we are witnessing is not the consequences of fact/value *dichotomy* (and its privileging of observation statements over "ought" judgments), but rather the *paradiastolic inversion* of "fact" and "value" as meaningful reference points. By this I mean, for example, the *rhetorical* recasting anthropogenic climate change as "liberal dogma," and the redemption of racist paranoia (e.g., "President Barack Obama is a foreign-born Muslim") as suppressed "truth." On the recent blossoming of anti-science, anti-authority sentiment within Republican politics, see: Theda Skocpol and Vanessa Williamson, *The Tea Party and the Remaking of Republican Conservatism* (Oxford University Press, 2012); and Chris Mooney, *The Republican War on Science*, New York: Basic Books, 2005).

2 Although Putnam inserts a further caveat that Quine went too far in dismissing the practical relevance of "trivial" truths and also failed to appreciate the blending of facts and values within "non-trivial" truths (Putnam 2002, 12).

3 "The word 'cruel' obviously—or at least it is obvious to most people, even if it is denied by some famous defenders of the fact/value dichotomy—has normative and, indeed, ethical uses" (2002, 34).

4 "[N]evertheless, to show that . . . the foundations of the idea that there is no rationality beyond purely instrumental rationality are in trouble, *may help* to combat that instrumentalisation and that manipulation." (Putnam 1994, 151, emphasis added).

5 See Hilary Putnam and Ruth Anna Putnam, "Dewey's Logic: Epistemology as Hypothesis," (Putnam 1994, 201–5).

6 Lionel Robbins, *An Essay on the Nature and Significance of Economic Science* (London and New York: Macmillan, 1932), 16.

7 See *The End of Value-Free Economics* (eds.) Hilary Putnam and Vivian Walsh (London and New York: Routledge, 2012).

8 See Martha Nussbaum, *Creating Capabilities: The Human Development Approach* (London and Cambridge, MA: The Belknap Press of Harvard University Press, 2011), 33–45.

9 Most of the initial series of exchanges are collected in: T.W. Adorno, et al., *The Positivist Dispute in German Sociology*, trans. Glyn Adey and David Frisby (London: Heinemann, 1976).

10 See Popper, *The Logic of Scientific Discovery* (New York and London: Routledge, 1959), 12–14.

11 "[A]lthough it is impossible to separate scientific work from extra-scientific applications and evaluations, it is one of the tasks of scientific criticism and scientific discussion to fight against the confusion of value-spheres and, in particular, to separate extra-scientific evaluations from *questions of truth*" (Popper 1976, 96–7).

12 "The split which exists in present social life between ideas and emotions, especially between ideas that have scientific warrant and uncontrolled emotions that dominate practice, the split between the affectional and the cognitive, is probably one of the chief sources of the maladjustments and unendurable strains from which the world is suffering. I doubt if an adequate explanation upon the psychological side of the rise of dictatorships can be found which does not take account of the fact that the strain produced by separation of the intellectual and the emotional is so intolerable that human beings are willing to pay almost any price for the semblance of even its temporary annihilation." (John Dewey, "Theory of Valuation," in *International Encyclopaedia of Unified Science (Volume II, No. 4)*, ed. Otto Neurath, et al. (Chicago: The University of Chicago Press, 1939), 65.)

13 See C. S. Peirce, *Collected Papers*, ed. Hartshorne and Weiss (Cambridge, 1960), Vol. V: "Lectures on Pragmatism," (5.27) and "How to Make Our Ideas Clear," (5.398).

14 Habermas, *Knowledge and Human Interests*, trans. Jeremy J. Shapiro (Boston: Beacon Press, 1971), 195–213. The definitive introduction to Habermas' social theory leading up to *The Theory of Communicative Action* (which he also translated into English) remains: Thomas McCarthy, *The Critical Theory of Jürgen Habermas* (Cambridge and London: The MIT Press, 1978).

15 See, for example, Cristina Lafont, "Moral Objectivity and Reasonable Agreement: Can Realism be Reconciled with Kantian Constructivism?" *Ratio Juris* 17 (2004); and Gordon Finlayson, "Habermas's Moral Cognitivism and the Frege-Geach Challenge," *European Journal of Philosophy* 13:3 (2005).

16 See Habermas, *The Philosophical Discourse of Modernity*, trans. Frederick G. Lawrence (MIT Press, 1990).

17 "Philosophy can rely neither on a knowledge of salvation nor on clinical knowledge and therefore cannot provide 'advice for how to live' [*Lebenshilfe*] in the way that either religion or clinical psychology can. In the form of ethics, it can offer guidelines for how to attain a reasonable understanding of one's identity, of who one is and who one would like to be. However, today the 'therapeutic' role of philosophical ethics consists at best in encouraging people to lead their lives *consciously*. Philosophical 'advice' remains ascetic when it comes to demands for 'making sense of life'; the responsibility for reflecting on the meaning of a person's life has to remain with that person" (2003, 289).

18 See, for instance, the example of workers on a construction site having various aspects of their social hierarchy and job security rendered problematic through the force of their own interactions: Habermas, *The Theory of Communicative Action Volume II: Lifeworld and System: A Critique of Functionalist Reason*, trans. Thomas McCarthy (Boston: Beacon Press, 1987), 121.

19 See Habermas, "From Kant's 'Ideas' of Pure Reason to the 'Idealising' Presuppositions of Communicative Action: Reflections on the Detranscendentalized Use of Reason," (Habermas 2003, 83–130).

20 "[A]t bottom [Habermas'] desire to treat all value discourse outside the narrow limits of discourse ethics as mere negotiation of differences between "life words" [*sic*], and also the reason that he fears conceding any objectivity that goes beyond this to such value discourse—namely that such a concession would not be compatible with "modernity" (meaning here the modern suspicion of everything that is supposed to be "metaphysical")—are, at bottom, positivistic desires and reasons" (Putnam 2002, 133)

21 Most recently discussed in Habermas, "European Citizens and European Peoples, The Problem of Transnationalizing Democracy," in *The Lure of Technocracy*, trans. Ciaran Cronin (Cambridge: Polity, 2015), 29–46.

22 See Habermas, "1989 in the Shadow of 1945: On the Normality of a Future Berlin Republic," in *A Berlin Republic: Writings on Germany*, trans. Steven Rendall (University of Nebraska Press, 1997), 164.

23 Putnam, "The Meaning of 'Meaning'," *Mind, Language, and Reality* (Cambridge University Press, 1975), 225–9.

24 Danielle Allen has suggested something along these lines through her interpretation of Aristotle, which focuses on the distinction he makes in *Nicomachean Ethics* between the comprehension of "the student" and the comprehension exercised by "the judge." Whereas the former inevitably involves *leading* others towards "truth," with the latter we get a rare glimpse of pedagogical restraint:

"Comprehension and prudence are not the same thing . . . for prudence is characterised by the giving of commands: its end is what one ought or ought not to do. But comprehension is characterized by judgement [*kritikē*] . . . And comprehension is neither the possession of prudence nor the gaining of it. Rather, just as "learning" is said to be comprehending, whenever it makes use of science, so comprehension is said to consist in *making use of opinion to render a decision about what someone else says* . . . . and rendering such a decision nobly. For doing something well is the same as doing it nobly. And from this, the name "comprehension"—in reference to which we speak of those who are of good comprehension—has arisen, namely, from the comprehension involved in learning. For we often say "learning" when we mean comprehending." [Aristotle, *Nicomachean Ethics*, trans. Robert C. Bartlett and Susan D. Collins (The University of Chicago Press, 2011), Book VI: 10, 1143a.] This approach to Aristotle would not be an attempt to usher an audience towards "truth" but rather to encourage the exercise of their own judgment. See Danielle S. Allen, *Talking to Strangers: Anxieties of Citizenship since Brown v. Board of Education* (University of Chicago Press, 2004).

References

Adorno, Theodor. W. (ed.). 1976. *The Positivist Dispute in German Sociology* Translated by Glyn Adey and David Frisby. London: Heinemann.

Albert, Hans. 1976. "The Myth of Total Reason." in T. W. Adorno, Dahrendorf, J. Habermas, H. Pilot, & K. Popper, eds., *The Positivist Dispute In German Sociology* (163–197). London: Heinemann.

Dewey, John. 1939. "Theory of Valuation." In *International Encyclopaedia of Unified Science*, Vol. II, No. 4, edited by Otto Neurath, et al. Chicago: The University of Chicago Press.

Habermas, Jürgen. 1971. *Knowledge and Human Interests.* Translated by J. Shapiro. Boston: Beacon Press.

Habermas, Jürgen. 1973. "Some Difficulties in the Attempt to Link Theory and Praxis." In *Theory and Practice*, translated by John Viertel (1–40). Boston: Beacon Press.

Habermas, Jurgen. 1976. "A Positivistic Bisected Rationalism", in T. W. Adorno, Dahrendorf, J. Habermas, H. Pilot, & K. Popper, eds., *The Positivist Dispute In German Sociology* (198–225). London: Heinemann.

Habermas, Jürgen. 1990. "Philosophy as Stand-In and Interpreter." In *Moral Consciousness and Communicative Action.* Cambridge: Polity Press.

Habermas, Jürgen. 1992. *Postmetaphysical Thinking.* Translated by William M. Hohengarten. Cambridge, MA: MIT.

Habermas, Jürgen. 1998. *The Inclusion of the Other.* Edited by Ciaran Cronin and Pablo DeGrieff. Cambridge, MA: MIT.

Habermas, Jürgen. 2003. *Truth and Justification.* Translated by Barbara Fultner. Cambridge, MA: MIT Press.

Nussbaum, Martha. 2011. *Creating Capabilities: The Human Development Approach.* London and Cambridge, MA: The Belknap Press of Harvard University Press.

Popper, Karl. 1959. *The Logic of Scientific Discovery.* New York and London: Routledge.

Popper, Karl. 1976. "The Logic of the Social Sciences.", in *The Positivist Dispute in German Sociology*, translated by Glyn Adey and David Frisby.

Putnam, Hilary. 1975. *Mind, Language, and Reality.* Cambridge: Cambridge University Press.

Putnam, Hilary. 1981. *Reason, Truth, and History.* Cambridge: Cambridge University Press.

Putnam, Hilary. 1994 *Words and Life.* Edited by James Conant. Cambridge and London: Harvard University Press.

Putnam, Hilary. 1999. "The Vision and Arguments of a Famous Harvard Philosopher." In *Key Philosophers in Conversation (The Cogito Interviews)*, edited by Andre Pyle, 44–54. London and New York: Routledge.

Putnam, Hilary. 2002. *The Collapse of the Fact/Value Dichotomy.* Cambridge and London: Harvard University Press.

Putnam, Hilary. 2003. "For Ethics and Economics without the Dichotomies", *Review of Political Economy*, 15(3): 395–412.

Putnam, Hilary. 2004. *Ethics without Ontology*. Cambridge and London: Harvard University Press.

Putnam, Hilary. 2016a. "The Fact/Value Dichotomy and the Future of Philosophy." In *Facts and Values: The Ethics and Metaphysics of Normativity*, edited by Giancarlo Marchetti and Sarin Marchetti. Routledge.

Putnam, Hilary. 2016b. "Realism", *Philosophy and Social Criticism*, 42(2): 117–31.

Putnam, Hilary and Walsh, Vivian (eds.). 2012. *The End of Value-Free Economics*. London and New York: Routledge.

Robbins, Lionel. 1932. *An Essay on the Nature and Significance of Economic Science*. London and New York: Macmillan.

Part II

Varieties of Entanglement

5 Reflections Concerning Moral Objectivity

Ruth Anna Putnam

For the last 25 years, or so, I have attempted, in various writings, to undercut the so-called fact-value dichotomy (Putnam 1985, 1987, 1998). In particular, I have claimed that the distinction between facts and values cannot bear significant ontological or epistemological weight, that, in an important sense, facts and values are equally objective.

At any rate, that is the view I meant to defend. However, it seems to me now that my arguments, in the first of those papers, could have been taken to defend, rather, the view that facts and values are equally subjective. For I was then trying to adapt Nelson Goodman's idea of world-making to the case of moral values (cf. Goodman 1978). Again, in the second of those papers, I argued that factual statements and value judgments are so intertwined in our practice, so dependent on each other that they form a seamless web. Here again, while I concluded that the acknowledged objectivity of factual statements entailed the objectivity of the relevant value judgments, a hostile critic could say that the subjectivity of value judgments infects our factual statements as well. Only the third of the papers mentioned seems to me immune to this inversion of its argument. Since it seems to me to be of great importance, for reasons I shall mention in due course, to defend the objectivity of (some) moral judgments, I shall, in this paper, make another attempt to do so.

1

I shall avoid speaking of facts and values; more precisely, I shall avoid speaking of facts. The notion of a fact, or the associated notion of a factual statement, is hopelessly fuzzy, quite aside from any possible relation to values. For example, facts are contrasted with theories. But where are we to draw the line between a theory and the facts it explains? Those who defend the teaching of neo-Darwinism in US high school biology classes sometimes say, "Evolution is a fact," while their opponents will counter with, "Evolution is a (mere) theory." Someone else might say that evolution is a fact that is explained by neo-Darwinian theory. Yet one could also say that fossil remains are a fact that is explained by evolution, or that to call certain objects fossils is to explain what they are. "Fossil" is a theory-laden term. Whether we classify an item as fact or as theory depends on our purposes in doing so as well as on our confidence in the item's truth.

The facts that scientific theories explain are, as C.S. Peirce pointed out, not particular events occurring at a particular place and time but what he called "phenomena," generalizations concerning repeatable events. Neither Newton's theory of gravitation nor Einstein's explains why a particular apple fell from a particular tree at a particular time; they explain "things falling." Yet when a judge admonishes a witness to report only the facts, the facts to be reported concern the particulars of a particular event in so far as the witness has direct knowledge of them. What the witness must not do is to present her inferences from or her attitudes to what she has observed. Facts in this sense are observed particulars.

Again, while the facts are what they are regardless of what we believe them to be, factual statements are what we say/believe about the facts. Even the simplest perceptual judgment—one kind of paradigmatic factual statement—already involves what Peirce called an abductive inference; even the simplest perceptual judgment is a hypothesis concerning future percepts. Or consider this. At a social occasion people speculate whether Susan has been offered a certain job and whether she would accept the offer were it forthcoming. Someone joins the group and says, "In fact, Susan got the offer but declined it." Here facts are opposed to conjectures. No doubt other examples could be found. I draw no conclusions from these examples; I merely suggest that it might be wise to eschew talking about facts or factual statements as if these notions were clear. I fully acknowledge that I myself have been guilty of doing just that.

In contrast, the notion of value is enormously rich. Various kinds of things (objects, states of affairs, character traits, etc.) have value, are valued, are evaluated, etc. Value may be monetary, moral, social, epistemic, aesthetic, etc. None of these lists are exhaustive. My aim here is not to defend the objectivity of all value judgments; as will be apparent in what follows objectivity is an aim to which we aspire and for which we strive but which we often fail to attain. Moreover, while objectivity is itself something we often value, there are occasions when striving for objectivity would be boorish, insensitive, or downright cruel. For example, one should not judge a young child's drawing by the standards of a work of art. Finally, I am concerned here only with moral value. I shall attempt to defend the view that moral judgments can be objective. But just what does that mean?

There are, I think, two ways to clarify the claim. One might choose some field of knowledge, claim that it provides a paradigm of objectivity, try to become clear on the sources or nature of that objectivity, and then examine other fields of (purported) human knowledge, in this case morality, to see whether they match the paradigm, or, if not, how they fall short. Since Descartes, one such paradigm has been provided by scientific theories, in particular by physics. Another paradigm, which may have been in the back of the minds of philosophers as diverse as Plato and Kant, is mathematics.

The second approach, the approach that I favor, begins by asking why we want objectivity. And here the answer may depend upon what area of human activity we have in mind. Notice that I said, "activity," whereas the first approach spoke of fields of knowledge. Knowledge, that is to say, belief, is involved in most human activities. Moreover, some activities involve the development (discovery or creation) of knowledge as their immediate aim, and some of that knowledge is only indirectly relevant to other human activities. However, in the vast majority of human activities, belief and action cooperate in bringing about some state of affairs other than knowledge. To be sure, the search for knowledge may play an important role on the way as well as knowledge being a byproduct of the process. In so far as that is the case, the pragmatic approach that I favor agrees that inquiry in the empirical sciences provides us with a paradigm of objectivity.

As an example of the sort of human activity and accomplishment I have in mind let us take the eradication of smallpox. The story is fascinating. Sometime in the latter part of the eighteenth century a surgeon's apprentice named Edward Jenner overheard a milkmaid saying, "I shall never get smallpox because I have had cowpox." Indeed, that cowpox provided immunity from smallpox was widely known among the rural population. It was also widely known that those who survive smallpox become immune to it. By the latter part of the eighteenth century it was not uncommon to seek protection against the horrible disease by deliberately infecting oneself with a mild case, a procedure called variolation, although the risks were not negligible. The remark of the milkmaid set in motion the events that led ultimately, after nearly 200 years, to the eradication of the disease. In 1796 Jenner vaccinated a boy with cowpox and later with smallpox. The boy showed no reaction to the latter. Vaccination using cowpox fluid spread rapidly throughout Europe and North America. As early as 1802 Jenner foresaw the ultimate eradication of smallpox as the result of vaccination. The World Health Organization declared on May 8, 1980 that the disease had been wiped out. [1]

2

Fascinating as this story is, what interests me here is that Jenner demonstrated, using scientific methods, the soundness, or we might say, the objectivity, of the folk belief in the protective power of cowpox. I could have said that he verified the folk belief, or more cautiously and accurately, that his experiment with the boy corroborated the folk belief, and that the subsequent history provided more than ample verification. We are certainly entitled to say that. But I want to look at the situation from a slightly different perspective. When one speaks of a folk belief, one suggests one of two things, either folk wisdom (a good thing, a useful thing, something one should take seriously for it is based on long experience) or old wives' tales (superstitions, something not to be taken seriously). The question confronting Jenner was whether the widespread belief in the protective power of cowpox was the former or the latter. A belief, as Peirce said crediting Bain, is something one is prepared to act on. In one sense, Jenner was prepared to act on the folk belief; he was willing to risk a boy's life on the experiment inspired by the milkmaid's remark. In another sense, Jenner was not prepared to act on the milkmaid's belief. It was not yet his belief; he was not prepared to vaccinate himself with cowpox fluid and then take himself to be safe from smallpox. The experiment with the boy changed that. Thereafter, he actively sought volunteers, persuaded other physicians to vaccinate their patients, etc. The status of the belief had changed; Jenner was convinced of its truth, as were many other people, because it had been scientifically tested.

Because we act on our beliefs, I am suggesting, we want our beliefs to be true; we want, to use an expression of William James's, that they carry us prosperously to where we want to go (James 1975, 34). And we realize that certain ways of arriving at our beliefs, of fixing them, as Peirce said, make it more likely that our beliefs are true than do other ways. Beliefs arrived at by such methods are objective. We take it for granted that the methods used in the various sciences are such methods. In so far as we arrive at or justify beliefs in everyday life in similar ways, we take such beliefs also to be objective. What are these methods or ways?

In "The Fixation of Belief," Peirce characterized the method of science as a method "by which our beliefs may be determined by nothing human, but by some external permanency—by something upon which our thinking has no effect." It must be something capable of affecting any person. "And," Peirce continues, "Though these affections are necessarily as various as are individual conditions, yet the method must be such that the ultimate conclusion of every man shall be the same" (Peirce 1931–5, 5.384).[2] In "How to Make our Ideas Clear," Peirce tells us that all scientific investigators are "animated by a cheerful hope" that scientific investigation pursued long enough will lead to one answer for each question, and that "This great hope is embodied in the conception of truth and reality. The opinion which is fated to be ultimately agreed to by all who investigate, is what we mean by truth, and the object represented in this opinion is real" (Peirce 1931–5, 5.407).

Although Peirce did not speak of objectivity, he suggests certain characteristics of inquiries that promise to lead to objective beliefs. Our belief must be determined by "an external permanency," and that is glossed as "nothing human" and as something "upon which our thinking has no effect" and "such that the conclusion of every man shall be the same."

I want to comment briefly on the last point. What is the connection between intersubjectivity and objectivity? The first thing to note is that the mere fact that a belief is widely shared is no guaranty of its truth or of its objectivity, for the belief may be due to widespread, successful indoctrination. Peirce had already warned against this mode of fixing belief; he called it the method of authority. Now, being concerned to note the salient features of the scientific method, he draws attention to intersubjectivity as a necessary condition. In the sciences, this is exemplified by the requirement that an experiment be repeatable. Any person with the requisite training should be able to reach the same experimental result and thus the same conclusion. It is more difficult to say what this requirement comes to in everyday life. To be sure, we distrust the accounts of witnesses to an event, if these accounts diverge. Yet we also know that such divergence is a common occurrence. We distrust accounts that are highly emotional, for we know that an observer who does not share the emotion may give a quite different account. Yet, a failure of empathy may also result in an inadequate account. In any case, Peirce makes a different point; he affirms his cheerful hope that inquiry sufficiently diligent, sufficiently prolonged will lead to intersubjective agreement. The crucial notion here is inquiry, not agreement but using a method that will lead to agreement. He says "fated" to lead to agreement just as we are fated to die. What is that method?

We have beliefs; many, perhaps most, of these beliefs are not chosen by us; we might say that they are thrust upon us. Our interaction with our environment causes us to have beliefs. I believe that we contribute to the content of these beliefs, and over that contribution we have some control. Precisely because we have some control over the content of our beliefs, the question of objectivity arises. We want our beliefs to be objective (maybe not all the time, but for the most part) because we act on them, and we do not want to experience unpleasant surprises. Because of the manner in which we arrive at them, objective beliefs are more likely to lead us where we intend to go.

I said that we contribute to the content of our beliefs, and that does not seem to fit Peirce's remark that we want our beliefs to be determined by nothing human. We contribute to the content of our beliefs in two ways: our interests determine what we investigate, and, within limits, we choose how to understand, how to think about, what we experience, that is, we choose, within limits, our concepts. When I say that our interests determine what we investigate, and hence what we come to know, the "we" in question may be as individual as the speaker, as when one chooses a research project. On the other hand, because of the way scientific research is funded, "we" may refer to funding agencies, whose interest or lack thereof determines the course of scientific research. Or "we" may refer to human beings in general, to the nameless millions who have shaped human civilization by their inquisitiveness and inventiveness. What I have just said is, I believe, entirely compatible with Peirce's demand that our beliefs be determined by nothing human. Our interests legitimately determine the subject matter of our beliefs—what we investigate—but not the result of those investigations.

Although Peirce called himself a scholastic realist, he did not think that one finds concepts as one finds pebbles on a beach. In some sense human beings "make" or develop concepts. And we find that some conceptual schemes work better than others. William James pointed out that the commonsense conceptual scheme of time, place, cause, thing, etc., has been developed by humanity through the ages and survives because it continues to be useful. But James thought that we might have developed a different scheme, that the commonsense scheme, the various scientific schemes, and perhaps yet other schemes are each good for its purpose, but that none has a claim to be how the world really is. Peirce, I believe, thought that scientific investigation indefinitely prolonged will lead to a unique conceptual scheme. Peircean Thirds—his term for universals that correspond to this scheme—would then be among the external permanencies that determine our beliefs, if we use the scientific method.

I have failed to mention a crucial aspect of the methods of inquiry—I use the plural deliberately—that lead to objective beliefs. These methods are all fallibilist. Beliefs are caused by our interaction with our environment, that is, by experience; beliefs so caused are subject to revision and even outright rejection by subsequent interactions with the environment. That is the basis of Peirce's cheerful hope. No matter how deeply you and I may disagree, if we both revise our beliefs in the light of our experiences, then, since we both experience the same world, sooner or later (perhaps only much later) we, or our descendants, will come to a shared belief. That cheerful hope will have to be stated in a somewhat attenuated form if one accepts Jamesean conceptual pluralism. But such niceties need not concern us now. Finally, I need to mention that there are many kinds of belief. To achieve objectivity in, say, historiography, is more difficult than it is in physics. Even the notion of objectivity relevant to accounts of the past may be somewhat different from that relevant in, say, engineering. But to discuss these matters would carry us too far afield. It is time to apply what we have learned about the objectivity of scientific beliefs to value judgments.

3

If beliefs are what we are prepared to act on, valuations are what guide our activities, what determine together with other factors where we intend to go. Consider again the case of Jenner. Everyone in those days was afraid of smallpox; anyone would have been interested in hearing how one could become protected against the dread disease. Jenner planned a career in medicine and had already been taught to think in terms of experiments. In other words, Jenner feared smallpox, as did everyone; unlike many of his contemporaries, he thought scientifically. The negative value of smallpox, his scientific mindset, and the urgency of the situation (the ever present threat of smallpox and its high fatality rate) prompted Jenner's experiment with the boy and his later activities. Quite early on he foresaw as a desirable ultimate result the total eradication of smallpox. That goal, or value, was to guide the actions of vast numbers of people.

I shall use the awkward term "valuation" rather than the familiar term "value" to make clear that it is the values one has, what one values, that guides one's action, not some value existing somewhere, say, in some transcendent realm. Among the objections Aristotle raised against Plato's doctrine of the Form of the Good was the following: it is not the sort of good a human being can pursue in action or possess; but that is just the sort of good we are looking for in our present inquiry (Aristotle 1985, 1096b, 12). I take it that Aristotle meant that only what *can be* an internal reason can be a subject of ethics. I want to emphasize that only what *is* an internal reason, what is one's own valuation, can guide one's action.

Some of our valuations are fleeting, enabling us, for example, to decide what meal to order in a restaurant on a particular occasion. Some of our valuations are norms that constrain our behavior again and again; among these are the moral principles by which we attempt to live. Some of our valuations shape large parts of our lives; one's choice of a career or of a life partner may serve as examples. From a slightly different angle, we note that some of our valuations are immediate affective responses, while others emerge after careful reflection. Finally, to repeat, there are many kinds of values, though the defenders of the fact-value dichotomy had mainly moral values in mind. On the other hand, those of us who deny that there is a dichotomy tend to emphasize the role played by judgments of epistemic value (coherence, relevance, reliability, objectivity, etc.) guiding the pursuit of scientific knowledge.

One is tempted to say that we act on our beliefs but are guided by our valuations. But valuations may be stated as propositions that we believe. For example, I believe that the economic disparities between United States citizens are unconscionably large. Because of this belief, I favor certain policies and oppose others, or, again, one might say that I believe that certain policies will move the country to greater economic justice while others will lead to yet greater injustice. Finally, acting on these beliefs, or guided by these valuations, I will vote for this candidate rather than that one. If the belief/valuation distinction mirrors the fact/value distinction, then, once again, one sees that the distinction, though a useful rhetorical device, will not bear much weight.

Several points need to be made clear before I move on to the question of the objectivity of moral judgments. First, while it is generally true that one's valuations will show themselves in one's behavior, this is not always the case. Of course, if I say that I love lobster, but I never eat it, you will conclude that I was joking or lying. But you may discover that I have decided to obey the dietary restrictions of my religion and that I therefore refrain from eating lobster although I remember how much I enjoyed it in the past. Even a sincere commitment to a moral ideal, say, economic justice, may not express itself in any behavior that differs from that of an adherent of laissez faire capitalism, either because one is too weak-willed, or because one thinks that one's efforts would be in vain, or because no opportunity presents itself. Nevertheless, generally our actions are guided by (some of) our beliefs and by (some of) our valuations. Generally, one's behavior is considered more rational the more it is so guided, the more one's beliefs, one's valuations, and one's actions form a coherent whole.

Second, while value judgments can be formulated as propositions, many valuations are expressed in actions without ever being entertained in propositional form. The infant seeks the breast; suddenly assailed by a foul smell one attempts to hold one's breath; amused, one stops to watch a kitten at play. Compassion on the one hand and moral outrage on the other may show themselves in actions that appear instinctive and may indeed prove to have been unwise because they were literally thoughtless. It is, however, worth remembering that we have a rich vocabulary in which to express our enjoyments and sufferings, our positive or negative reactions. We can and do think about these matters, and therein lies the possibility of objectivity.

Finally, following John Dewey, I want to distinguish between valuing and evaluating. Having stopped to watch a kitten at play, I remark that one ought more often to take the time to enjoy such happenings. I have evaluated my valuing (stopping to watch) the kitten's play. Or, we might say not only that I was amused but also that the kitten's play is amusing. The latter, we might say, evaluates my amusement, finds it appropriate. "The kitten's play is amusing," is a value judgment. We can ask whether it is objective. What would that mean? I am not sure; I am not sure the claim that a kitten's play is amusing is sufficiently contentious or sufficiently important in the larger frame of our morality to warrant speaking of objectivity. Let us turn, therefore, to weightier matters.

Why do we want our value judgments to be objective, and what process leads us to regard a value judgment as objective? There is no simple answer, for there are many different kinds of value judgments. But there is one uncontroversial case. We want our judgments of instrumental value to be objective, and non-cognitivists of all stripes agree that judgments of instrumental value have truth values, hence, like other beliefs, they can be objective. Like other beliefs we want them to be objective because we act on them, and we want our actions to promote our goals.

The judgment that cowpox protects against smallpox is a judgment of instrumental value, and as we have seen, it became objective when tested. I am not identifying truth with objectivity. It was always true that cowpox protects against smallpox, but folk wisdom was not sufficiently objective to base the treatment of fine gentlemen and ladies upon it. On the other hand, a belief that has all the earmarks of objectivity may turn out to be false. Fallibilism—preparedness to revise one's beliefs in the light of contrary evidence and, on appropriate occasions, the deliberate search for such evidence—is fundamental to all methods of inquiry that lead to objectivity.

Judgments of instrumental value, for example, that inoculation with cowpox is advisable, may be understood in two ways. When even non-cognitivists admit that such judgments have truth values, they read them as incomplete conditionals. "If one wants immunity from smallpox, one ought to be inoculated with cowpox." Or as simple causal statements, "inoculation with cowpox provides immunity from smallpox." But, so understood, judgments of instrumental value do not seem to be "action guiding," though one may of course act upon them, as one may act upon any belief.

But a judgment of instrumental value may be understood in a more interesting way, namely as claiming that something is a means to an end that is good. Where the goodness of the end is uncontroversial, the end is often not mentioned. Thus, the elderly are advised to get flu shots. That health, and specifically not getting flu, is good is not mentioned; it is understood. But if judgments of instrumental value are understood in this second way, their objectivity depends, in part, on the objectivity of the judgment that the end in question is good.

4

I believe that the twentieth-century obsession with metaethics, the nearly universal acceptance of the fact-value dichotomy—note that it is accepted by non-naturalist moral realists as well as by non-cognitivists—reflects the assumption that judgments concerning the value of ends are not known by empirical methods. Hence arises the question whether value judgments concerning ends can be objective.

I shall approach my answer somewhat indirectly. Following John Dewey, I shall defend the view that we can and do reason about ends. Thus, it is at least in principle possible that a judgment concerning the value of an end is objective. I am, of course, aware of the standard view, the view that knowledge of instrumental goods is entirely different from knowledge of things good-in-themselves, if there is such knowledge. The standard view often appeals to Aristotle's claim that we do not deliberate about ends. But the *Nicomachean Ethics* is a sustained argument for a certain kind of life—*eudaimonia*—as one's ultimate end. Moreover, Aristotle was well aware of the fact that the life worth leading is made up of many things, each by itself worth having or doing though perhaps also conducive to other goods. In living one's life, one has to balance these against each other, since, as William James points out, in this poor world we cannot realize all our ideals, all our hopes. Again, Aristotle's practice and, more importantly, *our* practice belie the claim that we do not deliberate, that is, reason about ends. And when we reflect upon such reasoning, we see that considerations of means and of ends, Dewey preferred to speak of "ends-in-view," influence each other. He also pointed out that one's ends are not endings. Once achieved, one's goals become part of the conditions of the rest of one's life.

Earlier I mentioned intersubjectivity as a necessary though not a sufficient condition of objectivity. How is it with judgments concerning ends? Certainly, we want some of our ends to be shared, many important ends—peace, economic justice, protecting the environment, and so on—require the cooperation of many, of virtually all, human beings. Sharing such ends means judging them to be worthwhile, to be attainable through cooperative efforts. It means widespread agreement concerning the means by which these ends are to be achieved. It means seeing each one of these ends as parts of an overall ideal that we may call, with Aristotle, human flourishing. Agreement on all these points and many more that I have not mentioned is based on beliefs arrived at by methods of inquiry that are, as already discussed, based on experience and are, speaking somewhat loosely, democratic. For in these cases, benefits and burdens will be distributed. Unless they are distributed fairly, agreement will not be stable. But the distribution will not be fair unless all voices are heard. William James wrote, "there can be no final truth in ethics any more than in physics, until the last man has had his experience and said his say" (James 1979, 141). The reference to physics, and thus to Peirce's definition of truth as the final opinion, reminds us that in ethics as in physics, we are to be fallibilists. We are to be prepared, as James says later in the same essay, to listen to the "cry of the wounded," to modify our ideals if, in their pursuit, we find ourselves creating havoc, sacrificing too many other things we or others also value. But in the case of ethics, or politics, I hear in that phrase "the last man" in particular a reference to the poor and the powerless, to those whose voices tend to go unheard. James warns not only against fanaticism but also against arrogance. When agreement concerning shared ends is the result of democratic deliberation based on objective information, those ends are objectively good. That is to say, the revisable judgment that they are good is objective.

There are, however, ends that do not fit so easily into the pattern of objectivity derived from scientific beliefs. I do not want everyone to love the man I love. Of course, I want my friends, people whose judgment I respect, to think well of him, but the opinions of the rest of humanity do not concern me. When one chooses a career, one wants one's choice to be realistic given one's talents; one does not want everyone to choose the same career nor even that everyone approve of one's choice. When one plants a garden, one hopes passers-by will find the sight pleasing, but one's main concern is to please oneself. There is, of course, room for objectivity, for one's deliberations about such ends involve considerations of resources and obstacles, of costs and benefits, and of future consequences, of how this end fits into the pattern of one's life. But, I am suggesting, the question, "Is this objectively good?" or anything like that question strikes me as misguided.

Of course, the man I love may turn out to be a scoundrel, or a bore, or I may simply cease loving him, or he may not return my love. Such tragedies may indicate that I am a bad judge of men, or of myself. I may learn from my experience. I may say that love blinded me to the man's faults and that now I see him objectively. But as a philosopher, I find such talk misleading for two reasons. First, as Iris Murdoch points out, it takes a loving gaze to see a person's characteristics clearly (Murdoch 1971, 17ff). Of course, she did not mean romantic love, but romantic love may lead to the love that enables one to see clearly. Second, when it comes to human relationships, or to the choice of a career, or even to the planting of a garden, one must, as William James points out, act in advance of the evidence. One must make a commitment and act on that commitment. To wait until one can judge the choice to be objectively good or wise is to forego the chance to make that choice. The man will find another woman to love; one will have missed the chance to acquire the knowledge or skills necessary for that career; summer will have passed and the garden been overrun by weeds.

I am suggesting that deliberation itself has costs, that at a certain point—and this is equally true of the social, or moral, goals I mentioned earlier—we must act, or forego the opportunity to pursue that goal. But that does not mean that, at that point, we must become fanatics. A deep and abiding commitment is compatible with, and should be accompanied by, clear sightedness. Our goals are not laid up in heaven, nor do they await us at the end of much effort just as we had imagined them at the beginning. Our ends-in-view change as we struggle to realize them—one may find out that one is better suited to a somewhat different career, or that the soil or climate is not conducive to growing this plant but excellent for that other one. Even if one is satisfied with one's accomplishment, life continues and sets new tasks, new ends-in-view.

Here someone may object that I have said nothing about ethics properly so called, although I have gestured at times in that direction. What matters, the objector may say, is not whether value judgments in general are objective; there are, indeed, many different kinds of value and nothing very helpful can be said about them as a class. What matters is whether moral judgments are objective.

We need moral norms, norms of behavior that are taken seriously, for we are social animals. We can only be fully human living with other humans. They nurture us when we are young and helpless; they give us language and they socialize us, as we in turn nurture our offspring and socialize them. But precisely because we are aware of the social origins of our own morality, we are inclined to raise questions. Is this really the best way to behave? Is this the kind of person I want to be? (Prone to these emotions, cultivating those virtues, doing this sort of thing). Can we do better as a society, coming closer in reality to the ideals we profess? History teaches us that norms change; anthropology teaches us that different societies differ in many of their mores. Yet there seems to be a common set of concerns, and in today's world no culture is radically cut off from the influence of others.

We cannot, we should not wish to avoid making moral judgments. Our own flourishing and that of those belonging to us in ever wider, ever more attenuated circles depends on our not sticking blindly to the morality we were taught. Moral progress as well as the very fitness of our morality to our situation depends on our willingness to reflect. And, indeed, even the most conservative and even the most fanatic (not the same thing) adherent of a system of morality cannot avoid making judgments. Because they concern us so intimately, we desperately want our reflective moral judgments to be right, yet for the same reason, it is difficult to avoid the personal perspective. Hence philosophers have suggested various methods by which to assure that our moral judgments are objective. Kant demands that we ask ourselves whether we can universalize the maxim of our action, or whether we respect the dignity of each human being. Rule Utilitarianism exhorts us to count each individual for one as we seek rules that, if generally obeyed, will maximize happiness. John Rawls proposes that we seek a reflective equilibrium between our principles and our considered judgments in particular cases. These methods, and others, were meant to be, and I believe can be, aids toward adopting a point of view that others can share.

We need a point of view that others can share, for the point of morality is to enable us to flourish jointly. "We," as I already suggested, sometimes refers to a small circle of our dearest friends and closest relatives, at other times it includes all humanity, and it often falls somewhere in between. Imagining the goal of human flourishing requires a wider perspective than the point of view of a single individual, and a lone human being cannot flourish. Moreover, the wider perspective from which we must reflect on moral questions cannot be a purely intellectual one. John Dewey pointed out that moral judgments are a species of value judgments, that moral value is found in "objects and activities which develop out of natural desires and normal social relations in family, neighborhood, community" (Dewey 1985, 263). He warned that we must not rely on immediate emotional responses or so-called "intuitions." He objected that intuitionism tends to be allied to conservatism, and insisted that even the strongest conviction concerning one's duty needs to be "revitalized by consideration of its meaning—that is, of the consequences which will accrue from acting upon it." (Dewey 1985, 268). Nevertheless, he concluded,

> To put ourselves in the place of others, to see things from the standpoint of their purposes and values, to humble, contrariwise, our own pretensions and claims till they reach the level they would assume in the eye of an impartial sympathetic observer, is the surest way to attain objectivity of moral knowledge. Sympathy is the animating mold of moral judgment not because its dictates take precedence in action over those of other impulses (which they do not do) but because it furnishes the most efficacious *intellectual* standpoint . . . Through sympathy the cold calculations of utilitarianism and the formal law of Kant are transported into vital and moving realities.
>
> (Dewey 1985, 270)

"Sympathy is the animating mold." Intersubjectivity turns out, as in the case of beliefs, to be a necessary condition of objectivity. But in the moral case, it plays a more central role. Once again others must be able to go through the same process of reasoning that we have gone through arriving at the same conclusion. But in this case, in order for others to be able to do that, our reasoning must already take their interests seriously as *their* interests, not merely as potential resources or obstacles to achieving our aims. This is what sympathy enables us to do.

I began by saying that undercutting the fact-value dichotomy, in particular defending the objectivity of moral judgments, has wide-reaching practical consequences. Fanaticism and the arrogance of power both are challenged by the understanding that one can reason, hence should reason together, about moral, which includes social and political, issues.

Dewey concluded the discussion of moral knowledge to which I referred several times with these remarks. "The need for constant revision and expansion of moral knowledge is one great reason why there is no gulf dividing non-moral knowledge from that which is truly moral." An example would be discovering the possibility of inoculation against smallpox, turning provision of such inoculation into a matter of moral concern. Because of the connection of moral and non-moral knowledge, Dewey writes, "each generation, especially one living in a time like the present, is under the responsibility of overhauling its inherited stock of moral principles and reconsidering them in relation to contemporary conditions and needs." This does not mean, he hastens to add, that moral principles are so relative that they have no binding force in any conditions. Rather, we are to discover which principles are relevant in our social condition. "Since the social condition is a fact, the principles which are related to it are real and significant" (Dewey 1985, 282–3).

Dewey emphasized the revisability of moral principle and the need to revise them in light of changing conditions. Earlier I suggested that fallibilism was an important aspect of any method for fixing objective beliefs. The collapse of the fact-value dichotomy brings in its train the collapse of the ethics/metaethics distinction. Because the willingness to revise one's beliefs is one condition of their objectivity, the (metaethical) claim that moral judgments are objective implies the (ethical) obligation to re-examine them when conditions change. Precisely because intersubjectivity is a necessary though not sufficient condition of objectivity, sympathy must be the animating impulse of moral reflection.

Notes

1 (Information taken from Nicolau Barquet, MD, and Pere Domingo, MD, "Smallpox: The Triumph over the Most Terrible of the Ministers of Death." *Annals of Internal Medicine*, 15 October, 1997, 127: 635–42. Taken from www.acponline.org/journals/annals/15oct97/smallpox.htm)

2 References to this edition are given by volume and paragraph, separated by a period.

References

Aristotle. 1985. *Nicomachean Ethics*. Translated by Terence Irving. Indianapolis and Cambridge: Hackett Publishing Company.

Dewey, John. 1985. *Ethics*. In *The Later Works, 1925–1953, Vol. 7*. Edited by Jon A. Boydston, Carbondale –Edwardsville: Southern Illinois University Press.

Goodman, Nelson. 1978. *Ways of Worldmaking*. Indianapolis and Cambridge: Hackett Publishing Company.

James, William. 1975. *Pragmatism: A New Name for Some Old Ways of Thinking*. Cambridge, MA and London: Harvard University Press.

James, William. 1979. "The Moral Philosopher and the Moral Life." In *The Will to Believe and Other Essays in Popular Philosophy* (141–162). Cambridge, MA—London: Harvard University Press.

Murdoch, Iris. 1971. *The Sovereignty of Good*. New York: Schocken Books.

Peirce, Charles S. 1931–35. *The Collected Papers of Charles Sanders Peirce*. Edited by Charles Hartshorne and Paul Weiss, vols. I–VI. Cambridge, MA: Harvard University Press.

Putnam, Ruth A. 1985. "Creating Facts and Values", *Philosophy*, 60: 187–204.

Putnam, Ruth A. 1987. "Weaving Seamless Webs", *Philosophy*, 62: 207–20.

Putnam, Ruth A. 1998. "Perceiving Facts and Values", *Philosophy*, 73: 5–19.

6 On Mattering

Naomi Scheman

> In other words, these various senses of mattering—ontology, epistemology, and ethics—are not separate/separable considerations. Given that you asked me about ethics, let's focus on the way in which the fundamental notion of responsibility is queered. According to agential realism, "responsibility" is not about right response, but rather a matter of inviting, welcoming, and enabling the response of the Other. That is, what is at issue is response-ability—the ability to respond. The range of possible responses that are invited, the kinds of responses that are disinvited or ruled out as fitting responses, are constrained and conditioned by the questions asked, where questions are not simply innocent queries, but particular practices of engagement. So the conditions of possibility of response-ability include accountability for the specific histories of particular practices of engagement (Barad 2012, 81).

If one is committed to distinguishing between facts and values, a claim that something *matters* would seem to fall on the value side—a "matter of concern," in Bruno Latour's terms, rather than a "matter of fact" (Latour 2004). I want to suggest that thinking about mattering destabilizes the distinction between matters of fact and matters of concern, between facts and values, by helping to dislodge a picture of "pure facts" as motivationally inert but doxastically potent—that is, as rationally compelling belief so long as untainted by the value judgments that would lead us to act on them. Mattering also challenges a picture of values and valuing as distinctively human: the "entanglement" of facts with values that Hilary Putnam (2002) argues is endemic to facts as we encounter them is, I will suggest, operative as well in the non-human world.

The entanglement of facts with values that Putnam argues for is part of what undermines the supposed motivational inertness of facts, making some apparently straightforward, bare facts (such as those marshaled by the Black Lives Matter movement, as I will discuss below) motivationally potent, a potency (perversely) revealed by some people's refusal or inability to recognize or acknowledge them, a phenomenon explored by the emerging field of epistemologies of ignorance (Mills 1997, Ortega 2006, Sullivan and Tuana 2007). On the other hand, while entanglement with values can lend motivational potency to facts, such entanglement has often been taken to undermine facts' doxastic potency, that is, their supposed ability—with respect to rational subjects—to compel belief. Motivational inertness and doxastic potency are connected: as rational believers, we are expected to accept demonstrable facts, while as free agents, we are allowed and expected both to have our own opinions about whether things *ought* to be as the facts tell us they are, as well as to choose how to act in the face of those facts. I want to suggest that thinking about how it is that value-entangled facts—facts that matter—are *both* motivationally and doxastically potent can help us think about the ethics and politics of belief.

Mattering, I will go on to suggest, has ontological as well as epistemological consequences: certain matters of fact (concerning race, for example) are constituted as matters of concern: that is, they are socially constructed in value-laden ways, as things or kinds of things that matter—but they are not for that reason unreal. To exist because of mattering is, I will urge, to really, robustly, objectively exist, a claim I want to speculatively connect with alternative conceptions of matter, emerging in a range of different fields and harkening back to Spinoza, according to which matter *matters*: the non-human, even non-living, world is characterized by what I call non-indifference. Mattering—that quintessentially evaluational affair—is, I will suggest, metaphysically fundamental: without it there would be no things, and no facts.

1 Facts are Not Motivationally Inert

> We must do away with all *explanation*, and description alone must take its place . . . The problems are solved, not by giving new information, but by arranging what we have always known.
>
> (Wittgenstein 1953, §109)

> In March 2003 . . . Donald Rumsfeld engaged in a little amateur philosophizing about the relationship between the known and the unknown: "There are known knowns . . . There are known unknowns . . . But there are also unknown unknowns . ." What he forgot to add was the crucial fourth term: the "unknown knowns," things we don't know that we know—which is precisely the Freudian unconscious, the "knowledge which doesn't know itself" (Žižek 2009, 457).

> This essay is really an exercise in archaeology, an excavation of important texts that somehow have become ruins, forgotten at the very same time that they are viewed and repeatedly brought to light. Perhaps it is true

> that sometimes the hardest thing to see is that which is in front of us, which should be the most visible (Ortega 2006, 58).

> for here there is no place that does not see you. You must change your life (Rilke 1995).

Some of the remedying of ignorance has to do specifically with *reminding* people of things that—in some sense—they already know, or of drawing their attention to facts that are, as we say, "staring them in the face," an ordinary but odd locution: what does it mean to see oneself and one's world "in the light of" certain facts, and why might one be motivated to resist doing so? The lines from Rilke describe the experience of feeling seen by a statue—a headless block of inanimate stone—"in light of" which one feels reproached, shamed. Žižek's point is a similar one, but with respect to facts rather than to works of art: there are facts that we cannot or will not face, in part because of what is revealed about us when we come face to face with them, when they "stare us in the face." I want to make a case for the radical and revelatory power of the description of facts (in contrast not to explanation, as Wittgenstein has in mind, but rather to explicit evaluation).

Think, in particular, about descriptions of what is, in some sense, plainly there, obvious, on the surface, not hidden or buried or disguised—but nonetheless overlooked or unnoticed or denied, at least by some of us. Phrases such as, "you know full well . . ." or, "you know as well as anyone . . ." signal this sort of "unknown known"; though we might alternatively use phrases such as, "you ought to have known . . ." or "how could you not have known?" It can, that is, often be unclear whether the failure is best described as our not knowing what we ought to know or as our not knowing (believing, acknowledging) that we actually do know. But in either case my concern is with what we are being charged with missing: the acknowledgment of the aptness of a particular description of what is going on, one that is on the face of it merely, straightforwardly factual, not explicitly evaluative, but that has the power to unsettle our sense of the world and of our place in it, and that reveals our ignorance as motivated.

Descriptions, I want to suggest, are actually never "mere," in part because they are never from nowhere; and their situatedness—it is from somewhere in particular that the world appears like that—is one important way in which facts are "entangled" with values. We could not possibly register all the facts that are out there, even those that are in plain sight: attention is always, necessarily, selective. An apparently trivial but striking demonstration of attentional selectivity is the invisible gorilla experiment: study subjects are directed to watch a video of two teams of people passing a basketball between them and to count the number of times members of the white-clad team pass the ball. While this is going on, a person in a gorilla suit enters the scene, pauses, beats his or her chest, and leaves, having been on screen for nearly one-third of the time. Queried afterwards, fewer than half of the subjects reported having seen the "gorilla": paying careful, close attention to one thing rendered something else—even something one would think striking and certainly worthy of attention—effectively invisible (Simons and Chabris 2009).

Less well known is a follow-up experiment conducted by Aneeta Rattan and Jennifer Eberhardt, in which before watching the video European American subjects were given the task of rating the popularity of a list of names, either stereotypically African American or stereotypically European American. Those who were thereby primed to be thinking of African Americans were significantly more likely to notice the "gorilla." The epistemic liability of narrowly focused attention was overcome by the force of the racist association between African Americans and apes, a chilling enough conclusion; but the authors go on to speculate that "social meaning might work to increase what we perceive, but constrain the range of objects deemed worthy of attention, 'fix[ing] and retain[ing] what the ordinary laws of association bring "before the footlights" of consciousness.'" (Rattan and Eberhardt 2010, 1087, quoting William James.)

There is no question here about the facts themselves: likely to be noticed or not, a person in a gorilla suit strode through the scene of the video. But if facts are to matter, to be appealed to—in order, for example, to settle empirical disputes—questions about their accessibility, about what makes them obvious or salient or even discernible, are clearly germane; and our ability to take in even those facts that are right in front of us is shaped by factors that are clearly value-laden. The irony of the Rattan and Eberhardt study is that activating socially harmful attitudes conveyed an epistemic advantage, allowing the subjects to see what otherwise would have passed unnoticed. What I want to suggest is that attentional selectivity can function to keep us from noticing facts that, were we to attend to them, would unsettle us, facts that would be clearly—without any additional valuational input—motivating, and that it is that motivational import that largely accounts for our attentional failures. We fail to notice what, being noticed, would tell us that we must change our lives.

Thus, efforts to motivate people to act against climate change or genocide or (the example I will consider below) systematic attacks on Black Americans typically consist in presentations of facts. It would weaken, not strengthen, the force of such appeals were those presentations to be accompanied by explicit value statements, since that force rests largely on its being presumptively obvious that the facts are horrific and call for action. It is true, of course, that the facts carry the force that they do against a background of shared values (about what is horrific, what calls for action); so, one might point out, the bare facts on their own might well be motivationally inert. But what is striking is that the relevant value claims are quite platitudinous, while the relevant facts are, when confronted, powerfully striking; and—most centrally for my argument—it is the facts, not the values, that are frequently denied, disputed, doubted, ignored, or (not-so-simply) unnoticed. I am, that is, perfectly happy to admit that motivating facts may be so only because they are in some sense entangled with values—either being themselves covertly evaluational or being taken in against a backdrop of unspoken values—since both these features are ubiquitous. Out of all the goings on around us, what presents itself as a fact is a matter of what is striking, salient, noticeable—what matters to us; and our failures to notice, to be struck by—to acknowledge—certain facts are frequently owing to their carrying with them motivational force we find it difficult to accept.

The emerging field of epistemologies of ignorance takes on issues such as these. The ignorance in question is not simply the absence of knowledge: it is motivated and resistant, and it is frequently ignorance about matters of fact that are in plain view, that ought to be obvious, or that, while perhaps relatively hidden to those who are ignorant of them, are obvious to others, whose testimony is systematically discounted. In particular, what Miranda Fricker theorizes as "testimonial injustice" allows some of us to dismiss the testimony of others who are describing our world in ways it would unsettle us to credit (Fricker 2009). The Black Lives Matter movement in the United States has as one of its aims making visible to white people the experiential world of Black Americans, a world about which most white people are ignorant (Sharpe 2010 and 2014; Bierria 2014; Garza 2014; Rankine 2014; Coates 2015). White ignorance about race is the central example around which the epistemologies of ignorance have been developed. It is, as Charles Mills puts it, "a particular pattern of localized and global cognitive dysfunctions (which are psychologically and socially functional), producing the ironic outcome that whites will in general be unable to understand the world they themselves have made" (Mills 1997, 18).

Sometimes the realities of race are there to be seen, by anyone, but sometimes its manifestations are not actually directly observable to most white people. The phenomenon of "driving while Black," for example, tends to be observable only by Black drivers or those who are in the car with them when they are pulled over by the police and treated in ways that range from rude to deadly. But even in such cases, the realities are hardly hidden: they are reported (though not fully enough in mainstream media) and spoken about: especially these days it probably takes more effort *not* to know than to know. The easiest way to accomplish the not-knowing is by discounting the testimony of those who do know. Even sympathetic, conscientiously anti-racist white people will sometimes respond to the accounts given by Black people of their experiences by acknowledging that, yes, this is how things are experienced by African Americans and that having such experiences certainly makes it rational that they—African Americans—have the feelings and beliefs, the fear and sadness and anger, that they do, but resist taking the next step: acknowledging that those accounts are accounts of the world that *we*—white Americans—live in, that those accounts are presenting us with facts about our world, facts "in the light of which" we are exposed.

Facts are not, of course, the sorts of things to literally return our gaze or to shed light, but then neither is the headless statue in the presence of which Rilke describes the experience of being transfixed. In that case it matters that the statue is a powerful work of art, one that embodies an ideal of human corporeality, in light of which the poet sees himself and feels called upon to change, to live up to what the stone holds out as a human possibility. The facts I'm interested in come to stare us in the face rather differently, showing us in their light not what we might become but what we already are or have been, what we have done or allowed to be done—all things we have reason to avoid acknowledging. While the statement that Black lives matter is explicitly evaluational, the power of the words and images the movement marshals lies in their factual content: we don't need to be told that what we are reading about or seeing is wrong, and it would undercut the power of those words and images were we told what to think and feel about them. The point is to *just look*: this is happening here and now; this is real; this is the everyday reality of Black people's lives—how could you not have known?

Insofar as white people are being addressed by the Black Lives Matter movement the demand is for acknowledgment. In the first instance what needs acknowledging is that in one sense white people and Black people (and, differently, other people of color) live in different worlds, face different dangers and challenges, benefit or suffer differently from structures of power and privilege. It is a demand for acknowledgment of these differences and how they shape one's way of being in the world, of the legitimacy of how the world appears to those whose lives have been shaped in particular by anti-Black racism. But it doesn't stop there: it is often not enough to have one's subjective take on one's own world acknowledged as rational. There is also a need for acknowledgment as a teller of truths, as a testifier, about the one real, shared world, and, in particular, as someone who can shed light on the world as white people experience it. I am not the sole, privileged authority about what it is like to be me, or about how the world appears from where I am located in it: my subjective impression of what the world looks like from here can be inaccurate. It's not just that the world looks different when viewed from *there*, but it only seems to me to look like *this* from *here* because of what I am not allowing myself to notice. Work in the epistemology of ignorance aims to show just how active, motivated, and intransigent ignorance can be—ignorance even (or especially) of facts that are staring me in the face, facts in the light of which I am not the person I take myself to be (Ortega 2006; Pohlhaus 2012). The world in question—the world of racialized people and racially inflected interactions—is a social construction, and, as such, as I will argue below, robustly real. Simply acknowledging difference, even rationally comprehensible difference, overlooks the relationality that informs world-construction, and the ways in which how the world appears (subjectively) to me might be a mirage, screening from my view what you are able to see, what is staring me in the face though I am culpably unable to notice it.

A central focus of the Black Lives Matter movement has been specifically on Black children: on the precariousness of their lives, in part because, although as a matter of fact they are children, they are not seen as children (Tilton 2010; Goff et al. 2014), and on the heart-breaking anxieties of parenting them, of teaching them about all the ordinary, ubiquitous things they need to fear, about the self-effacing lengths they need to go to attempt to appear harmless, attempts that can never be enough to truly protect them from those who see "black skin as a weapon" (Rankine 2015), and what they need to know and to do to turn self-effacement into self-respect, in a world where self-respecting demeanor on the part of Black people is seen by those with or without badges but with the power to kill them as a sign of predatory dangerousness (Coates 2015). These facts and the facts about how Black women and men are routinely treated, especially by police, are facts about the world that all Americans—not just African Americans—live in, facts that white Americans are strongly motivated not to acknowledge, in part because of how we appear in their light. Since the facts about race and racialization are socially constructed around white supremacist values (see the following section), we can expect that such facts will be difficult for white people to recognize—even when those facts are "staring us in the face." This is the point of epistemologies of ignorance: as a white person I am indicted by such facts, even if I do not endorse the values on which they rest, and my (laudable!) refusal to endorse those values can actually motivate my resistant ignorance of the facts they inform.[1]

When Wittgenstein says that "everything lies open to view" (Wittgenstein 1953, § 126), he clearly doesn't mean *everything*: many things are hidden, from some or all of us; many things are simply unknown; many things need to be discovered or uncovered. His remark is directed specifically to philosophers: philosophy should not be in the business of discovering or uncovering new or hidden truths, or of probing for the deep reality underneath the misleading surface. As he puts it, "the work of the philosopher consists in marshalling reminders for a particular purpose" (Wittgenstein 1953, §127). His call for description, specifically description that arranges—and thereby reminds us of—what we have always known, is meant as an alternative to explanation, to the demand that what really matters lies hidden, underneath the "merely superficial." That demand goes back at least to Locke and is central to the epistemology of modern science, according to which surface appearances are not reliable guides to the real natures of things. When it comes to the sorts of questions philosophy poses, Wittgenstein wants us to reject this picture: philosophy's questions arise from what we do, not from some buried essence, and what is called for is attention to our practices, not disdainful dismissal.

There are similarities between Wittgensteinian exhortations to "don't think, but look!" (Wittgenstein 1953, §66) and what is going on with Black Lives Matter and other efforts to get the privileged to see what is staring them in the face, similarities that cluster around the notion of reminders. Reminding someone of something always risks the charge of presumptuousness: there is the presumption that those we are reminding know, have known, or at least could and ought to have known, what we are reminding them of, as well as the presumption that they have forgotten it or failed to keep it in mind. There is also the presumption that we are in a position to be reminding them—typically either in some position of authority (as your teacher, I remind you that your paper is due tomorrow; as your supervisor I remind you of the report that needs to be on my desk in the morning) or in a relationship of some intimacy (I remind you of our up-coming anniversary or of the first time we met).

On Wittgenstein's conception philosophers are similarly presumptuous—claiming to speak for all of us out of what can seem like an inappropriate assertion of authority but is, I think, better thought of as a presumption of intimacy or at least entangled relationship, even if all(!) we're entangled in is a language or a form of life. When I speak as a philosopher, I don't speak for *them*, whoever they may be; I speak for *you*, and the presumption behind my words is that there is a "we" that you and I comprise (Cavell 1996; Scheman 2011; Bauer 2015). As problematic as this presumption can be, I don't think that it's optional, and, although I think it's a distinctively philosophical activity, not only philosophers engage in it. The activity I have in mind is that of helping someone to notice what is staring them in the face, to put the pieces together, reminding them of what they (ought to) know, and challenging them to acknowledge how they themselves appear in light of those facts. And, of course, as with Wittgenstein's interlocutor, in many cases the "someone" we address needs in the first instance to be ourselves, lest we be not just presumptuous but also arrogant.

Both Wittgensteinian philosophical therapy and the efforts of Black Lives Matter in relation to white people can seem problematically presumptuous because those doing the reminding are in positions neither of authority nor of intimacy relative to those they are reminding. They have no claim on their interlocutors' attention prior to gaining it: the legitimacy of their presumption can be secured only through securing acknowledgment of that to which they are drawing attention. We need to be persuaded to *just look*, and through looking to truly see, and to see ourselves in light of those facts. The images circulating of Black men, women, and children being brutalized, even killed, are meant to be heart-breaking—that is, meant to break open closed (white) hearts, to, in George Yancy's term, "unsuture" us, break down the resistance to seeing ourselves in the light of those facts. We need, as Yancy puts it, to "lose our way," to no longer be protected by the soothing ignorance that masquerades as knowledge and erects a shield against the vulnerability of facing and being faced by, revealed in the light of, the brutal facts on which our privilege rests (Yancy 2014).

2 Values are not Ontologically Undermining

> Why can we never discover the same stubbornness, the same solid realism by bringing out the obviously webby, 'thingy' qualities of matters of concern? Why can't we ever counteract the claim of realists that only a fare of matters of fact can satisfy their appetite . . . ?
>
> (Latour 2004, 237)

> So, I think my problem, and "our" problem, is how to have simultaneously an account of radical historical contingency for all knowledge claims and knowing subjects, a critical practice for recognizing our own "semiotic technologies" for making meanings, and a no-nonsense commitment to faithful accounts of a "real" world, one that can be partially shared and that is friendly to earthwide projects of finite freedom, adequate material abundance, modest meaning in suffering, and limited happiness (Haraway 1988, 579)

In his 2004 paper, "Why Has Critique Run out of Steam: From Matters of Fact to Matters of Concern," Bruno Latour laments the increasing popularity of a dismissively critical attitude toward well-founded factual claims, whether about anthropogenic climate change or the attacks on the World Trade Towers. While not dismissing the role of cynical manipulators of public opinion,[2] Latour is issuing a challenge to himself and others to shift the target of critical theorizing away from undermining the reality of matters of fact and toward shoring up the reality of matters of concern. The claim of realists that their appetite could be satisfied only by matters of fact—conceived of as independent of our concerns, uncontaminated by values—constitutes a demand that Latour and others in science studies and feminist philosophy of science have long argued cannot be met. But what Latour wants to draw our attention to is the peculiarity of its seeming to follow from pointing out that facts become facts through our being entangled with them—through their mattering—that they are therefore less than real, less consequential, less deserving (and in need) of attention and respect.

I want to take up a particular manifestation of the ontological discounting of matters of concern by thinking about social constructions, whether of particulars or of kinds. On a standard conception of realism in terms of mind-independence, there would seem to be a problem with socially constructed entities; and many theorists have taken social constructionism to be a form of anti-realism (Mason 2016). Along with others, I have argued for a realist understanding of social constructions, and I want to continue that line of argument by thinking about specifically value-laden constructions—things (particulars or kinds) that exist in part because of shared concerns, attitudes, and evaluative stances. Thus, for example, races arguably came into existence through the workings of fundamentally racist European projects of self-constitution and global colonization, and persist through the complex interactions of continuing racism, on the one hand, and, on the other, commitments to resistance grounded in shared identities.[3] I want to suggest that being in this way mind-dependent and value-laden doesn't detract at all from being real.

One way of getting at this question is by considering Sandra Harding's conception of "strong objectivity," which "requires that the subject of knowledge be placed on the same critical causal plane as the objects of knowledge" (Harding 1993, 69). For traditional accounts of objectivity (what Harding calls "weak objectivity") the idealization of the (objective, disinterested, unimplicated) knowing subject has been demanded as a way of securing objective knowledge of a presumptively independent—hence, real—world; and it is this demand that strong objectivity rejects, making the case that the world is better known when knowers take themselves and are taken by others to be located in the social world, with distinctive epistemic advantages and liabilities. This epistemological thesis has ontological implications. From the perspective of strong objectivity, mind-dependence is revealed as a real phenomenon, just one of the many ways in which things in the world are causally related to each other, and is no more ontologically deflationary than, say, trees' dependence on carbon dioxide.

This point can be clarified by looking at Stuart Hampshire's essay "Sincerity and Single-Mindedness" (1971), in which he discusses the relationship between states of mind, such as emotions, and an agent's own beliefs about them. Hampshire is concerned to argue against an ideal of sincerity that cautions against reflection, that characterizes as sentimental a concern with scrutinizing one's own mental states. But what I find most useful in his discussion is his noting that our beliefs about and attitudes toward our own states of mind are constitutive of those states of mind: disavowed jealousy is different from jealousy we acknowledge; if we are conflicted about what we feel, those feelings themselves are conflicted.

That is, our takes on our states of mind help to make those states of mind what they are: they are not just mind-dependent (as I assume states of mind necessarily are), they are dependent on explicitly value-laden judgments about them—the dependency is part of the phenomenon.

Similarly, the ways in which social constructions are entangled with value-laden beliefs and attitudes do not detract from their reality, since those entanglements are themselves real. Once we give up on the epistemic ideal of the detached observer—once we take knowers to be as real as their objects of knowledge and coming to know to be a real relationship involving a range of influences and dependencies—there is no reason to discount the reality of facts that are constituted in part by values. As Sally Haslanger puts it, "[t]he only kind of mind-dependence is ordinary mind-dependence in which the mind in question is an ordinary human mind that exists in the natural world, interdependent with other natural things" (2012, 213).

We can, of course, be mistaken about what sort of thing something is, and we can take something that is dependent on us—on our social practices and meanings—to exist independently of those practices and meanings—as, to take the most discussed examples, we can mistakenly take race and gender to be "natural," meaning something like biological. That can be an important mistake to make, in part because it can lead us to misdirect efforts at effecting change: in Haslanger's words, "if we are going to change the world, we need to know how the problematic parts are created and maintained" (2012, 215). But being value-laden, mind-dependent, and socially constructed is a way of being "created and maintained"—i.e., of being real.[4]

Feminist and critical race theoretic metaphysicians such as Haslanger have focused on the ontological status of kinds, notably gender and race, and much less, if at all, on the ontology of particulars. But precisely analogous considerations hold. One way of framing a central question about the ontology of particulars is what van Inwagen calls the "special composition question": under what conditions do some objects compose some further thing? (van Inwagen 1990; Ney 2014, 103–14). The claim that there are socially constructed particulars rests on noting that many things that we ordinarily take to exist, such as nations, corporations, labor unions, and universities, have among their composition conditions socially salient relationships (Scheman 2000). People or things belong to or are parts of one of them because of how they are meaningfully related to other people or things: independently of social practices—of what matters to us—none of these things would exist, but that fact does not make them unreal, any more than socially constructed kinds are unreal. Those social practices are typically the enactments of shared values, thus making it the case that not only facts but also the things that the facts are about are saturated with values. Such things exist as robustly real—causally implicated in the world—but only because they matter to us.

Recognizing that and how they matter may well lead us to lament their existence or at least to regard their continued existence as a matter of controversy, but we cannot just think them away: being socially constructed is not being imaginary.

3 Mattering is Not a Human Prerogative

> The ubiquitous puns on 'matter' do not, alas, mark a rethinking of the key concepts (materiality and signification) and the relationship between them. Rather, it seems to be symptomatic of the extent to which matters of "fact" (so to speak) have been replaced with matters of signification (no scare quotes here). Language matters. Discourse matters. Culture matters. There is an important sense in which the only thing that does not seem to matter anymore is matter.
>
> (Barad 2003, 801)

Since Karen Barad's 2003 paper, and partly because of her work, matter has come to matter across a wide range of fields and theoretical approaches in the humanities and social sciences.[5] Most obviously, this is a claim about what matters to theorists, about what people choose to think, talk, and write about. But in Barad's work and that of diverse others we also see arguments about mattering as not just a human affair. Rather, mattering, I want to suggest, (in)forms everything. That is, insofar as mattering is on the "value" side of a distinction between facts and values, valuing isn't something that comes into the world just with us, or even with some other privileged animals; it is, rather, ontologically fundamental.

A wide range of theoretical perspectives are converging on what might be thought of as a resistance to anthropocentrism. My own route to this complex space was idiosyncratic: a combination of my earlier Wittgensteinian account of the social construction of mental phenomena and indigenous accounts of the wisdom of stones. The thesis of social construction, concerning both kinds and particulars, maintains that things that we had taken to be in some sense "natural" do not pre-exist our ways of talking and thinking. In particular, I argued that emotions, beliefs, desires, and so on cannot, given their causal roles, be physical events, states, or processes, since those causal roles depend on the social salience and intelligibility of the phenomena of mentality. Abstracted from social meaning, whatever might be taken to comprise those phenomena disaggregates—a thesis Patrick Laine (forthcoming) has helpfully named the "amorphism" of the mental, to contrast it with Davidson's "anomalism," which maintains an untenable commitment to token identities (Scheman 2000). While I still believe that this account is right, as are the many arguments for the socially constructed nature of kinds such as race, gender, and sexual identity, I became troubled by a problematic anthropocentrism. This troubling is, of course, odd: social constructionism is obviously anthropocentric: the claim is precisely that some things that we had taken to be in the world prior to and independent of certain forms of human social activity are in fact dependent (conceptually, not just causally) on that activity. Non-anthropocentric social constructionism doesn't make sense. Or does it? This is where the other line of thinking comes in.

In explaining the Western, scientific worldview to a working committee concerned with bridging that worldview with that of indigenous peoples, especially those who were the original inhabitants of the state I lived in and whose way of life continues to be under threat,[6] Paul Schultz, an Ojibway elder from White Earth in Minnesota, sketched a pyramid with, at the bottom, stones: dumb and senseless; then plants, with a bit more on the ball; then non-human animals: sentient, moving; then, at the pinnacle of creation, humans: the smartest things around. By contrast, he sketched a pyramid representing an indigenous worldview: stones, having been around the longest, know the most; then plants; then other-than-human animals; then, finally, humans: mere babies, ignorant, needing to learn from our elders. As an analytically-trained philosopher with a family ethos of atheism and scientific rationality, I couldn't make literal sense of this account, as appealing as I found it: how could stones *know* *any*thing? It helped to shift from thinking about *knowledge*, especially understood as (something like) "justified true belief that p", to thinking about *wisdom*, and, specifically, about how it might matter to have been in the world for a very long time. One thought is that it matters only if what you lived through mattered (made a difference) to you, if you were shaped and changed by it.

Following on this thought, I came to realize how profoundly wrong it is to think of the non-human (or even non-sentient, or even non-alive) world as a world in which nothing matters, as *indifferent*. Nothing is indifferent; specifically, no *thing* is indifferent. To be indifferent is to fail to distinctively shape, and to be distinctively shaped by, the world around you. But by that measure stones are hardly indifferent. They respond in distinctively stony fashion to some things, and not at all to others; and they respond differently depending on what sort of stone they are and what has previously happened to them. They are narratives of the times they have passed through: the classifications of sedimentary, igneous, and metamorphic are genres, types of stories; and particular stones and rock formations bear the traces (in important ways just *are* the traces) of more particular, distinctive responses. And things around them similarly bear the traces of what those stones have done. Similarly, think about what can (and cannot) be learned from the rings of trees, or about the co-evolution of predator and prey species: in all these cases what something is now is a record of what has happened to it and how it has related to other things around it.

Narrative, most fundamentally, is space and time made salient: from space-time points to here and there, now and then, once upon a time, long ago and far away. And being a *thing* is a matter of saliencies, of non-indifference: things are emergent loci of difference in webs of relationship, a matter of mattering, and of persisting through time and into the future, resisting disaggregation. (Spinoza called this *conatus*.) Part of what matters is being recognized by your surroundings as the sort of thing that you are. Workers, for example, may share a sense of solidarity and may even act in concert, but something ontologically significant happens when they organize and achieve certification as a union. Similarly, something ontologically significant happens when couples marry—witness the importance attached to the achievement of marriage equality for same-sex couples. In both cases, being that particular kind of thing (a labor union or a married couple) means being recognized by the world as having distinctive causal powers and vulnerabilities, supported by laws. Laws of nature work similarly, shaping vulnerabilities and capacities of things because of the kinds of things they are. Think, for example, about a stone vs. a pile of sand. A stone has a comparatively robust narrative—how it came to be the size and shape that it is, to have the composition it does—and a comparatively robust proclivity to continue in existence as the thing that it is. It also acts as a whole in causing events and absorbing effects. By contrast, the pile as such has a relatively shallow story (I just scooped it up from the beach) and relatively little tendency to continue as the thing it is (I toss it and it scatters). If the pile is on the ground and I kick it, it doesn't move as one, nor does the mass of the pile itself bruise my toes—and, unlike stones, piles of sand can't break windows.

As I suggest above and as I and others have argued, social constructionism is not ontologically deflationary. It is precisely as socially constructed that particulars such as nations, corporations, labor unions, and mental phenomena, or kinds such as races, genders, and sexualities, are real, that is, causally implicated in the world: their causes and effects are inflected by their social meanings and cannot be accounted for if they are understood in nonsocial terms—as, e.g., land masses, brain states, or biological kinds. But, further, I want to suggest that social construction is a special case of a more general feature of reality. A socially constructed thing or kind is what it is (is any*thing* at all) because of how it fits into an ineliminably social set of narratives, shaped by saliencies, by perspectivity—by mattering. But while mattering, along with narrativity, salience, and perspectivity are typically thought of as part of what we, as humans, bring to the world, they are, I want to suggest, ontologically more fundamental: mattering, valuing, is at the heart of thing-ness.

4 Conclusion

The themes of attention and motivated inattention, of the power of "just looking," and of the subjectivity of the non-human come together in thinking about photography. In addition to verbal descriptions of what it is like to be Black in the United States, the Black Lives Matter movement puts into the public realm photographic and video images, many of which have become iconic. Despite the well-known possibilities of altering photographic images, especially digitally, photographs and videos carry the power of factual documentation—thus, for example, the omnipresence in some cities of video camera surveillance and the push for police to wear body-cams to record their interactions with those they stop. But especially when we look at intentionally taken photographs, we are seeing someone's perspectival framing of a scene. What we see is a complex entanglement of the quality of their attention and the authority of photographic facticity; and often the force of their attention is to compel our own, in part by placing us in the position of the photographer, seeing through their eyes, registering the fact that what we are seeing matters and that we are complicit in—entangled with—it.

I started thinking seriously about photography in relation to the work of a friend, the late photographer Lynne Cohen, who was married to Andrew Lugg, a Wittgensteinian philosopher. Cohen's last book has the Wittgensteinian title *Nothing Is Hidden* (2012). Throughout her career she used an 8"x10" view camera, and over the years her prints grew from contact size to extremely large. All the images are indoors, in highly visually intentional spaces, and there are never any actual people in them. They are, nonetheless, charged with human intentionality, including with the force of Cohen's steady, attentive gaze: none of the spaces she photographs are, at the time she is photographing them, actually unpeopled, since she herself is there; and though we never see her, it is through her eyes that we see what we see. And what we see is strange and estranging, some places we might not ordinarily have access to and some we pass through without noticing, but all of them part of the world we inhabit, accept, and make—without attending to.

In a conversation that Cohen and Lugg recorded, reflecting on their professional practices and aesthetic and political motivations and choices, Cohen explained, in a way that brings together the various strands of my discussion:

> Another thing I'd add is that in the early days we were very much taken by Francis Ponge's phrase '*le parti pris des choses*'. I like the idea of taking the side of things and giving them a voice rather than lording it over them. The public life of objects is important to me since it says something about how we want ourselves, collectively, to be seen. Perhaps this is peculiar but I'd like to be regarded as treating objects with fairness and respect. That's not the same as a concern with social justice but it's linked to it. By presenting objects non-hierarchically and avoiding the trap of telling the viewer how to think, I hope to convey something about how the world could be different. I'm not neutral about what I photograph. I have opinions and

wouldn't pretend for a minute that a military installation is the same sort of place as a health spa. The seeming neutrality of my pictures is a false clue. I present a spa pool and a practice range similarly, trusting viewers will reflect on what is being done in our name (See Cohen and Lugg 2009 and 2013).

Notes

1 Petra Green (forthcoming) discusses this phenomenon in relation to the persistence of sexist attitudes among young heterosexual Swedes in part because of—not despite—the deeply ingrained national self-image of gender equality.
2 Such manipulation is especially evident concerning climate change. See Oreskes and Conway (2011).
3 I have found the perspective of "post-positivist realists" especially suggestive. See Moya and Hames-Garcia (2000) and Mohanty et al. (2006).
4 Having read Hilary Putnam's recently published paper on realism (2015), which exhibits his characteristically expansive, pluralist, pragmatist intelligence about the matter, I was especially looking forward to discussing these issues with him—an additional reason, were one needed, to be saddened by his death.
5 The literature is vast. I have been particularly inspired by Barad (2003), Bennett (2010), and Cohen (2015), by Actor Network Theory, e.g., Latour (2005), and by work in material feminism, e.g., Alaimo and Hekman (2008).
6 For background on this work, see: Nibi-Manoomin Bridging Worldviews Committee. https://www.cfans.umn.edu/wildrice, accessed 6 April 2016. The following paragraphs are drawn from Scheman (2015).

References

Alaimo, Stacy and Hekman, Susan (eds.). 2008. *Material Feminisms*. Bloomington: Indiana University Press.

Barad, Karen. 2003. "Posthumanist Performativity", *Signs*, 28: 801–31.

Barad, Karen. 2012. "Interview by Adam Kleinmann." Special DOCUMENTA (13) Issue of *Mousse Magazine* (Milan, Italy). Accessed on-line 30 March 2016. https://www.academia.edu/1857617/_Intra-actions_Interview_of_Karen_Barad_by_Adam_Kleinmann_.

Bauer, Nancy. 2015. *How to Do Things with Pornography*. Cambridge, MA: Harvard University Press.

Bennett, Jane. 2010. *Vibrant Matter: A Political Ecology of Things*. Durham, NC: Duke University Press.

Bierria, Alisa. 2014. "Missing in Action: Violence, Power, and Discerning Agency", *Hypatia*, 29: 129–45.

Cavell, Stanley. 1996. *A Pitch of Philosophy: Autobiographical Exercises*. Cambridge, MA: Harvard University Press.

Coates, Ta-Nehisi. 2015. *Between the World and Me*. New York: Spiegel and Grau.

Cohen, Jeffrey J. 2015. *Stone: An Ecology of the Inhuman*. Minneapolis: University of Minnesota Press.

Cohen, Lynne. 2012. *Nothing Is Hidden*. Göttingen: Steidl. http://www.lynne-cohen.com/. Accessed 6 April 2016.

Cohen, Lynne, and Andrew Lugg. 2009. Edited and revised, 2013. "Two Feet on the Ground". Unpublished ms.

Fricker, Miranda. 2009. *Epistemic Injustice: Power and the Ethics of Knowing*. Oxford: Oxford University Press.

Garza, Alicia. 2014. "A Herstory of the #BlackLivesMatter Movement", *Proud Flesh: New Afrikan Journal of Culture, Politics, and Consciousness*, 10: 103–8.

Goff, Phillip Atiba, Matthew Christian Jackson, Brooke Allison, Lewis Di Leone, Natalie Ann DiTomasso, and Carmen Marie Culotta. 2014. "The Essence of Innocence: Consequences of Dehumanizing Black Children", *Journal of Personality and Social Psychology*, 106: 526–45.

Green, Petra. Forthcoming. "Intimate Epistemologies: Knowledge and Ignorance under Conditions of Inequality" (working title). PhD diss, Umeå University.

Hampshire, Stuart. 1971. "Sincerity and Singlemindedness." In *Freedom of Mind and Other Essays*, 232–56. Princeton, NJ: Princeton University Press.

Haraway, Donna. 1988. "Situated Knowledges: The Science Question in Feminism and the Privilege of Partial Perspective", *Feminist Studies*, 14/3: 575–99.

Harding, Sandra. 1993. "Rethinking Standpoint Epistemology: What Is Strong Objectivity?" In *Feminist Epistemologies*, edited by Linda Alcoff and Elizabeth Potter, 49–82. New York: Routledge.

Haslanger, Sally. 2012. *Resisting Reality: Social Construction and Social Critique*. Oxford: Oxford University Press.

Laine, Patrick. Forthcoming. "A Story about Minds, or: A Mind Is Not a Thing, But It's Not Nothing Either." PhD diss., University of Minnesota.

Latour, Bruno. 2004. "Why Has Critique Run out of Steam? From Matters of Fact to Matters of Concern", *Critical Inquiry*, 30: 225–48.

Latour, Bruno. 2005. *Reassembling the Social: An Introduction to Actor-Network-Theory*. Oxford: Oxford University Press.

Mason, Rebecca. 2016. "Are Social Kinds Mind-Dependent?" Draft presented to WOGAP, Cambridge, MA. Unpublished ms.

Mills, Charles. 1997. *The Racial Contract*. Ithaca, NY: Cornell University Press.

Mohanty, Satya P., Paula Moya, Linda Martín Alcoff, and Michael Hames-Garcia. 2006. *Identity Politics Reconsidered*. New York: Palgrave-Macmillan.

Moya, Paula M. L. and Hames-Garcia, Michael. 2000. *Reclaiming Identity: Realism and the Predicament of Postmodernism*. Berkeley: University of California Press.

Ney, Alyssa. 2014. *Metaphysics: An Introduction*. New York: Routledge.

Oreskes, Naomi and Conway, Erik M. 2011. *Merchants of Doubt How a Handful of Scientists Obscured the Truth on Issues from Tobacco Smoke to Global Warming*. New York: Bloomsbury Press.

Ortega, Mariana. 2006. "Being Lovingly, Knowingly Ignorant: White Feminism and Women of Color", *Hypatia*, 21: 56–74.

Pohlhaus, Gaile, Jr. 2012. "Relational Knowing and Epistemic Injustice: Toward a Theory of Willful Hermeneutical Ignorance", *Hypatia*, 27: 715–35.

Putnam, Hilary. 2002. *The Collapse of the Fact/Value Dichotomy and Other Essays*. Cambridge, MA: Harvard University Press.

Putnam, Hilary. 2015. "Naturalism, Realism, and Normativity", *Journal of the American Philosophical Association*, 1: 312–28.

Rankine, Claudia. 2014. *Citizen: An American Lyric*. Minneapolis: Graywolf Press.

Rankine, Claudia. 2015. "The Condition of Black Life Is One of Mourning." *New York Times*, 22 June.

Ratan, Aneeta and Jennifer, Eberhardt. 2010. "The Role of Social Meaning in Inattentional Blindness: When the Gorillas in Our Midst Do Not Go Unseen", *Journal of Experimental Social Psychology*, 46: 1085–8.

Rilke, Rainer Maria. 1995. "Archaic Torso of Apollo." In *Ahead of All Parting: Selected Poetry and Prose of Rainer Maria Rilke*, translated by Stephen Mitchell. New York: Modern Library. Accessed online 6 April 2016. https://www.poets.org/poetsorg/poem/archaic-torso-apollo.

Scheman, Naomi. 2000. "Against Physicalism." In *Cambridge Companion to Feminism in Philosophy*, edited by Miranda Fricker and Jennifer Hornsby, 49–67. Cambridge: Cambridge University Press. Reprinted in Scheman, Naomi. 2011. *Shifting Ground: Knowledge & Reality, Transgression & Trustworthiness*, 81–96. Oxford: Oxford University Press.

Scheman, Naomi. 2011. "A Storied World: On Meeting and Being Met." In *Stanley Cavell and Literary Studies: Consequences of Skepticism*, edited by Richard Eldridge and Bernie Rhie, 92–105. New York: Continuum Press.

Scheman, Naomi. 2015. "Discombobulation and Epistemologies of Ignorance: On Having Your Conceptual Schemes Up-Ended." *APA Newsletter on Indigenous Philosophy*, 9–11.

Sharpe, Christina. 2010. *Monstrous Intimacies: Making Post Slavery Subjects*. Durham, NC: Duke University Press.

Sharpe, Christina. 2014. "The Lie at the Center of Everything", *Black Studies Papers*, 1: 189–214. Web. Accessed 6 April 2016. http://nbn-resolving.de/urn:nbn:de:gbv:46–00103785–17.

Simons, Daniel and Chabris, Christopher F. 2009. *The Invisible Gorilla: How Our Intuitions Deceive Us*. New York: Random House.

Sullivan, Shannon and Tuana, Nancy. 2007. *Race and Epistemologies of Ignorance*. Albany: SUNY Press.

Tilton, Jennifer. 2010. *Dangerous or Endangered? Race and the Politics of Youth in Urban America*. New York: NYU Press.

van Inwagen, Peter. 1990. *Material Beings*. Ithaca: Cornell University Press.

Wittgenstein, Ludwig. 1953. *Philosophical Investigations*. Edited by G.E.M. Anscombe and Rush Rhees, translated by G.E.M. Anscombe. Oxford: Blackwell.

Yancy, George. 2014. "Introduction: Un-Sutured." In *White Self-Criticality beyond Anti-Racism: How Does It Feel to Be a White Problem?* edited by George Yancy, xi–xxvii. Lanham MD: Lexington Books

Žižek, Slavoj. 2009. *In Defense of Lost Causes*. London: Verso.

7 Change in View

Sensitivity to Facts and Prospective Rationality

Carla Bagnoli

Rational agents often make progress by revisiting their previous judgments about what to believe and what to do. In fact, practical reasoning in general may be thought to be a complex activity by which we bring what matters into view. On this construal of practical reasoning, the process of revision takes center stage, and it often includes (even though it is not limited to) rethinking and re-describing the facts of the matter. Sensitivity to facts is thus an important aspect of practical and theoretical rationality. However, it is far from obvious what "sensitivity to facts" consists in, and what sorts of capacities it requires rational agents to exercise. I will argue that among these capacities, emotional engagement figures prominently. This occurs when agents are actively—though emotionally—involved with aspects of the scenario they are thinking about. Emotional engagement importantly contributes to practical reasoning in general insofar as it contributes to changing view and revising judgment and decision. In particular, emotional engagement with the circumstances of action is a crucial component of deliberation. The upshot of this argument is that to account for the impact of emotional engagement and, consequently, to make sense of ordinary functions of reasoning, one has to overcome the sharp distinction between facts and values.

In this chapter, I offer a constructivist account of practical reasoning as an activity that is transformative, taking up the plea for the study of reasoning as an activity of revision and change in view, argued by philosophers as diverse as Iris Murdoch and Gilbert Harman. Within this context, I account for the role of sensitivity to facts, claiming that sensitivity to facts, understood as emotional engagement, is partially constitutive of facts. I consider both the epistemological and ontological aspects of this claim. From the epistemological point of view, the claim is that the agents identify the relevant facts through emotional engagement. From the ontological point of view, the claim is that by emotionally engagement, agents can modify the objects of their emotions, hence shaping the facts and bringing what matters into view. To some important extent, then, the facts are not fully separable from what are the concerns of the agents in their perspective. This is not to say that there are no facts of the matter, or that facts are mere projections of the agent's sensibility onto the cold, colorless, and valueless reality. Rather, the claim is that the ontological boundaries between facts, their normative status and significance are internal to the practical standpoint, which is defined in terms of the agents' concerns within rational practices of mutual recognition.

The most relevant implications of the constructivist view have to be sought at the level of prospective rationality, that is, the temporal dimension of rationality. Emotional engagement produces changes in view which are perspectival, and thus take into account the agents' place in their circumstances. Nonetheless, they exhibit some sort of objectivity and authority. By examining the dynamics of change in view, we can make sense of moral progress, development, and decadence over time. Thus, these notions provide us with theoretical tools to make sense of the perspectival dimension of practical reasoning as ordinary agents perform it.

1 Facts of the Matter and Matters of Value

Rational agents often make progress by revisiting their previous judgments about what to believe and what to do. The process of judgment revision often includes, even though it is not limited to, taking into account or reassessing the facts of the matter. It is a basic requirement of rationality that the agents ought to take facts into account. Failure to do so undermines the normative status of one's judgment and demands that it be corrected. In some cases, such a failure also undermines the status of rational agents, e.g., when they appear to be impervious to evidences and reasoning, ignorant of relevant facts, careless or negligent in gathering information, unresponsive to the demands of rational justification to others. Sensitivity to facts is thus an important quality of the rational agent. Conversely, practical and theoretical irrationality often takes the form of insensitivity to facts, e.g., denying, resisting, or discounting their normative relevance or evidential import. Again, irrationality may take different forms, including a lack of curiosity that reveals epistemic dogmatism, or a sort of callousness that prevents the agent to appreciate the salience of some moral and political implications of his inquiry, or indifference to facts that concern others and that would affect an impartial reasoner.

While sensitivity to facts is an important dimension of practical and theoretical rationality, it is far from obvious what "sensitivity to facts" precisely consists in and what kind of constraints it puts on practical reasoning. Answers to these questions largely depend on the alleged ontological and epistemological status of facts, whether an inquiry about facts is wholly independent of values, and what we take reasoning to be. This is the cluster of issues that concerns us in this chapter, and here is a way to approach them. Drawing on a very simplified contrast between the non-cognitivist and the cognitivist views,[1] we may say that the non-cognitivist view is that matters of facts and matters of value radically differ in nature. Values can be anchored on to facts, inquiry about facts may be driven by values, but facts and values are sharply separated.[2] By contrast, the cognitivist view is that values are identical with or reducible to natural facts, even though they have some special trait, or that they are so entangled with facts that they cannot be separated from their factual basis. The possibility of separating fact and value partly rests on other dichotomies that are likewise problematic, such as the descriptive/prescriptive distinction in meaning, the analytic/synthetic distinction in judgment, and the thin/thick distinction in concepts.[3] In what follows, I will focus on the impact of the fact/value dychotomy on practical reasoning.

2 Reasoning as Inquiry

Cognitivist and non-cognitivist models issue very different accounts of practical reasoning. For non-cognitivists, reasoning is about facts, and value merely attaches to them, in more or less complicated ways, granting their efficacy and motivational force. Facts are neutral, and the presence of values explains their normative significance and relevance to action. How practical reasoning should look in the cognitivist model depends on how values are supposed to relate to facts. If values simply *are* facts, then there is no difference whatsoever between reasoning about facts and reasoning about values. This is the route pursued by naturalism.[4] On the other hand, if values are *special kinds of* facts, or facts are *value-laden*, it is important to clarify what kinds of objectivity is attainable. To avoid the charge of arbitrariness, one is committed to explain further what reasoning is, if not an inquiry into facts of the matter driven by (special, personal, or communal) values.

The problem of the dichotomy (or entanglement) of fact and value is pervasive. It concerns all the domains of human inquiry. Scientific inquiry is in no safer place. Hilary Putnam has forcefully argued that in the practice of science, which many suppose to be neutral and freestanding, inquiry is driven by values and virtues (Putnam 2002).[5] Since science is considered the bastion of objectivity and yet it is not value-neutral, there seems to be a good case for thinking that objectivity does not require neutrality.[6] In what follows, I track a line of criticism whose consequences about the standards of rational action are large and span from ethics to politics and economics. Putnam holds that the main motivation for thinking that science is a value-neutral inquiry about matters of fact is that it deploys *formal* methods (Putnam 2002).[7] I am concerned with the extension of Putnam's claim originally limited to scientific inquiry to the larger domain of rational inquiry.

Most (economic and philosophical) approaches to rational decision share the presumption that a method is objective when it is formal. As a consequence, their aim is to identify an algorithm or function that determines what is rational to do, given the circumstances and the facts of the matter. The subject's deliberative job consists in selecting a formal procedure and supplying the bare facts of the matter. Utilitarianism is certainly the easiest case at hand, but the critique can plausibly be extended to other ethical and economic theories of rational action that are thoroughly procedural. It appears that the fact/value dichotomy supports a view of rational deliberation whose objectivity is warranted by the formality of the procedure and the factuality of contents. The selection of the algorithm and the criteria of relevance of facts are a matter of choice, which may suggest that they are idiosyncratic and arbitrary, or perhaps constrained only by the limitations imposed by nature.

On this very point, it is instructive to recall the early controversy in which Iris Murdoch engages non-cognitivism—not to show how deeply Putnam is indebted to it, but rather to take stock of an important line of argumentation, which is easily lost on our contemporaries. In this controversy, Murdoch's antagonist is Richard M. Hare, but the same objections she moves to Hare can be extended to contemporary metaethics, and especially to those metaethical theories that fixate on the so-called logic of moral language, disregarding the broader dimension of practical reason and rationality. Hare and his followers purport to show that practical reasoning is firmly grounded on the logical features of moral concepts, because of their descriptive meaning (cf. Hare 1952 and 1963, Ch. 1). On this view, the logic of moral language, together with the relevant facts of the matter, generate a framework of reasoning that should sort out all the questions of ethics. All we need in order to solve moral questions is the formal devices that competent users of moral language have, plus the bare facts of the matter.[8] I name this view of reasoning "procedural," because it treats deliberation as a procedure, which comprises of a function and the relevant facts. I would like to suggest that there is a connection between the claim that there is a fact/value dichotomy, and the claim that practical reasoning is reducible to a procedure, which is put to work when the relevant facts of the matter are added. For instance, Hare holds that the facts are neutral and selected according to normative criteria of relevance.[9] One might say that relevant facts acquire value and normative weight through the application of a procedure, but they are always in principle separable from value-concepts. Value-concepts attach to matters of facts, and criteria merely apply to facts without altering them, as rings that are super-imposed onto reality.

Thus far the debates have taken advantage of Murdoch's explorations about facts and values only to a very limited extent. They have privileged one aspect of the dispute, which revolves around the contrast between thick and thin concepts. Instead, I would like to develop her insight further to bear more deeply on a larger issue concerning the scope and form of practical reasoning. Putting Murdoch's critique of the fact/value dichotomy in this perspective alerts us about the risks of focusing too narrowly on the semantics of value-concepts. The impact of the collapse of the fact/value dichotomy can be fully appreciated by considering how it reshapes the account of practical reasoning. Note the following implications. First, contrary to non-cognitivist views, we do not learn much about the nature and purpose of practical reasoning by studying of the logic of moral concepts. The semantic features of moral discourse do not generate all there is to know about the structure of practical reasoning. Second, and in a sense to be clarified, reflection is not governed by the principles of logic. Thus, the life of the mind ought to be investigated by means that are different than the rules of inference. Third, and as a consequence, practical reasoning cannot be reduced to gathering the relevant facts and assessing them by some evaluative standards, but it importantly focuses on re-describing what matters and bringing it into view.

3 Against the Proceduralist Model of Reasoning

Murdoch rejects a very common way of conceiving of the objectivity of facts and values, which she finds misguided and misleading. Part of her argument is that by endorsing this widespread account of the fact/value divide, one is bound to mischaracterize the ordinary activities that constitute the life of the mind, and also the standards of objectivity.[10] The proposal, as I take it, is that we should think of ethical objectivity as a moral achievement and thus account for its historical dimension.

A first step in this direction is to stress the continuity between the executive and non-executive aspects of deliberation. By the "non-executive aspects" of deliberation I mean a broad category of activities that include mental processes like rethinking, reconsidering, and imagining (Bagnoli 2011, 205). The proceduralist model of reasoning privileges the executive aspects and ignores the non-executive aspects of deliberation stands. Murdoch articulates her critique in broad strokes, by opposing the metaphor of vision to that of choice and movement. She defends the former as superior to the latter, which suits the behaviorist model of agency and is often endorsed by non-cognitivist accounts of ethics. Murdoch's oppositional use of these metaphors is instrumental to her argument for rebuilding ethics on a psychological footing. Her conviction is that ethical theory cannot do without a thorough study of the mind,[11] and that metaethics cannot be confined to the study of semantics. The alternative she envisions does not reduce practical reasoning to a thoroughly empirical inquiry, however. In fact, her more ambitious project is to provide a philosophical exploration of the mind, which describes ordinary ways of thinking and normative standards of being moral. The aim must be descriptively plausible in order to be of practical relevance. The practical efficacy of reasoning is not ensured by juxtaposing the arbitrary criteria of choice, or super-imposing abstract moral ideals, such as the moral law. It comes about by repeated and individual acts of focused attention. Murdoch's argument is based on the case in which attention is re-focused, that is, the case of a change in view. I would like to emphasize that this is nothing less than a proposed method of inquiring: to study the mind by studying its alterations, e.g., conflicts and changes. In what follows, I endorse this method and bring it to bear on the topic I announced at the outset.

4 Perspectival Alterations of the Mind

Focusing broadly on the alterations of the mind may seem to be a false start because in standard cases of reasoning, the division between matters of fact and matters of value is straightforward. In this section, I offer some examples in which the mind alters in order to adjust to new findings and new concerns. Only some such alterations count as change in view, and in the next section I will show how such cases undermine the fact/value dichotomy.[12]

Suppose Ori discovers that there is a convenient morning train to St. Petersburg and decides to cancel his afternoon reservation. This is a case in which the agent acquires more information about the relevant facts, which substantially alters the deliberative scenario, canceling the normative force of previous considerations. Such considerations simply cease to be reasons, because of the acquisition of new relevant facts. The relevant epistemic change is explained by the availability of more information. This sort of reasoning modifies the means to be taken in the execution of a plan: it issues a *revision* of belief (e.g., regarding the best way to reach one's destination).

In other cases, however, sensitivity to facts requires something more and different than gathering and acquiring information. Here is a simple illustration. Saddened by the decay of a jasmine on his terrace, Fabien regrets not to have planned carefully his summer vacation, during which the jasmine was not watered. His regret is retrospective, and it corrects his judgment about what it takes to plan well one's vacation. There is a sense in which we want to say that Fabien knows more after this experience. He knows more about what happens to jasmines during a hot summer, if one does not water them. He is also more knowledgeable about the sorts of things one should arrange in view of taking vacation, and also about his own vulnerability to regret. The change that Fabien undergoes over time cannot be characterized precisely in terms of increased information. The example illustrates the case in which an agent corrects his judgment and gains a better understanding of what matters to him, e.g., through the mild pain of regret. To fully capture Fabien's look back on his past experience is better to say that he has put matters into view. Insofar as this transformation is not merely a revision of previous beliefs, it qualifies as a very basic case of *change in view*.

For the sake of the argument, I restrict the term "change in view" to cases in which there is more is at stake than acquiring new information.

Clearly, the acquisition of new facts, which are relevant to settle the case, or the correction of mistaken beliefs are important ways in which we modify and improve our epistemic position. However, these revisions do not count as changes in view.[13]

What is interesting about this example is that Fabien's transformation cannot be easily factored out in a change in beliefs on the one hand, and change in values on the other. Furthermore, the change is not based on acquisition of new facts of the matter, nor is it a requalification of information Fabien had, e.g., about the plants needing water or the effect of dry weather. Finally, Fabien does not passively undergo a change of heart. It would be ridiculous to suggest that he is overcome by the passion for jasmine. Rather, he has developed an emotional attachment, which puts matter into a new perspective. Perhaps, this gives him new reasons and goals. The nature of the object of attachment, a plant, is such that the normative echo of introducing this new goal is rather limited. The example is meant to show that there are trivial and mundane cases of change in view that cannot be treated in terms of matter of facts plus function of value. If these relatively simple experiences of change resist such a treatment, there is reason to think that the non-cognitivist formula will be inadequate also in more complex cases.

How is a complex case of change in view more complex than Fabien's case? My suggestion is that we consider changes where the descriptive dimension is profoundly altered in a given temporal trajectory. Suppose that, in a communist country, the parents of a talented boy named Kai give their child away to the state, so that he can be trained properly. In exchange, they receive enough support to provide for their other two children. They know that they are taking away the boy's childhood, and sacrifice him to support the other two. They do not know that his career prospects will make up for this sacrifice. These parents think that they are choosing between two evils, and suffer severe guilty feelings for sending their boy away. The child feels wronged. Time goes by, and Kai becomes a virtuoso, profoundly devoted to music and completely satisfied with what he has become. His attachment to music annuls his resentment for having been wronged. The parents feel proud. The son's success cancels the parents' sense of guilt. They think they have done what was best for Kai: they have believed in him and they have overcome their natural but egoistic attachment toward him in order to increase his otherwise meager prospects of leading a good life.

The temporal trajectory reveals a complex multi-dimensionality of the change in view. Differently than in the previous case, the characters of the story not only bring what matters into view, but they also positively change the facts of the matter and the normative valence of what they did and was done to them. The locus of this complex change is the description of what happened, but such a description is not merely about facts.

The way in which each of the characters tells the story from beginning to end reflects a profound revision of what matters.

The case is complex also because the driving force of the change is complex, indeed more complex than in Fabien's case. All characters in the example revise the *description* of the context of choice because of their present emotional involvement. Emotional attachment is not a brute force and, unlike the previous case, it is part of the complex network of personal relations and normative expectations. Kai adjusts to the expectations and demands of others (e.g., to develop his talent and do well). The parents also adjust to the normative demands and expectations of others (e.g., to provide for their children, and give them the best chances of a good life). Eventually, both the mature musician and Kai his parents agree that the decision to send him away was right while at the time the decision was taken all agreed that Kai was wronged.

Furthermore, emotional involvement changed the focus of the description. Initially, the relevant matter is whether it was right to deprive the child of his childhood in order to gain support for all other children. In the end, the relevant matter is whether sending the child away was the best thing to do for bettering his prospects.[14] By calling attention onto this shift, I do not mean to suggest that the characters of the example have merely developed a new evaluation that attaches to the facts of the matter. Nor is it that they are merely wording and articulating the same crude facts in more positive terms. Rather, the point is that their emotional involvement makes the characters of the story sensitive to different aspects of the situation, so that they relate different facts. Supposedly, this is not because time heals, but because such characters have *reasoned through* what happened to them in new ways, until their description of the initial situation altered. They have engaged emotionally with the events, their choices, and their relations.

My claim is twofold: first, the experience of reasoning entails emotional engagement by which the facts of the matter are altered; and, second, an important form of practical reasoning consists in the active alteration of the description of the facts that matter. One may find both claims too radical. The first implies that our emotional engagement has an effect on what we take the facts of the matter to *be*. It also implies that one can change one's past by emotionally relating to the past, and literally not metaphorical. This is a rather awkward position, which violates an obvious claim about the irreversibility of the past. The second claim may seem unnecessarily restrictive, especially because it is contestable that a change in view is the appropriate outcome of practical reasoning. I will provide further considerations about the legitimate aims of reasoning in the two final sections of the chapter. I now turn to the defense of the first claim about emotional engagement.

5 Why a Change in View is Not a Reassessment

In commenting the example of the talented musician I have argued that it counts as a process marked by a historical trajectory. The co-protagonists give the trajectory a direction, by taking chances and making decisions. While they certainly do not control all aspects of the story, they are not merely affected by external changes. These agents exercise their agential role partly by engaging emotionally with the circumstances of action.[15] This claim will prove important in arguing further that the exercise of reasoning is not disjoint from the exercise of some emotional capacities, and that marks the distinction between facts and value. Emotional attachment serves important evaluative functions, and it is not a blind force driving people in their reasoning. In fact, the previous example of the musician shows that because of the impact of emotional attachment, the protagonists change the facts of the matter over time.

A critic might object that this strong claim is unnecessary to make sense of the example above. He may say that the same point can be made without admitting any alteration in the facts of the matter, simply by noting that the characters in the story have acquired new information, by living. At the time the parents chose, they hoped but did not know for sure that sending away the child would turn out to be for his best—nor did the boy know that music will become the main reason of his life in the end, it all turned out well. One may have qualms about whether the original decision was a lucky guess, or else well supported by a sophisticated probabilistic calculus, but the new description is due merely to the acquisition of facts that were not available beforehand. They all have learned by experience new facts of the matter, which provides the newly furbished descriptive basis for a different normative assessment.

Compare the previous reconstruction of Kai's case with the following example. Suppose Yael loses interest in the *Book of J*, which he had cherished almost as sacred in his youth; it seems inadequate to say that he grew more acquainted with the book, and as a consequence his attitudes toward it changed. It is also insufficient to say that Yael grew less attached to the book, and became disinterested in it, despite that she judges the contents of the book and all the relevant facts about it in the same way. If, for instance, Yael's loss of interest in the book is due to the fact that Yael does not believe that the book has been written by God, this is hardly something she has learned by new acquisition. The fact that Yael does not believe the book is written by God is not due to her discovery that J is not God, or that J has not written that book. Perhaps, Yael encountered some philosophical argument she did not know before and found it persuasive. More likely, however, it is the case that she does not believe any longer that the book contains the sort of guide she was expecting from God. Hard to say whether this "discovery" falls on the side of the facts or on the side of value. But the process leading to Yael's disbelief is certainly not a reassessment of facts.[16]

There is another consideration that counts against reducing changes in view such as Kai's to reassessment of the facts of the matter. It concerns examples in which the direction of revision is problematic. On the assumption that facts and values are sharply separated, and the change in view is reducible to a normative assessment of facts, the question about the direction makes sense. In some cases, a change in view is a revision in beliefs due to the acquisition of new facts or correction of mistaken beliefs, as it happens when considering how best to reach a destination. The new reassessment is based on a change in belief. In other cases, the direction of revision is the opposite. For instance, Fabien acquires a new taste about jasmines, and then some facts become salient in his perspective.

My argument is not that such there are no interesting cases of (practical) reasoning that move from acquisition of facts newly available (e.g., facts that others have construed through their inquiry, such as the facts of science) toward a revision of previous value judgments; and, conversely, from the acquisition of new goals, interests, and values toward the examination and reassessment of facts that were previously available. On the contrary, I think these are two ways in which we ordinarily make progress in thinking about how to act. My point is that there are large varieties of (mundane and tragic) cases that do not fit this mold. In particular, there are cases in which the question about the direction of change (from facts to values or vice versa) does not admit of an answer because it *does not make sense*. I take this point to be supported by the following example.

In a memorable page of *Anna Karenina*, Karenina suddenly notices that her husband's ears are disturbingly inelegant. Here is Tolstoy's description of the event: "'Great heavens! What happens to his ears?' she thought, gazing at his cold and commanding figure, and especially to his gristly ears which now so strike her, pressing as they did against the rim of the hat" (Tolstoj 1995, 102). Does this reaction indicate that Karenina had not seen her husband's ears before, or that she never saw him really, for what he was? Is it sign that she does not love him any longer, or is it a consequence of her loss of interest in him? This line of questioning presumes that the only interesting matter is whether Karenina's reaction tracks a revised belief about her husband's appearance, where this is to be understood in terms of objective facts of the matter, or else it registers something else, a change of heart, perhaps. In sum, on the model under scrutiny, a change in view is either a change in the facts, or a change in the value, and the question is what generates what.

Tolstoy's description of Karenina's reaction and subsequent reasoning shows another dimension of investigation, which eludes the non-cognitivist model. First, we learn that Karenina feels an unpleasant feeling of dissatisfaction, as if she expected to find her husband different. Then we learn that she is particularly struck by the feeling of dissatisfaction with herself, which she experienced when she met her husband. This remark opens a new dimension of exploration of Karenina's mind. In fact, Karenina is portrayed as engaging in a second level of reflection, concerned with her self, and focused on the legitimacy of her feeling of dissatisfaction with her husband. During this painful exploration, she pays attention to her emotional attachments and aversions, to further understand the nature of her feelings. As we follow Tolstoy's insisted description of the event, we progressively lose track of the original facts of the matter, Karenin's gristly ears.[17] The subsequent re-descriptions of the event reveal the inward direction: the facts of the matter that triggered Karenina's reaction and her reflection were nothing but the occasional starting point of a self-exploration. The real task of deepening the description is to explore Karenina's mind. Past the relation between facts and values and the questions about the direction of fit, we are investigating the nature of the emotional dissatisfaction, which is the focus of Karenina's reflection. We learn that this is not a feeling unknown to her. On the contrary, "it was that ordinary well-known feeling, as though she was dissembling, which she experienced in regard to her husband, but formerly she had not noticed it, while now she was clearly and painfully conscious of it."

To make sense of Karenina's displeasure and her subsequent reflection about her self, and to understand the point and the direction of her reasoning, we have to appreciate the epistemic role of her emotional engagement.

6 Emotional Engagement

In the previous section, I have replied to the objection that the examples I related can be fully accounted in terms of revision of beliefs or normative reassessment. There might be other reasons to resist my claim that emotional involvement is part of the practical reasoning leading to a reasoned change of view. In particular, a worry might be that by crediting emotion with this crucial role, I thereby treat practical reasoning as a heteronomous process, driven by forces alien to agential activity. The worry is motivated by the presumption that emotions are blind forces, which threaten agential autonomy, that is, the autonomy that is distinctive of agents in acting on the basis of reasons.

What role emotions are said to play in practical reasoning importantly depends on what emotions are said to be. I take emotions to form a broad and heterogeneous category, thus I do not think there is a unique response to the worry I just reported. If emotions are taken to be mere sensations, their role in practical reasoning might be thought to be limited. By contrast, if they are (like) perceptions, then their epistemic function becomes central. The point I want to make bypasses the dispute about the definition of emotions. This is because in my view the key epistemic component of the change is emotional *engagement*.[18]

I take emotional engagement to signify the sort of active involvement that agents ordinarily pursue when they feel something *and* they also feel under pressure to reflect on the nature of their feeling. Emotions generate some sort of affective involvement on the side of the subject. Indeed, this is what (at least some higher order) emotions do: They create ties between the subjects and their surroundings by focusing attention, and eventually trigger action.[19] I am interested in the particular case in which the subject is not just affectively involved in her emotion, but she also reflectively engages with her emotion in a way that leads to a change in view, e.g., by paying attention to the emotion she changes her mind about something. The case in point is that of a woman who engages with her unpleasant feeling of disgust for the appearance of her husband, she is displeased with herself for this feeling, and by focusing attention on the feeling itself, she starts to bring matters into view.

The change in view is brought about by the emotional engagement and consists in the fact that she engages fully with her feelings and faces her discomfort. Emotional awareness counts as a reflective achievement. But such emotional awareness is not the sole upshot of this reflective activity. There is a prospective deliberative component in this change in view. Aware of the depth of discomfort, Karenina appreciates her unhappiness and also the condition of entrapment: she prepares herself to search a way out. This is how emotional engagement interfaces practical reasoning, without undermining agential autonomy.

7 From Emotional Engagement to Reasoning

It is useful to clarify the import of this view for practical reasoning, by restating the distinction between emotional involvement and emotional engagement. To be sure, the mere presence of emotional involvement does not warrant that the agent learns something new, or something true about what matters. In fact, emotional involvement may work against the truth, rather than toward it. Traditional views of emotions as disturbances, external to practical reasoning, often relate this very point. But the ambiguous role of emotions may be noted independently of any specific view on the nature of emotions. It is often the case that rational agents fail to consider the evidential weight or the importance of some facts because they are involved emotionally in ways that impede them to see things aright. (This is just in a manner of speaking, which does not allude to any metaphysical view). Emotional involvement can cloud our judgment and interfere with our rational capacities. For instance, it may provide motives to positively resist evidence. This interference is not only registered at the level of motivation, however. It is often the case that agent's emotional attachments posit obstacles to revision and correction because they undermine the relevance of some salient facts. Their role is somewhat epistemic and normative, as it consists in discounting facts that in other circumstances the agent would consider salient in her assessment of the situation. Such emotional attachments may be deep enough to make the agent immune to guilty feelings in the case of transgression or violation of moral duties.

These phenomena where emotional attachments interfere with the good functioning of our complex rational capacities are significant, but they do not support the non-cognitivist view. Emotions themselves can be either appropriate or inappropriate, and there are normative criteria to assess the varieties of their epistemic role and importance. On the basis of such criteria, we can identify ways in which emotional attachments directly contribute to practical reasoning. Furthermore, and contrary to what the non-cognitivist insists, such roles are not thoroughly passive. For instance, in some cases, emotional involvement is sufficient to alter one's reasoning, by providing new objects of attention or, more generally, to give reasoning a point. For instance it is because Serena loves her son Elia that she gets interested in planning a summer vacation, which suits his needs, and starts gathering information about the seashore facilities for children, comparing prices, reviewing offers, and considering pros and cons. What would be the point of engaging in this reasoning, if she did not love him so?

For some philosophers, this is because emotions are like perceptions and thus fulfill an epistemic role. In my view, emotions may be modes of valuing, but I would not say that they are forms of moral discernment or that they track value; nor does my view imply that emotions straightforwardly provide normative reasons for action. They provide considerations that have normative significance and should be audited in the process of forming reasons. My view is that emotions acquire such a determinate normative status through the reflective engagement of the agent. It is through this engagement that the agent becomes aware of her emotional attachment and of its normative significance. I take this self-reflective aspect of engagement to be extremely important for it provides a starting point for practical reasoning, at least in some ordinary cases such as the ones discussed in this section.

By stressing the self-reflective aspect of the emotional involvement I want to indicate that it brings an awareness that is in itself a reflective achievement. However, emotional involvement intervenes at various stages in thinking about action. To some extent, its contribution is cognitive, even though not in the sense in which theories of moral perception say. The cognitive function of emotional engagement is not analogous to the evidential support: loving music does not bring evidence that music is lovable. Love is not a mode of discernment of the value of music. Nor does the presence of love guarantee that its role is transformative for the agent.

Rather, the musician's emotional engagement with his profession broadly identifies the practical standpoint from which the agent starts reasoning. Such emotional engagement constrains his considerations of what it takes for his life to be worth living: it gives his reasoning a point. In some cases, such reasoning may be deliberative, concerning what to do next. In this case, emotional attachments play a crucial role in identifying the circumstances of action. My conjecture is that this is because emotional attachments play a constitutive and hermeneutical role in shaping the facts of the matter and bringing what matters into view. We identify, classify, and interpret the facts of the matter through our emotional involvement with them. This conjecture runs against the alleged fact/value dichotomy, and carries epistemological and ontological consequences. From the epistemological point of view, the claim is that the agent identifies the relevant facts through emotional assessment. From the ontological point of view, the claim is that we can modify the objects of our emotions, hence the facts of the matter, by emotional engagement. To some extent, then, the facts are not invulnerable to judgment and evaluation.

This view has implications for the prospective dimension of rationality, that is, rationality as it unfolds in time.

8 Objective Standards for Prospective Rationality

As mentioned in section 1, a problematic aspect of theories that admit or, rather, defend the entanglement of fact and value is that they seem to lose objectivity. This is a serious problem, especially for a study dedicated to reasoning. To reject this conclusion, prominent philosophers, such as Putnam, have argued that one should not equate objectivity to description, as if "description" meant to convey a neutral representation of mind-independent objects. There might be objectivity without objects, and since descriptions are about objects, there might be objectivity without description.[20] A further point is that there might be descriptions that contain thick evaluative concepts, whose descriptive or factual meaning is significant, and yet they perform an evaluative function (cf. Putnam 2002, 34–8). It is in this connection that the pioneering work of Murdoch is generally mentioned.

However, it seems to me that the most interesting aspect of Murdoch's work goes in the opposite direction; that is, in the direction of *reinforcing* the role of description. In fact, her main claim is that description is where the efforts of the mind concentrate. Objectivity is an achievement rather than conformity of the intellect to the object. But this achievement is obtained by re-describing the facts as they come into view, hence visible in descriptions. Ordinary agents make progress by taking notice of the object of emotional engagement, e.g., the other as a target of our loving attention. Thus, describing is a key activity, rather than a neutral and valueless representation of objects. It is something in which agents engage actively. Precisely because this sort of engagement is a privileged locus of moral activity, it comes with distinctive responsibilities. This is the sense in which moral activity does not take the form of a procedural deliberation, but it is exercised in many ways, which engage emotion, attention, and imagination.[21]

9 Reasoning as a Transformative Activity

I share Murdoch's view that objectivity is an achievement, and also the claim that it is a personal achievement. Differently than Murdoch, however, I argue that this is an achievement brought about by practical reasoning.[22] To vindicate this claim, it is important to discard the view that practical reasoning is procedural and algorithmic, a merely formal scaffold by which we organize factual matters. The sort of cognitive operations required in the application of a procedure are mechanic and routine, hence do not require any significant contribution from the agent. Of course, the agent may intervene at the very start by picking the decision procedure, but for this choice to be reasoned, rather than arbitrary, we still need an account of reasoning. To some extent, reaching a formalization of a decision procedure is the result of reasoning, not the way to do it. Thus, my argument does not deny that it is possible and rationally justified to construct, and then rely on decision procedures. What I dispute is the claim that reasoning amounts to applying a procedure or a function.

On the basis of what I have argued thus far, reasoning should be treated as an activity that produces a reasoned alteration, not necessarily a revision of belief or a change in concern, but a change in view. Reasoning is an activity in which we engage because of our concerns, and starting from our personal perspective. The examples I offered are cases of change in mind in which emotional engagement is crucial, and emotions are distinctively perspectival.[23] Nonetheless, the descriptions that one achieves as a result of focused emotional engagement can be more or less perspicuous and objective. In these final two sections, I provide a constructivist account of the "rational activity" of the mind, and show that such activity is not unconstrained, even though perspectival.[24]

To some extent, the constructivist conception of reasoning matches the definition of reasoning as "reasoned change in view" proposed by Gilbert Harman (1986).[25] Like Harman's, it starts with the acknowledgment that reasoning is something people *do*. Constructivism conceives of reasoning as an activity, rather than as the product of reasoning; it highlights that reasoning itself understood as the way in which reasoners produce a change in view is not linear, and it is to be distinguished from arguments and proofs. The "change in view" is a complex perspectival change, which is the result of the networking of discursive and emotional capacities.[26]

Thus, I agree with Harman on a fundamental point, which concerns the form that reasoning must take. The conception of reasoning as an activity is in tension with the standard view that practical reasoning is either deductive or inductive, and governed by the rules of logic, hence takes the form of practical inference and syllogism.[27] This may come as a surprise because Kantian ethics is often associated with deductive reasoning. As I have argued elsewhere, this interpretation rests on a misconception of the role of universal principles in the Kantian account of practical knowledge. The constructivist account of practical reason has the advantage of clearing the ground from this mistake, by pointing to universality as the structure of rational activity. Reasoning as a revision is productive and generative because it deploys principles as tools for construction, rather than as material first points. Constructivists themselves have not been clear enough on this issue. Current formulations do not fully capture the real difference between classical and constructivist accounts of rational justification.[28] Rather than reference to formal procedure, a more robust conception of construction is needed for constructivism to qualify as a distinct form of rational justification, genuinely alternative to practical inference and other standard logical forms of reasoning. On this more robust conception, a construction of reason is generative, that is, it produces something that was not in place before: a reasoned change in view.

In support of this robust understanding of construction, I offer a more discriminating characterization of practical reasoning, which I take to be peculiar and distinctive of (Kantian) constructivism.[29] In contrast to reasoning understood in terms of arguments and proofs, I conceive of reasoning as an activity of reason. Activities differ from processes in that they require the active engagement of agents. Thus, practical reasoning is an activity guided by structural principles of reason. In order to issue authoritative judgments, agents engaging in practical reasoning ought to be guided by such principles. A Kantian claim is that being guided by the principles of reason is a structural warrant of autonomy. This is to say that the sort of autonomy that reason provides is a minimal requirement for being authoritative judges and agents. Of course, this minimal requirement does not guarantee that such reasoners are autonomous in more mundane and ordinary senses. Thus, more needs to be said about whether a given subject entering reasoning is autonomous. The definition above fixes the formal condition for his activity to be autonomous, not a threshold for subjects to have personal autonomy.

The distinction between "activity" and "process" brings to light another difference in the way Harman and I conceive of reasoning. For

Harman reasoning can be unconscious, while Kantian constructivism takes reasoning to be a conscious, self-reflective activity. This is not to deny that there are operations that we make unconsciously, and which may underlie our reasoning. However, that agents are conscious of their reasoning is a condition for their reasoning to be subjectively authoritative, that is, authoritative in the first person.[30]

Finally, the constructivist characterization allows for a finer distinction between practical and theoretical reasoning. Harman assumes that one's reasoning can be "separated into distinct segments of practical and theoretical reasoning," even though he recognizes that any given segment of reasoning can affect beliefs and intentions (Harman 1986, 1).[31] He also assumes that theoretical and practical reasoning concerns beliefs and intentions, respectively; even though he recognizes that intentions may affect beliefs, and beliefs may affect intentions. By contrast, Kantian constructivism is interested in the distinction between theoretical and practical reasoning that is best captured in terms of first-person vs. third-person authority (cf. Bagnoli 2013). This is because the constructivist view is concerned with subjective authority, rather than with the objects or final ends of reasoning. I propose a general account of reasoning, which is held together by principles.[32]

Is this theory descriptive or normative, given that it deploys structural norms? Structural norms tell us how we ought to think about what to do. Harman says it is both descriptive and normative, since it starts with what we do, but it contains abstractions and idealizations that are necessarily normative. The appeal to principles is meant to guarantee the coherent structure of our thought, but also a coherent account of deliberation over time. The rational principles that guide our thought and action are principles that vulnerable, interdependent agents, endowed with similar emotional capacities, could share. The constraint that principles put on our practices makes a constitutive reference to others, not as representative of an abstract community, but as our interlocutors and co-legislators.[33] Our freedom to change view is constrained by principles, hence by others. Reasoning about outward objects and inward reasoning are governed by the same principles, and likewise are made possible by shared practices of validation and mutual recognition.[34]

Notes

1 "A Cambridge Philosopher once said that, since the use of the definite article implies that there is only one thing to refer to the use of the definite article in the expression 'The Oxford Movement" is an indication that Oxford moved only once. Whatever truth there is in this allegation, It would certainly be a mistake to suppose that there is but one emotive theory of ethics, because we speak of the emotive theory of ethics. There are in fact as many such theories as there are philosophers whose view can be so described" (Urmson 1968, 7).

2 As examples of this view, one may take Hare (1952, Ch 7.1 ff); Hare (1963, Ch: 1–3); Edwards (1955); Stevenson (1963); Urmson (1968); and Mackie (1977).
3 Hare's reply to Foot is that for such thick concepts "it is possible to accept that an act satisfied the descriptive conditions for being called 'rude' without being committed to evaluating it adversely, even though 'rude' is normally an adjective of adverse evaluation" (Hare 1981, 74). This is because the thick concept 'rude' can be deployed purely descriptively, which indicates that its normative or prescriptive meaning is separable from its descriptive meaning. For Hare the standards of objectivity are thus rooted in the descriptive meaning, which is separable from the evaluative meaning. By contrast, many think that thick concepts are not factorable into a purely descriptive component and an attitudinal or prescriptive component, see McDowell (1985) and Williams (1985).
4 There is an interesting contrast between Humean and Aristotelian naturalism, see McDowell (1995) and Foot (2001, 6ff). Emotivists appeal to Hume for an original epistemological argument, see Urmson (1968, 19); and Hare (1963). Humean naturalism tends to support non-cognitivism, broadly considered, because it denies that there are moral facts. By contrast, Aristotelian naturalism is a form of cognitivism, because it recognizes something like natural normativity, see Anscombe (1957) and (1958); and Foot (2000).
5 There seems to be a fairly large agreement that virtues and values are important also in the practice of science and drive scientific inquiry. However, philosophers arguing for this claim also generally hold fast to a distinction between epistemic and moral virtues and values. I doubt that we can make sense of the distinction. Take, for instance, the virtue of accuracy; it is unclear why one should insist that it is an epistemic rather than moral virtue. On the role of emotions in service of cognition, see Sheffler (1991, Ch. 1).
6 "If objectivity of normative claims is to be impugned, then it must be on other grounds" (Elgin 1997, 185).
7 This claim is profoundly related to Quine's attack on the synthetic-analytic distinction; see Quine (1951). The focus of my argument, however, is the formal character of the method. This reconstruction is not too distant from Urmson's view of the rise of emotivism, as originated in "epistemological despair," once naturalism is found to fail, and non-naturalism is considered abhorrent (Cf. Urmson 1968, 19).
8 There might be complications, but these are generated by motivational hindrances or mere ignorance. If the agent does not follow up on her reasoning, this can be explained by insincerity or misunderstanding, see Hare (1952 and 1963); and Murdoch (1997, 315–16).
9 Hare soon became keenly aware of the limitation of this way of dealing with moral problems, insofar as he recognized that there are conflicts that cannot be solved by appealing to the procedure generated by the logic of moral discourse, see Hare (1963, 137–85). However, in the subsequent elaboration of his ethical theory, he did not abandon the claim that there is such a procedure. On the contrary, the development of preference Utilitarianism rests on the claim that all moral problems can be treated by a preference Utility function, see Hare (1981).
10 I have discussed Murdoch's argument in the context of her exploration of moral life in Bagnoli (2011, 205 ff). See also Diamond (1996) and Millgram (2005).
11 This might be thought to be in line with Anscombe's famous remark that there cannot be an ethics without a moral psychology, see Anscombe (1958). However, Anscombe defends a classical view of practical inference, see Anscombe (1957).

12 I endorse only partially Baier's definition of change in view, even though I agree that not all alterations of the mind count as a change in view. In my view, Baier overemphasizes the active aspect of the change in view, as if no other alteration of the mind contained any cognitive effort on the side of the agent. For instance, she writes: "Not every alteration of "cognitive state" or "registration," or "Bayesian belief" is a change of mind. Acquiring new information which pushes back the frontiers of one's ignorance is not, for us or for our half-wanton ancestors, a change of mind, nor is the mere updating of old information, letting one's mind reflect the changes in the observed world around one. A change of mind is a correction, but not all cases of mistake-correction, and no cases or ignorance-remedying, count as such changes" (Baier 1979, 54). I think this is an exaggeration, which does not fit all the ways in which ignorance is remedied and mistakes corrected, as Fabien's case shows.

13 I agree with Baier that not all revisions are changes in view, and I also agree that an important feature of the change in view is the activity of the agent (cf. Baier 1979, 55–66). However, I do not fully endorse Baier's definition of change in view.

14 Recently, Wallace has argued that present attachments make us immune to regret, see Wallace (2013) and Bagnoli (2016). In this essay, I focus on description in order to press the case that we are completely passive neither in regard to our present emotional attachments, nor in regard to our past commitments and choices. This is because of the effect of emotional involvement and engagement with our past and present choice, as I will argue in sections 6–7.

15 I am working out something similar to what McDowell calls "acting in the light of fact", but I insist that the fact that matters is shaped by emotional engagement, compare McDowell (2013).

16 It may be inappropriate to count it as a "change in view" in case this happens as a result of a radical change of framework. Baier reserves the term change in view to short or long-term changes in goal, and excludes that conversions may be counted as changes in view. However, Yael's change may not be the result of any major revolution in values. In fact, Yael may hold the very same values and beliefs as before concerning how to lead her life and what makes it good; except, she now does not think all that in terms of following God's command.

17 Tolstoj seems to me to use a particular form of internal analepsis, where the narrator goes back repeatedly to the same event, every time uncovering some new detail, and opening a new narrative dimension. It is not the usual external flashback, but an insisted re-description that moves inward (Tolstoj 1995, 102).

18 This is not to say that all emotions can be the focus of emotional engagement; some emotions may not be. However, typically, emotions generate involvement: the opposite case it generally categorized as alienation or lack of authenticity.

19 Perhaps it would be more prudent to restrict this consideration to so-called higher order emotions. However, I am not convinced that even the most basic emotions, such as fear, are not capable of triggering self-reflective processes. While walking, fear of losing track of one's path may be so confusing that one loses track of one's location; but it may also be so acute that it alerts the agent to effective counter-measures to adopt, so that he is more carefully and attentive to details than ordinarily is.

20 "What I am saying is that it is time to stop equating *objectivity* with *description*. There are a lot of statements [. . .] that are not descriptions, but that are under rational control, governed by standards appropriate to their particular functions and contexts" (Putnam 1989, 33).

21 Millgram notices that changes in view take time because they involve redirecting attention: "On Murdoch's account, moral progress is a matter of gradually redirecting one's attention; this happens incrementally, and can be assisted, but not compelled, by providing better objects on which to fix one's attention. So we should not expect an argument, even a novelistic argument, that is intended to force an abrupt turn in one's moral life" (Millgram 2005, 195, n. 31). Murdoch says that the "constructive activity" of imagination "introduces value into the world which we confront." Her claim is that "We have already partly willed our world when we come to look at it; and we must admit moral responsibility for this fabricated world" (Murdoch 1997, 201). See also Murdoch (1997, 331–2); and Bagnoli (2011, 208–9). Imagination is a mode of attention, see Lovibond (1983, 41–4, 46, 48).

22 In this respect, I differ from Murdoch, because she does not trust reason to guide all interesting forms of reflective change. I think this difference is partly due to her narrow understanding of reasoning, which she associates with proceduralism, as I show in Bagnoli (2011).

23 This is feature of emotions that I consider decisive for the functioning of practical reasoning. See also Greenspan (1988). On the relation between objectivity and perspectival judgment, see Elgin (2010).

24 The conclusion I reach in these sections are close to the sort of constructivism that Caterine Z. Elgin defends about science: "If scientific facts can be relative and value-laden, there is a strong *prima facie* case for saying that relativity and value-ladeness do not undermine objectivity" (Elgin 1997, 185). My argument will be on entirely different grounds, however, and it is to be situated in a larger picture of practical knowledge centered on autonomy, see Bagnoli (2013).

25 See, in particular, the first programmatic chapter, interestingly titled "A Plea for the Study of Reasoning."

26 Harman stresses the significance of emotions in changing view. See also Greenspan (1988 and 2011).

27 I have argued for this claim in Bagnoli (2016). Harman takes this stand, on the basis of different considerations. "Obviously, there is deductive argument, but it is not similarly obvious that there is deductive reasoning. Again, it is not clear that there is such a thing as inductive reasoning [. . .]. Analogous remarks also apply to the suggestion that there is such a thing as practical syllogism" (Anscombe 1957, 57). A syllogism is a form of argument, and although there is practical reasoning, there is not obviously any such thing as practical argument or logic and so not obviously any such thing as a practical syllogism [. . .] It may be a mistake to expect principles of reasoning to take the form of a logic" (Harman 1986, 6). Harman concludes that it is unclear how the formulation of logics of thought might contribute to the study of reasoned revision.

28 This is true of definitions of constructivism in terms of procedure, such as Korsgaard's, as well as of those characterizations that use the notion of entailment confined to practical standpoint, such as Street. It is not obvious that rules of inference, implication, and entailment have anything to do with practical reasoning.

29 There are different varieties of constructivism and Kantian constructivism. For the latest statement of my variety, a cognitivist but irrealist account of constructivism, based on the claim of autonomy, see Bagnoli 2013.

30 Arguably, it is also a condition for moral responsibility and imputability, even though it might be that, in a different broad sense that does not include imputability, agents are morally responsible also for aspects of reasoning of which they are not aware, insofar as such aspects are expressive of character

31 Harman distinguishes between two phases of reasoning, the first consisting in the process of reflection, and the second consisting in the actual revision of one's view. Correspondingly, he identifies maxims of reflection and principles of revision. This distinction is not relevant to my purposes here.

32 In contrast to Harman, I do not think that the principle that we ought to be coherent is defeasible, compare Harman (1986, 11). On the contrary, I take it to be constitutive of reasoning. However, the difference is likely to be less sharp than it appears because we disagree on the status of the principle. On the Kantian view, coherence is not defended as a *logical* principle.

33 According to Baier, "the liberty to change one's mind importantly depends on the validation and recognition of others" (Baier 1979, 66). However, she insists on the conventional practice of "second-thinking."

34 I would like to thank Lorenzo Greco, Elijah Millgram, and Robert Schwartz for comments on previous drafts.

References

Anscombe, G. E. M. 1957. *Intention*. Cambridge, MA: Harvard University Press.

Anscombe, G. E. M. 1958. "Modern Moral Philosophy", *Philosophy*, 33(124): 1–19.

Bagnoli, Carla. 2011. "The Exploration of Moral Life." In *Iris Murdoch, Philosopher*, edited by J. Broackes, 193–221. Oxford: Oxford University Press.

Bagnoli, Carla. 2013. "Constructivism about Practical Knowledge." In *Constructivism in Ethics*, edited by Carla Bagnoli, 153–82. Cambridge: Cambridge University Press.

Bagnoli, Carla. 2016. "Rooted in the Past, Hooked in the Present: Vulnerability to Contingency and Immunity to Regret", *Philosophy and Phenomenological Research*, 92 (3): 763–770.

Baier, Annette. 1979. "Mind and Change of Mind", *Midwest Studies in Philosophy*, 4(1): 157–76. In (1985). *Postures of the Mind*. Minneapolis: Minnesota University Press.

Diamond, Cora. 1996. "We Are Perpetually Moralists: Iris Murdoch, Fact and Value'." In *Iris Murdoch and the Search for Human Goodness*, edited by Maria Antonaccio and William Schweiker, 79–109. Chicago: University of Chicago Press.

Edwards, Paul. 1955. *The Logic of Moral Discourse*. Glencoe, IL: Free Press.

Elgin, Catherine Z. 1997. *Between the Absolute and the Arbitrary*. Ithaca: Cornell University Press, 176–91.

Elgin, Catherine Z. 2010. "Keeping Things in Perspective", *Philosophical Studies*, 150(3): 439–47.

Foot, Philippa. 2001. *Natural Goodness*. Oxford: Oxford University Press.

Greenspan, Patricia S. 1988. *Emotions and Reasons: An Enquiry into Emotional Justification*. London: Routledge.

Greenspan, Patricia. 2011. "Craving the Right: Emotions and Moral Reasons." In *Morality and the Emotions*, edited by Carla Bagnoli, 39–61. Oxford: Oxford University Press.

Hare, Richard M. 1952. *The Language of Morals*. Oxford: Clarendon Press.

Hare, Richard M. 1963. *Freedom and Reason*. Oxford: Clarendon Press.

Hare, Richard M. 1981. *Moral Thinking: Its Levels, Method, and Point*. Oxford University Press.
Harman, Gilbert. 1986. *Change in View: Principles of Reasoning*. Cambridge, MA: MIT Press.
Lovibond, Sabina. 1983. *Realism and Imagination in Ethics*. Oxford: Blackwell.
Mackie, John. *Ethics: Inventing Right and Wrong*. London: Penguin Books.
McDowell, John. 1985. "Values and Secondary Qualities." In *Morality and Objectivity*, edited by Ted Honderich, 110–29. London: Routledge.
McDowell, John. 1995. "Two Sorts of Naturalism." In *Mind, Value and Reality*, 167–98. Cambridge, MA: Harvard University Press, 1998.
McDowell, John. 2013. "Acting in the Light of a Fact." In *Thinking about Reasons: Themes from the Philosophy of Jonathan Dancy*, edited by David Bakhurst, Margaret Olivia Little, and Brad Hooker, 13–28. Oxford: Oxford University Press.
Millgram, Elijah. 2005. *Ethics Done Right: Practical Reasoning as a Foundation of Moral Theory*. Cambridge: Cambridge University Press.
Murdoch, Iris. 1997. *Existentialists and Mystics*. New York: Penguin.
Putnam, Hilary. 2002. *The Collapse of the Fact/Value Dichotomy and Other Essays*. Cambridge MA: Harvard University Press.
Quine, Willard V. O. 1951. "Two Dogmas of Empiricism", *Philosophical Review*, 60(1): 20–43.
Sheffler, Israel. 1991. *In Praise of Cognitive Emotions*. London: Routledge.
Stevenson, Charles L. 1963. *Facts and Values*. New Haven: Yale University Press.
Tolstoj, Leo. 1995. *Anna Karenina*. Edited by E.B. Greenwood. Hertfordshire: Wordsworth.
Urmson, James O. 1968. *The Emotive Theory of Ethics*. Oxford: Oxford University Press.
Wallace, Jay. 2013. *The View From Here: On Affirm*ation, *Attachment, and the Limits of Regret*. Oxford: Oxford University Press.
Williams, Bernard. 1985. *Ethics and the Limits of Philosophy*. London: Fontana Press.

8 Normativity without Normative Facts? A Critique of Cognitivist Expressivism*

Alexander Miller

1 Introduction

Can we do without normative facts? Morality is the paradigm example of a normative area of thought and talk, and a prominent theme in twentieth and twenty-first century metaethics has been the attempt to account for our everyday moral practices without relying on the notion of a moral fact. Philosophers have been suspicious of moral facts for a number of reasons, but one influential line of argument goes as follows. Let a *natural* fact be a fact that is part of the subject matter of the natural or social sciences (cf. Moore 1903, 92). If moral facts are held to be non-natural, suspicions arise about their nature (they seem to be very different from the familiar facts of science or commonsense), and about how we come to have knowledge of them (the familiar sorts of knowledge-gathering activities deployed by science or commonsense seem to provide knowledge only of natural facts). Non-natural moral facts are thus suspicious on *metaphysical* and *epistemological* grounds. So could moral facts be natural facts? If so, the metaphysical and epistemological worries are perhaps more tractable, but we now face difficulties in the area of moral *semantics*: for one thing, moral judgments don't seem like judgments about natural states of affairs since the former seem to be intrinsically motivational in a way not shared by the latter, and Moore's open-question argument suggests that moral judgments have a normative force not possessed by naturalistic judgments.

Historically, considerations like the above have prompted many philosophers to eschew moral facts, either via error theories or via forms of non-cognitivism or prescriptivism. Error theorists (e.g. Mackie 1977) attempt to avoid the semantic worries facing moral naturalism by construing the contents of moral judgments in terms of non-natural facts, but avoid the metaphysical and epistemological worries facing non-naturalism by asserting that there are no moral facts, the upshot being that atomic, positive moral judgments are systematically false. Non-cognitivists and prescriptivists agree that there *are* no moral facts, but avoid the implication that moral judgments are systematically false by denying that they are in the market for truth and falsity, either because they express non-cognitive sentiments or desires (Ayer 1936; Stevenson 1937) or because they are more like orders or commands than attempts to describe the world (Hare 1952).

Although the error-theoretic and non-cognitivist or prescriptivist traditions have distinguished descendants in contemporary metaethics (e.g., see Joyce 2001 for the former, Blackburn 1984, 1993, 1998, 2002; and Gibbard 1990, 2003 for the latter), they are not the main topic of this chapter. Instead, I want to look in a little detail at a form of opposition to moral facts held by its proponents to avoid the standard worries faced by error theories or descendants of non-cognitivism and prescriptivism. This is the "cognitivist expressivist" account of moral judgment, developed in an ingenious series of works by Terence Horgan and Mark Timmons (I'll focus mainly on Horgan and Timmons 2006a and 2006b, though I'll refer to other of their papers as we proceed).[1] In sections 2 and 3 I will give a brief outline of the cognitivist expressivist position. In sections 4 and 5 I will argue that cognitivist expressivism, as developed by Horgan and Timmons (hereafter H&T), can be construed either as a view akin to the "minimalist" account of moral discourse developed by Crispin Wright (1992, 1996) or as a view akin to the "quasi-realist" position advocated by Simon Blackburn (see the works cited above). I will argue that because Wright's minimalist position and Blackburn's quasi-realism are incompatible, H&T cannot have it both ways, so that cognitivist expressivism is forced to go down one or other of the minimalist or quasi-realist routes. I'll conclude that whichever of the two routes H&T eventually opt for, the cognitivist expressivist position as they adumbrate it is underdeveloped and problematic.

2 Cognitivist Expressivism

Following H&T (2006a), let *descriptivism* be the view that moral judgments are in the business of representing reality, that they express *descriptive beliefs*. Let *expressivism* be the view that moral judgments express motivational states, states distinct from descriptive beliefs. Let *factualism* be the view that there are moral facts, and let non-factualism be the view that there are no moral facts.[2] Then we can identify some familiar metaethical positions: realism, relativism, and constructivism are forms of *descriptivist factualism*, and error theories are forms of *descriptivist non-factualism*. An emotivist view would be a form of *non-factualist expressivism*: on this view, there are no moral facts, but an error theory is avoided because moral judgments are viewed as expressing sentiments or feelings rather than beliefs. In denying that moral judgments express beliefs, emotivism is a form of *non-cognitivism*. H&T argue that there is space for a further position under the banner of non-factualist expressivism: *cognitivist expressivism* holds that moral judgments express *nondescriptive beliefs*, beliefs whose primary function is motivational rather than representational.

H&T argue in favor of cognitive expressivism by arguing that it can capture the characteristics of moral thought and talk better than its meta-ethical competitors. The characteristics to be captured include (H&T call this "moral phenomenology" (2006a, 223)):

MP1: moral judgments appear to express genuine beliefs and make genuine assertions
MP2: moral judgments appear to be truth-apt and (in some cases) true
MP3: there can be genuine, deep moral disagreements
MP4: moral judgments are action-guiding

and others.

H&T suggest (they call this the "coherence criterion") that a meta-ethical theory should capture the above features in a way that coheres with plausible theories in philosophy and other areas of inquiry (including the sciences), so that the two main tasks facing them are:

Task 1: Give an account of belief that allows us to say that moral judgments express genuine—but non-descriptive—beliefs.
Task 2: Argue that cognitivist expressivism can plausibly capture moral phenomenology.

At various places, H&T mention Blackburn's quasi-realism as having affinities with their cognitivist expressivism (2006b, 256–7, 2015 passim).[3] Elsewhere, they also mention Wright's (1992) minimalism as having affinities with cognitivist expressivism (2006b, 277, n. 15, 2015, 207 n. 8, Horgan 2001, 77, n. 11).[4] After outlining the bare bones of H&T's approach (to Task 1 in particular) in section 3, I'll go on in the remainder of the chapter to argue that since Blackburn's quasi-realism and Wright's minimalism are incompatible, and because H&T apparently fail to appreciate this, their cognitivist expressivism turns out to be neither fish nor fowl: whatever way we take it, H&T's position fails to convincingly execute crucial tasks that Blackburn-style and Wright-style positions, respectively, need to address.

3 Belief and Assertion

H&T construe beliefs as attitudes towards actual or potential states of affairs, and also as psychological commitment states of which there are two main types, *is-commitments* and *ought-commitments*. For example, the belief that John gave Mary the money he owed her is an *is-commitment* with respect to the state of affairs in which John gives Mary the money he owes her. Given H&T's repudiation of moral facts, it is important for them that there are no actual or potential moral states of affairs. Thus, a moral belief like the belief that John ought to give Mary the money he owes her is an *ought-commitment* with respect, not to a (moral) state of affairs in which John ought to give Mary the money he owes her, but rather to the (non-moral) state of affairs in which John gives Mary the money he owes her.

H&T argue (2006a, 232–3) that ought-commitments are genuine beliefs, since they possess the generic features that are characteristic of beliefs: their contents are declarative, these can figure as constituents in logically complex judgments and logical inferences, and they can combine with other beliefs to yield new beliefs in an appropriate manner, and so on. For example, the content of Alison's belief that John ought to pay Mary the money he owes her is given by the declarative "John ought to pay Mary the money he owes her." This can figure as a constituent in the logically complex "Either John has paid the money he owes to Mary or he ought to do so," and can figure as the conclusion in an inference from "John hasn't paid Mary the money he owes her" and "If John hasn't paid the money he owes to Mary then he ought to do so." Likewise, Alison's belief that John hasn't paid Mary the money can combine with the relevant logically complex belief to yield the belief that John ought to give Mary the money.

It is important to note that when H&T refer to "declarative contents" they mean contents possessed "simply as a result of grammatical form" (2000, 122). E.g., the example they use to show that the contents of ought-commitments can figure as constituents in logically complex judgments is "Either John has paid the money he owes to Mary or he ought to do so." As noted in the previous paragraph, at the level of underlying logical form there are—according to H&T—*no* moral states of affairs to act as the contents of ought-commitments. Hence, there is no possible state of affairs in which John ought to give Mary the money he owes her, whereas this is the second disjunct of the example H&T use to illustrate the point about logical complexity.[5]

Since, moral judgments—ought-commitments—possess the various features listed as generic characteristics of belief they are, according to H&T, genuine beliefs. (This allows cognitivist expressivism to capture MP1).[6] Moreover, "the job of moral beliefs is not descriptive but evaluative: one evaluates some potential state of affairs positively or negatively. In doing so, one's primary purpose is oriented towards action"(2006a, 233), and "moral judgements directly dispose us toward appropriate action, independently of our pre-existing desires" (2000, 126). So moral beliefs are distinct from descriptive beliefs (whose job is representational rather than action-guiding). Moral judgments are thus ought-commitments, and these are nondescriptive beliefs. Hence we arrive at cognitivist expressivism. H&T think that because cognitivist expressivism eschews moral facts, it is metaphysically and epistemologically solvent, and also that because it views moral judgments as expressive of belief, it avoids the standard problems faced by non-cognitivist views like emotivism. They dub the resulting view as one on which we have "Morality Without Moral Facts."[7]

4 Wright's Minimalist Account of Truth-Aptitude

According to the minimalist account developed in Wright (1992), a sentence is truth-apt iff (i) it possesses certain syntactic features (can appear in conditionals, can be negated, can be embedded in propositional attitude ascriptions, etc.) and (ii) it figures in an area of discourse that is disciplined.

Moral sentences satisfy both (i) and (ii). For example, the following are all syntactically well-formed: "If murder is wrong, then getting little brother to commit murder is wrong," "It is not the case that John ought to pay Mary the money," and "Alison believes that John ought to pay Mary back." So (i) is satisfied. Moreover, moral discourse is certainly disciplined (or as Wright puts it "subject to communally acknowledged standards of proper use"(1992, 35–6)): we can say things like "I thought the claim that James ought to pay the money back was warranted, but I've since discovered that the money was a gift, so I'm withdrawing the claim."

Thus, on Wright's account moral sentences are (minimally) truth-apt, and so, since belief comes in train with truth-aptitude, moral judgments—uses of moral sentences—express (minimal) moral beliefs.[8]

Suppose now that some moral judgments are true (since H&T reject error theories, they have no quarrel with this). In addition to the platitudes linking the notions of belief, assertion, and truth accepted by H&T—"To believe something is to take it to be true," "Sincere assertions express beliefs," and "Truth and falsity are properly predicated of beliefs and assertions" (2006a, 234)—Wright also accepts a platitude connecting truth and correspondence. According to Wright, it is a platitude that: "P" is true if and only if "P" corresponds to the facts. This can be derived in a manner that exhibits its platitudinous status as follows:

(i) "P" says that P
(ii) "P" corresponds to the facts iff things are as "P" says they are
(iii) "P" is true iff P
(iv) "P" is true iff things are as "P" says they are (from (i), (iii))
(v) "P" is true iff "P" corresponds to the facts (from (ii), (iv))

(i)—(iii) are platitudes, and (v) follows from them via (iv) and elementary logic: so for Wright, a true judgement—an assertoric use of a truth-apt sentence—is true if and only if that judgment corresponds to the facts.

However, the claims " 'P' corresponds to the facts" and "The judgment that P corresponds to the facts" are simply *paraphrases* of the LHS of (v) that import no "seriously dyadic" (1992, 83) relation of correspondence and hence no metaphysically weighty notion of a fact. So, the true moral judgment that you ought to be kind to children corresponds to the fact that you ought to be kind to children, expresses the (minimal) belief that you ought to be kind to children, and we can say that moral judgments aim to represent the (minimal) moral facts. Since facts are what true beliefs describe, moral judgments and moral beliefs are thus *descriptive*, though minimally so.

So what Wright's minimalism delivers isn't cognitivist expressivism, but rather cognitivist (minimalist) descriptivism. On this kind of view true moral judgments do "correspond to the facts," so that "Morality Without Moral Facts" would not be an accurate description of the resulting position. This suggests that contrary to what they occasionally suggest,[9] the position developed by H&T should not be viewed as in the spirit of Wright's minimalism. However, H&T say various things that appear to imply that they are aware of the fact that the position they advocate implies (so long as some moral judgments are true) the existence of (minimal) moral facts. For example, they frequently say that their position does not imply the existence of "in-the-world" moral facts (e.g. 2006, 235), suggesting that there is some metaphysically less substantial kind of fact or facts whose existence *is* implied by their position.[10] However, whereas Wright goes to great lengths to spell out potential differences between "merely minimal" facts and metaphysically heavyweight facts—in terms of his "realism-relevant cruces" of width of cosmological role, cognitive command, the Euthyphro contrast and so on (see 1992 ch. 3–5 as well as the discussion of "superassertibility" in ch. 2 together with the essays in Wright (2003)), H&T give no detail at all in their expositions of cognitive expressivism to explain the difference between the "in-the-world" moral facts which they eschew and the facts that correspond to merely minimally true moral judgments. [11]

Hence, if H&T are putting forward a view along the lines of Wright's minimalism, there is what appears to be a very serious lacuna in their position: they need to motivate a distinction, or set of distinctions, between "merely minimal" metaphysically lightweight facts and robust, metaphysically weighty facts, argue that the moral facts to whose existence they are committed fall within the former category, and that this blunts the sorts of worry they—H&T—raise for competitor metaethical positions that situate moral facts within the latter.

Moreover, if H&T were proposing a view akin to Wright's minimalism, their rather baroque account of how moral judgments can figure in logically complex judgments (see e.g. 2006b, 277–81 and appendix) would be otiose. On a minimalist view, the content of a logically complex judgment is given by its minimal truth-condition and in the case of, e.g., the judgment that either John has paid Mary the money he owes her or he ought to pay her back, the relevant truth-condition is a straightforward function of the truth-conditions of, respectively, the judgment that John has paid Mary the money back and the judgment that John ought to pay Mary back.[12] Again, the cognitivist expressivist position proposed by H&T appears to sit ill with Wright-style minimalism.

5 Blackburn's Quasi-Realism

Wright's "mere minimalism" about moral judgment—with its component minimalist view of truth-aptitude and belief—is flatly inconsistent with Blackburn's quasi-realism.[13] Whereas the minimalist secures truth-aptitude and the idea that moral judgments express beliefs on the basis of the relatively superficial conditions on syntax and discipline described in the previous section, Blackburn holds that serious philosophical work needs to be done to *earn the right* to think of moral judgments as truth-apt and expressive of belief: this is the point, e.g., of his various attempts to deal with the Frege-Geach problem. Blackburn thus rejects minimalism about truth-aptitude (though not, perhaps, minimalism about truth).[14] If, then, Wright's "mere minimalism" is not a good model for cognitivist expressivism, might Blackburn's quasi-realism be a more appropriate analog for H&T's position?

It is difficult to see how it can be. A key part of the quasi-realist strategy is earning the right to apply the concepts of truth and falsity to moral judgments. We can make two points about H&T's remarks on this matter.

First, there is a suggestion that the cognitivist expressivist can make progress in this regard via an account of truth-ascriptions as metalinguistic speech acts. H&T argue that since the truth-predicate operates in accordance with the T-schema

"S" is true if and only if S

"in sincerely thinking or uttering a statement of the form, 'S' is true, one is committed to affirming S and vice versa" (Horgan and Timmons 2006a, 234). Then

> When one predicates truth of a moral statement, one is engaged in an act of affirming 'metalinguistically' the first-order moral claim in question (that is, affirming the first-order judgement expressed by the statement one is calling true). Such an affirmation, done metalinguistically by employing the semantic concept of truth, is a morally engaged 'fusion' of semantic and moral appraisal (Horgan and Timmons 2006a, 234).[15]

This, however, falls prey to a point made by Frank Jackson, Graham Oppy and Michael Smith (1995) in the context of a critical commentary on Paul Boghossian's (1990) suggestion that minimalism about truth implies minimalism about truth-aptitude. Jackson, Oppy, and Smith focus on the following instance of the T-schema:

(1) "Torture is wrong" is true if and only if torture is wrong

They write: "Granted, *if* (1) gave the condition under which 'Torture is wrong' is true, it would follow that 'Torture is wrong' is truth-apt. But (1) only does that if its right-hand side is truth-apt, and *that* is the question" (1995, 289). Or, at least, the right to view the right-hand side of (1) as truth-apt is precisely what the cognitivist expressivist, if construed as a quasi-realist, is in the business of attempting to earn. If H&T are here attempting to earn the right to apply the concepts of truth and falsity to moral judgments, what they say fails to do so because "the fact that one way of giving the truth-conditions of a truth-apt sentence is to write down the sentence itself does nor per se tell us anything about which sentences are truth-apt" (Jackson, Oppy, and Smith 1995, 289).[16]

Second, another key part of the quasi-realist strategy of earning the right to apply the concepts of truth and falsity to moral judgments is the provision of a solution to the Frege-Geach problem on the basis of a fairly austere set of materials. Although H&T outline how their view purports to handle logically complex judgments with moral consistuents, I'll now argue that what they say seems completely inadequate if construed as an attempt to deal with the Frege-Geach problem.

First, a brief reminder of what the Frege-Geach problem consists in, using a simple form of emotivism as our stalking horse. Let's take our emotivist to propose B! (–Φ) as an account of the state of mind in terms of which the meaning (or logical form) of "It ought to be the case that Φ" is to be given. Now consider the argument

(1) Dick isn't cheerful.
(2) If Dick isn't cheerful then Harry ought to apologize.

Therefore:

(3) Harry ought to apologize.

For an emotivist, (3) is analyzed as *B! (Harry does not apologize)*, and the problem for the emotivist is that it is very difficult to see how the contribution of "Harry ought to apologize" to the meaning of the conditional in the "unasserted context" in premise (2) can be given in terms of the sentiment which gives its meaning in the "asserted" context (3). The emotivist seems forced to assign different meanings to "Harry ought to apologize" as it appears in the unasserted context (2) and the asserted context (3), so that the argument from (1) and (2) to (3) is guilty of a fallacy of equivocation. This is the Frege-Geach problem. In its full generality, the challenge posed by the problem can be stated as follows: give an account of how evaluative judgments contribute to the meanings of statements in which they appear in unasserted contexts (such as the antecedents and consequents of conditionals) in terms of the sentiments they express in simple asserted contexts, in a manner which preserves the validity of the intuitively valid inferences in which they appear.

It will be instructive to compare one of Blackburn's attempts to solve the Frege-Geach problem with what H&T offer on behalf of cognitivist expressivism. In his 1984 book, Blackburn analyzes the argument as follows:

(1*) —Cd
(2*) H! [—Cd; B! (—Ah)]
(3*) B! (—Ah)

Here, the conditional in premise (2) is analyzed in terms of a higher order attitude of approval (H!) towards moral sensibilities that combine the belief that Dick isn't cheerful with a sentiment of disapproval towards the state of affairs in which Harry does not apologize. According to Blackburn, then, someone who endorses (1) and (2) but not (3) believes that Dick isn't cheerful, approves of combining this belief with a sentiment of disapproval towards Harry's not apologizing, yet doesn't have this latter sentiment of disapproval. For Blackburn, then, someone who endorses (1) and (2) but not (3) has a "fractured sensibility" or a "clash of attitudes"(1984, 195), so that the intuitive validity of the argument is preserved.

Blackburn's "toolkit" is relatively austere: crucially, it avails itself of beliefs towards non-normative states of affairs (such as those in which Dick is cheerful or in which Harry apologizes) and sentiments of approval and disapproval towards non-normative states of affairs, but normative states of affairs (such as that in which Harry ought to apologize) play no part in the story. Blackburn's guiding thought is that if the Frege-Geach problem and the other problems facing emotivism can be dealt with on the basis of these relatively austere resources, we will have earned the right to treat moral judgments as truth-apt and expressive of belief. [17] It's worth noting that Blackburn's toolkit is similar—indeed, identical—to that deployed by H&T. As we saw in section 3, the is-commitments and ought-commitments deployed by H&T are attitudes towards non-moral states of affairs. Is-commitments are descriptive beliefs, and although ought-commitments are not reducible to desires, they are regarded as pro-attitudes with the distinctive characteristics of moral commitments.[18]

Blackburn's attempt to solve the Frege-Geach problem is a serious attempt at earning the right to view moral judgments as genuinely truth-apt, and the similarity between his relatively austere naturalistic toolkit and that deployed by H&T invites the question whether H&T have the resources to deal with the standard objections against Blackburn's attempt to resolve the Frege-Geach problem. The standard objection to Blackburn's 1984 response to the Frege-Geach problem is that it doesn't capture the idea that someone who endorses the premises but not the conclusion of a logically valid argument is guilty of a logical failing. The point is well-made by Crispin Wright:

> Anything worth calling the validity of an inference has to reside in the inconsistency of accepting its premises but denying its conclusion. Blackburn does indeed speak of the 'clash of attitudes' involved in endorsing the premises of the modus ponens example, construed as he construes it, but in failing to endorse the conclusion. But nothing worth regarding as inconsistency seems to be involved. Those who do that merely fail to have every combination of attitudes of which they themselves approve. That is a moral failing, not a logical one.
> (Wright 1988, 33)[19]

How do matters stand with H&T's comments on moral judgments and logically complex judgments in which they figure as constituents, such as the belief that either Dick is cheerful or Harry ought to apologize?

H&T introduce two "sentence-forming operators":

I [ ]: It is the case . . .
O [ ]: It ought to be the case . . .

These take non-sentential formulae ("that"-clauses) as arguments and yield whole sentences as outputs. For example, we can insert, respectively, "that Dick is cheerful" and "that Harry apologizes" as inputs and we get the sentences "Dick is cheerful" and "Harry ought to apologize" as outputs. Note that in line with the expressivist agenda H&T eschew moral non-sentential formulae, since this would entail helping themselves to distinctively moral contents or potential moral facts. So, the non-sentential formula "that Harry ought to apologize" is not available for use in the account of the logically complex belief we are considering since if this refers to anything, it refers to a distinctively moral content or potential moral fact.

So, we have I [Cd] and O [Ah] as formal versions of the sentences "Dick is cheerful" and "Harry ought to apologize,, each of which is built up from non-moral sentential formulae via the application of sentence-forming operators. But how about the complex sentence "Dick is cheerful or Harry ought to apologize"? H&T introduce an operator-forming connective designated by "o." This operator works on sentence-forming operators to yield the "complex, disjunctive sentence forming operator"

I [ ] o O [ ]

which takes pairs of non-sentential formulae as inputs and yields disjunctive sentences as outputs. In this way, through the application of sentence-forming operators, operator-forming connectives, and complex sentence-forming operators, H&T attempt to show how the complex sentence "Dick is cheerful or Harry ought to apologize" is a function of simpler constituents and, crucially attempt to do so without recourse to moral non-sentential formulae. (Compare this with an account that could be given by someone with no qualms about helping themselves to truth-evaluable moral sentences: "Dick is cheerful or Harry ought to apologize" could be regarded as a simple truth-function of "Dick is cheerful" and "Harry ought to apologize").

Does this style of account of the semantics of logically complex moral judgments provide the materials for a satisfying response to the Frege-Geach problem? I will now argue that it does not.

Consider the intuitively valid argument:

(1) Dick isn't cheerful,
(2) If Dick isn't cheerful then Harry ought to apologize,

Therefore:

(3) Harry ought to apologize.

In terms of the formal apparatus introduced by H&T, this argument can be represented as follows:

(1**) I [—Cd]
(2**) I [Cd] o O [Ah]
(3**) O [Ah].

The Frege-Geach problem challenges the expressivist to give an account of how the state of mind expressed by "Harry ought to apologize" as it appears in (3) plays a role in determining the contribution that that sentence makes to the state of mind expressed by the conditional (2) in which it features as consequent, and crucially, in such a way that the validity of the inference from (1) and (2) to (3) is preserved. We saw above that Blackburn's (1984) proposal appears to fail, because on that construal someone who accepts the premises of such an argument but not the conclusion appears to be guilty of a moral rather than distinctively logical shortcoming. How does H&T's account fare in this regard? What, on their account, is the state of mind exhibited by someone who accepts premises (1) and (2) but not the conclusion (3)? With regard to (1) they have the descriptive belief that Dick isn't cheerful, an is-commitment with respect to the content that Dick isn't cheerful, and with regard to (3) they do not have an ought-commitment with respect to the content that Harry apologizes to Dick. What about the conditional premise (2)? H&T write:

> Crucial and essential to such a logically complex commitment state [i.e.(2**)] . . . is its distinctive constitutive inferential role in the cognitive economy of a cognitive agent (*insofar as the agent does not exhibit lapses in logical competence*). Such a state is one that is poised to interact with other potential beliefs to inferentially generate yet further beliefs that are inferentially 'in the offing'. For instance, if Tom occurrently has the disjunctive belief about Dick and Harry, and Tom also has occurrently the belief that Dick is not cheerful, these two beliefs together should inferentially generate (*insofar as Tom does not exhibit a rationality failure*) the occurrent belief that Harry ought to apologise.
>
> (Horgan and Timmons 2006b, 278, emphases added)

It is difficult to see how this account of the state of mind expressed by (2) could contribute to a plausible attempt to tackle the Frege-Geach problem. There appear to be two possibilities.

The first possibility is that we suppress the references to logical competence and rationality in the above and view the state of mind simply as a disposition to have O [Ah] when one has I [—Cd]. In this case, though, we have no solution to the Frege-Geach problem because an agent who accepts premises (1) and (2) but not the conclusion (3) on this account simply fails to manifest a disposition they possess—but if this is a failing (and it's not clear that it is), it is certainly not the kind of failing one expects in the case of someone who accepts the premises but not the conclusion of a logically valid argument.

The second possibility is that we take seriously the references to logical competence and rationality (as emphasized in the quotation above). We can now perhaps generate the right kind of failing in an agent who accepts the premises but not the conclusion in that they fail to manifest a disposition they would manifest if they were logically competent or rational. But, of course, if we are in the process of developing a quasi-realist style account of moral judgment we have to *earn the right* to the notions of logical competence and rationality and cannot simply take them for granted in the way we could if we viewed the conditional premise as having a truth-condition that is a function of the truth-conditions of its antecedent and consequent. On this latter sort of view since the obtaining of the truth-conditions of the two premises logically ensures the obtaining of the truth-condition of the conclusion, someone who accepts the premises but rejects the conclusion has a logically inconsistent set of beliefs: but this sort of account is not available to the expressivist *in advance* of solving the Frege-Geach problem, since the expressivist's right to the relevant notion of logical inconsistency is supposed to be a *product* of his solution to that very problem and the various other problems facing emotivist-style accounts of moral judgment.[20]

Either way, it is difficult to find a plausible attempt at solving the Frege-Geach problem in H&T's remarks on logically complex moral judgments.[21]

6 Conclusion

Cognitivist expressivism can be viewed either as a Wright-style minimalist project or as a Blackburn-style quasi-realist project. It cannot be both as the Wright-style minimalist and Blackburn-style quasi-realist projects are mutually inconsistent. However, on either reading of the cognitivist expressivist project, there are substantial philosophical tasks that are inadequately addressed. If it is a form of Wright-style minimalism, we need a set of "realism-relevant cruces" of the sort developed by Wright together with an account of the metaphysically robust sorts of fact to which a moral realist is committed; if it is a form of Blackburn-style quasi-realism, we need a much better attempt to earn the right to apply the concepts of truth and falsity to moral judgments. As things currently stand, it is wishful thinking for H&T to claim that cognitivist expressivism can capture the characteristics of "moral phenomenology" outlined in section 2 above. So we are not yet in a position where we can lay claim to a metaethically satisfying account of morality without recourse to moral facts. Insofar as morality is the paradigm case of a normative practice, we are not yet in a position where we can lay claim to a philosophically satisfying account of normativity free of metaphysical commitments to a realm of normative facts.

Notes

* An early version of this paper was presented at the December 2014 New Zealand Association of Philosophy conference at the University of Canterbury, Christchurch. I'm grateful to the audience on that occasion for their helpful comments. Thanks, also, to Kirk Surgener and Joss Walker for useful feedback.

1 The need for a view like cognitivist expressivism is supposed to be a consequence of the attack on moral realism mounted in Horgan and Timmons's "moral twin earth" papers and in their related papers arguing that moral realism inevitably leads to moral relativism. See e.g., Horgan and Timmons (1992, 1996, and 2006c).

2 So "factualism" as H&T use the word stands for a metaphysical position, not (as is more common in the literature) for a view ascribing a fact-stating semantics.

3 They also see their position as belonging to the same metaethical tradition as Hare, Stevenson, and Gibbard (2006b, 256). See also (2000, 151 n. 25) and (2006c, 73–7).

4 See also Horgan and Timmons (2000, 146–7, 150).

5 I say "important" because this emphasis on surface syntactical features implicating the presence of genuine belief is one of the features of Wright's minimalism. See the next section.

6 They write: "The same general points can be made about linguistic speech acts in which one utters or writes sentences that express one's moral beliefs. Such speech acts are genuine assertions"(2006a, 233).

7 This is the title of Horgan and Timmons (2006a). For reasons of space, I won't run through all of H&T's arguments to the effect that on the basis of the considerations just mooted we can capture all of the facets of moral phenomenology in a way that respects the coherence criterion. We should note, though, that they hold that platitudinous linkages between the concepts of belief, assertion, and truth together with the considerations relevant to MP1 ensure that moral judgments are truth-apt (i.e., MP2).

8 See Divers and Miller (1995).

9 And as is suggested by the role H&T assign to surface syntactical features in their story about the generic characteristics of belief (see n.4 above).

10 Elsewhere they gesture in the direction of a similar distinction when they speak of some of their terminology needing to be taken with a "metaphysical grain of salt" (2006b, 261), of "*worldly* object-kinds and properties" (2006b, 264, emphasis added) and of "substantive and non-substantive truth-conditions" (2006b, 283). See also H&T (2015) where they speak of "descriptive states of affairs" (196) "ontologically robust facts" (194), "metaphysical facts" (194) and "metaphysically robust facts" (204). Also (2000, 123) where they mention a distinction between robust and minimal facts (2000, 124), where they speak of "way-the-world-might-be" content (2000, 147), where there is talk of "more or less robust metaphysical commitments," and (2006c 91 n.20) for "metaphysically robust truth-makers."

11 It is possible that the "contextual semantics" developed in a number of related works could be deployed to respond to this concern (see e.g., Horgan 2001, Horgan and Timmons 2002, 2015 section 6). It seems to me, though, that—in marked contrast with Wright's efforts in this area—H&T provide at most a *framework* within which the relevant distinctions might be developed and apparently make little attempt to spell out the relevant detail (in particular as it applies in the moral case). The detail can't be provided simply by raising one's voice ("it's a fact that torture is wrong but there are no moral FACTS!"). Matters are also complicated somewhat by H&T's misconstrual of Wright as "seriously flirting" with the idea that truth might be identified across the board with "superassertibility" (Horgan and Timmons 2006b, 277, n.15). This is a mistake. Wright (1992) suggests that where truth is epistemically constrained it can be identified with superassertibility, but whether truth *is* to be viewed as everywhere subject to epistemic constraint is not a question that Wright (1992) seeks to answer. Wright himself makes this point in a reply to Horgan (Wright 2003, 95–6).

12 For more on H&T's account of logically complex moral belief and judgment and how it fares with the Frege-Geach problem, see section 5 below. We should note that it has been argued that in the context of his pluralism about truth, Wright's minimalism fails to evade something like the Frege-Geach problem. For example, Christine Tappolet (1997) argues that mixed inferences will pose a problem for Wright: she argues that since truth is what is preserved in a valid argument, if there is not a single notion of truth applicable to all of the steps in an argument, we can't make sense of that argument's validity. If this were true, the argument of this section might perhaps be weakened: Wright's minimalist would not have such an easy way with

logical complexity. It seems to me, though, that Tappolet's argument is neutralized by Wright's observation that on his view, "true" is *univocal* across different domains: the pluralism relates to what *constitutes* truth from one domain to another rather than the sense of the word "true," so that there is no suggestion on Wright's part that "true" is lexically ambiguous (see e.g. Wright 2003, 58–9). But a full discussion of Tappolet's charge is not possible here.

13 See Divers and Miller (1994 and 1995) and Miller (2013b).

14 See, e.g., Blackburn (2002), where he explicitly repudiates minimalism about representation (i.e., truth-aptitude).

15 See also (2000, 146–7) and (2006b, 275).

16 H&T here appear, then, to confuse the question whether truth is minimal with the question whether truth-aptness is minimal, and the point made by Jackson, Oppy, and Smith is that these should be kept distinct: "One question is what makes for truth among the truth-apt, what divides the true from the false among the items that are apt for truth and falsity; the other question is what makes for truth-aptness among the whole range of linguistic items, only some of which are truth-apt" (1995, 290).

17 For an account of these other problems and Blackburn's attempts to deal with them, see ch. 4 of Miller (2013a).

18 Horgan and Timmons (2000, 134 and 2006b, 271).

19 See also Hale (1986).

20 Again, see Miller (2013a, ch. 4).

21 Similar remarks apply to the discussion in Horgan and Timmons (2000, 136–9) where H&T run the same line as that criticized in the text. Note that at (2000, 149, n. 23) H&T admit that their account assumes that there is a notion of logical consequence applicable to arguments involving moral judgments and beliefs, but appear unaware of the fact that this frustrates the idea that their view is in the spirit of broadly projectivist accounts of moral judgment.

References

Ayer, Alfred. 1936. *Language, Truth, and Logic*. London: Gollanz, (Second Edition 1946).

Blackburn, Simon. 1984. *Spreading the Word*. Oxford: Oxford University Press.

Blackburn, Simon. 1993. *Essays in Quasi-Realism*. Oxford: Oxford University Press.

Blackburn, Simon. 1998. *Ruling Passions*. Oxford: Clarendon Press.

Blackburn, Simon. 2002. "Quasi-Realism in Moral Philosophy: An Interview with Simon Blackburn", *Ethic@*, 1(2): 101–14.

Boghossian, Paul. 1990. "The Status of Content", *Philosophical Review*, 99: 157–84.

Divers, John, and Miller, Alexander. 1994. "Why the Expressivist about Value Should Not Love Minimalism about Truth", *Analysis*, 54: 12–19.

Divers, John, and Miller, Alexander. 1995. "Platitudes and Attitudes: A Minimalist Conception of Belief", *Analysis*, 55: 37–44.

Gibbard, Alan. 1990. *Wise Choices, Apt Feelings*. Oxford: Clarendon Press.

Gibbard, Alan. 2003. *Thinking How to Live*. Cambridge, MA: Harvard University Press.

Hale, Bob. 1986. "The Compleat Projectivist", *Philosophical Quarterly*, 36: 65–84.

Hare, Richard. 1952. *The Language of Morals*. Oxford: Oxford University Press.

Horgan, Terence. 2001. "Contextual Semantics and Metaphysical Realism: Truth as Indirect Correspondence." In *The Nature of Truth: Classic and Contemporary Perspectives*, edited by Michael Lynch, 67–96. Cambridge, MA: MIT Press.

Horgan, Terence, and Timmons, Mark. 1992. "Troubles for New Wave Moral Semantics: The 'Open Question Argument' Revived", *Philosophical Papers*, XXI: 153–75.

Horgan, Terence, and Timmons, Mark. 1996. "From Moral Realism to Moral Relativism in One Easy Step", *Critica: Revista Hispanoamerica de Filosofia*, 28: 3–39.

Horgan, Terence, and Timmons, Mark. 2000. "Nondescriptivist Cognitivism: Framework for a New Metaethic", *Philosophical Papers*, 29: 121–53.

Horgan, Terence, and Timmons, Mark. 2002. "Conceptual Relativity and Metaphysical Realism", *Philosophical Issues*, 12: 74–96.

Horgan, Terence, and Timmons, Mark. 2006a. "Morality without Moral Facts." In *Contemporary Debates in Moral Theory*, edited by James Dreier, 220–38. Oxford: Blackwell.

Horgan, Terence, and Timmons, Mark. 2006b. "Cognitivist Expressivism." In *Metaethics after Moore*, edited by Terence Horgan, and Mark Timmons, 255–98. Oxford: Oxford University Press.

Horgan, Terence, and Timmons, Mark. 2006c. "Expressivism, Yes! Relativism, No!." In *Oxford Studies in Metaethics* 1, edited by Russ Shafer-Landau, 73–98. Oxford: Clarendon Press.

Horgan, Terence, and Timmons, Mark. 2015. "Modest Quasi-Realism and the Problem of Deep Moral Error." In *Passions and Projections: Themes from the Philosophy of Simon Blackburn*, edited by Robert N. Johnson and Michael Smith, 190–209. Oxford: Oxford University Press.

Jackson, Frank, Oppy, Graham, and Smith, Michael. 1995. "Minimalism and Truth-Aptness", *Mind*, 103: 287–302.

Joyce, Richard. 2001. *The Myth of Morality*. Cambridge: Cambridge University Press.

Mackie, John L. 1977. *Ethics: Inventing Right and Wrong*. Harmondsworth: Penguin.

Miller, Alexander. 2013a. *Contemporary Metaethics: An Introduction* (Second Edition). Cambridge: Polity Press.

Miller, Alexander. 2013b. "Ethics and Minimalism about Truth." In *The International Encyclopedia of Ethics*, edited by Hugh LaFollette. Oxford: Blackwell.

Moore, George E. 1903. *Principia Ethica*, rev. edn, edited by Thomas Baldwin. Cambridge: Cambridge University Press, 1993.

Stevenson, Charles L. 1937. "The Emotive Meaning of Ethical Terms." In *Logical Positivism*, edited by Alfred Ayer, 264–81. Glencoe: Free Press.

Tappolet, C. 1997. "Mixed Inferences: A Problem for Pluralism about Truth Predicates", *Analysis*, 57: 209–10.

Wright, Crispin. 1988. "Realism, Anti-Realism, Irrealism, Quasi-Realism." In *Midwest Studies in Philosophy*, edited by P. French, T. Uehling and H. Wettstein XII: 25–49. Minneapolis: University of Minnesota Press.
Wright, Crispin. 1992. *Truth and Objectivity*. Cambridge, MA: Harvard University Press.
Wright, Crispin. 1996. "Truth in Ethics." In *Truth in Ethics*, edited by Brad Hooker, 19–34. Oxford: Blackwell.
Wright, Crispin. 2003. *Saving the Differences: Essays on Themes from Truth and Objectivity*. Cambridge, MA: Harvard University Press.

9 The Evolutionary Debunker Meets Sentimental Realism[1]

Mauro Rossi and Christine Tappolet

1 Introduction

Value Realism holds that there are evaluative facts that are objective, or stance-independent, in the sense that they exist independently of what we think or feel about them. On this view, facts such as pain being bad, knowledge being desirable, or helping others being admirable are as objective as facts postulated by the natural as well as the social and human sciences. But what would evaluative facts be? What, if anything, distinguishes evaluative facts from other kinds of facts? Is the value realist just adding more grim facts to our ontology or are evaluative facts distinct? If evaluative facts are distinct, what is the difference with natural facts? A useful way to frame the problem is to ask what values are, if one were to follow Hilary Putnam (2002) and accept that the fact/value dichotomy is a failure. Clearly, this question needs to be answered if we want to assess Value Realism.

Value Realism is opposed to Value Anti-Realism, the thesis that there are no objective evaluative facts. Both Value Realism and Value Anti-Realism come in very different versions, which can each be paired with a variety of stances regarding the epistemology of values, the semantic of evaluative sentences, and the moral psychology of evaluative judgments.[2] In spite of the large number of possibilities that results, some combinations are standard. Thus, Value Realism is generally associated with Rationalism, the view that the evaluative is grounded in reason, and which often takes the form of the epistemological claim that knowledge of what are conceived of as *a priori* evaluative truths can be gained through the exercise of reason. Quite generally, Rationalism is opposed here to Sentimentalism, the view that the ground of evaluative judgments lies in the sentiments or more generally in affective-*cum*-motivational states. Because they take affective and motivational states to be non-cognitive, purely motivational states, sentimentalists most often reject Value Realism (Nichols 2004; Prinz 2007).

In this paper, we propose a defense of Value Realism that relies on the unusual combination of Value Realism with Sentimentalism. What this account, which we call "Sentimental Realism," holds, in a nutshell, is that what makes evaluative facts special is their relationship to emotions. More precisely, Sentimental Realism claims that evaluative facts are fully objective facts, but that such facts are picked out by concepts that are response-dependent, in the sense that they are essentially tied to emotions. Our plan is as follows. We shall start with a presentation of Sentimental Realism and a discussion of its main virtues. On the basis of this, we shall discuss an objection to Value Realism that draws on evolutionary considerations, the Evolutionary Debunking Argument. We shall argue that Sentimental Realism safely escapes from this dilemma.

Before we start, let us clarify the scope of our discussion. The normative domain can be divided into what belongs to the evaluative, which concerns the good and the bad as well as the different ways of being good and bad, such as *admirable* or *courageous*, on the one hand, and the deontic, which is exemplified by concepts such as *ought*, *obligatory*, *forbidden*, and *permissible*, on the other. As should be clear from the terminology we used, our discussion focuses only on one part of the normative, namely, the evaluative. We have little to say in this paper about deontic facts, such as the facts that we ought to reduce pain or that it is forbidden to kill innocents.

2 Sentimental Realism[3]

What makes Sentimental Realism a form of Sentimentalism is its commitment to the thesis that evaluative concepts are essentially tied to affective states, and more specifically to emotions, such that such concepts count as being response-dependent. This claim is particularly plausible in the case of evaluative concepts such as *admirable, shameful*, or *amusing*, which wear their response-dependence on their sleeves. As many have highlighted, a plausible way to spell out the response-dependence of such concepts is as follows:

(S) By conceptual necessity, *x* is admirable/shameful/amusing/etc., if and only if admiration/shame/amusement/etc., is appropriate in response to *x*, if one were to contemplate x.[4]

There are very different ways to understand (S). A first question that arises is how one should understand the equivalence. According to one reading, (S) consists in a conceptual analysis, in which evaluative concepts are broken down to what is taken to be simpler conceptual elements, namely, affective concepts and the concept of appropriateness (e.g. Chisholm 1986, 58–60). The problem with this interpretation is that it is far from clear that affective concepts can be understood without appealing to evaluative concepts. How, for instance, could one explain the concept of admiration without invoking admirableness? In our view, it is better to understand (S) as offering a conceptual elucidation—as opposed to an analysis—in which none of the two sides of the biconditional has conceptual priority (Wiggins 1987). On such a no-priority view, the grasp of evaluative and affective concepts would be interdependent, such that none would be considered more fundamental. A second question concerns the notion of appropriateness, which can be taken to mean quite different things. According to some, such as Roderick Chisholm, appropriateness has to be explicated in terms of the deontic notion of requirement (Chisholm 1986, 58–60). Others, such as Thomas Scanlon, spell out the equivalence in terms of reasons for the responses at stake (Scanlon 1998, 95–100). We believe that a better way to understand such equivalences is to claim that being appropriate in this context is to be correct from an epistemic point of view.[5] The idea is that someone is admirable, for instance, just if feeling admiration towards her is correct, or accurate, from an epistemological point of view. Put differently, an appropriate emotion is one that represents the evaluative facts as they are.

The worry that will immediately arise here is that understood this way, (S) is much too circular to be of interest. By proposing to elucidate evaluative concepts in terms of responses that are appropriate just if their objects fall under the concepts in question, the equivalence draws nothing but an uninterestingly little circle. The circularity cannot be denied, but we believe that (S) is nonetheless of interest. As David Wiggins stressed when defending what he called "sensible subjectivism," the aim is to "elucidate the concept of value by displaying its actual involvement with the sentiments. One would not . . . have sufficiently elucidated what value is *without* that detour" (1987, 189). The point underscored by (S) concerns our epistemic "access" to values. To paraphrase Wiggins, the important point is that when we try to find out whether something is admirable or shameful, for instance, there is nothing more fundamental to appeal to than our responses of admiration and shame. What this means is that (S) makes a point that concerns the epistemology of values. Clearly, the expertise in affective concepts involves the ability to feel the relevant emotions. Thus, (S) points toward the epistemic indispensability of our emotional responses.

Given this, it is easy to see that this account of evaluative concepts is compatible with the claim that evaluative facts are objective, in the sense of being stance- or response-independent. The emotion of admiration toward someone can be appropriate only if in fact this person is admirable; and nothing prevents us from saying that the evaluative fact in question is objective. Indeed, on a natural understanding, it is because someone is objectively admirable that feeling admiration towards her is appropriate. Thus, what Sentimental Realism adds to (S) is the following claim:

(R) There are objective evaluative facts.

Now, here too, there are different ways to go. In particular, (R) allows for both naturalist and non-naturalist versions, depending on how the relationship between evaluative facts and natural facts is thought of. According to radically non-naturalist versions, evaluative facts are *sui generis* normative facts, which are distinct from natural facts. What is denied, more precisely, is that evaluative facts are identical or reducible to natural facts. What both non-naturalists and naturalists accept in general, however, is that evaluative facts supervene on natural facts, in the sense, roughly, that there is no evaluative difference without a difference at the level of natural facts. The view we find most plausible is that evaluative facts not only supervene on the natural, but are constituted by natural facts that involve disjunctive natural properties. Whether this view should count as a form of non-naturalism or not is better left open here.

Sentimental Realism, the conjunction of (S) and (R), is clearly a possible account. But why would one want to embrace it? Independently of the reasons to embrace both (S) and (R), which we will briefly discuss, we believe that the combination of the two claims makes for an attractive conception of values, which is in a position to answer important objections to Value Realism.

Let us start with the second conjunct, (R). Quite generally, as most would agree, (R) can be considered to be the default meta-normative position (Brink 1989). Indeed, this claim is even stronger in the case of values, where there is little motivation for the view that what we are concerned with are rules that are supposed to govern human interactions. Evaluative predicates, such as 'good' or 'admirable', behave like ordinary predicates. Thus, the structure of 'Sarah is admirable' appears in no way different from that of 'The ball is round.' Both types of sentences can be evaluated in terms of truth, for we can ask 'Is it true that Sarah is admirable?' just as we can wonder whether the ball is round. The two types of sentences appear to have cognitive contents that are genuinely truth-assessable. Similarly, when making evaluative judgments, we seem to express cognitive states that appear to be ordinary, truth-assessable beliefs. Indeed, both our evaluative sentences and our evaluative judgments seem to be about an objective evaluative reality, which exists independently of our own attitudes. This is evidenced by the fact that disagreement about value presents itself as genuine disagreement, in which at least one of the disagreeing parties makes a mistake.[6] According to the value realist, these considerations should be taken at face value. This means that, unless we have strong reason to think otherwise, we should admit that there really are objective evaluative facts and properties, which our evaluative judgments attempt to capture, and about which we disagree.

Let us turn to the other conjunct, (S). What supports (S) is that it captures the close relationship between the evaluative and the affective. Given the lexical proximity between terms such as 'admirable' and 'admiration,' it is difficult to deny that such concepts are closely connected to affective concepts. This observation might suggest that being admirable is nothing but to be admired by someone. But in contrast to this subjectivist thesis, (S) is in a position to explain why what we *de facto* admire is not always admirable, while what is admirable is not always what *de facto* causes admiration. What (S) holds, is that it is only when admiration towards someone is appropriate that this person is admirable. Correlatively, given that it does not aim at reducing normative concepts to natural concepts, (S) does not fall prey to the naturalistic fallacy, i.e., the error that consists in analyzing normative concepts in natural terms (Moore 1903).

In our view, an important reason to adopt Sentimental Realism is that this approach is grounded in an independently plausible account of emotions, namely, the Perceptual Theory. According to the Perceptual Theory, emotions are a kind of perceptual experiences, namely, perceptual experiences of evaluative facts.[7] The idea is that the admiration you feel towards someone represents that person as admirable, in the same way as color or shape experiences represent things as having color or shape properties.

The Perceptual Theory gains to be compared with its main competitors, Feeling Theories (James 1884; Lange 1885; Whiting 2011), Conative Theories (Frijda 1986; Scarantino 2014) and Judgmental Theories (Solomon 1976; Nussbaum 2001).[8] By contrast to Feeling Theories, which hold that emotions are mere feelings, and Conative Theories, according to which emotions are constituted by motivational states such as desires, the Perceptual Theory holds that emotions are intentional states that have representational contents, in the sense that they represent things as having certain evaluative properties. It can thus easily account for the fact that emotions not only have intentional objects, but are naturally taken to be assessable in terms of the nature of these objects, fear being appropriate when it is directed at fearsome and dangerous objects. In contrast to Judgmental Theories, according to which emotions are normative or evaluative judgments, the Perceptual Theory holds that it is not necessary to have the ability to make evaluative judgments and to possess evaluative concepts to feel emotions. Thus, the Perceptual Theory has no difficulty to account for the fact that non-human animals and infants can feel a number of emotion kinds, such as fear or anger.

This is not the place to spell out the full argument for the Perceptual Theory, let alone to consider the different objections that have recently been raised against it.[9] Let us simply note that the main argument for the Perceptual Theory is an argument by analogy. Emotions and sensory perceptual experiences share a number of striking features, the main of which are the following. Emotions and sensory perceptual experiences both have phenomenal qualities, in the sense that there is a way it is like to feel an emotion, just as there is a way it is like to have sensory perceptions; they both are characterized by a lack of direct control by the agent, in the sense that you cannot feel an emotion or have a sensory experience merely because you decide to do so; they both manifest "recalcitrance," such as when you fear something which you judge to be harmless, or you see something as bent while you judge that it is straight.

Sentimental Realism is a promising account of evaluative facts concerning the admirable, the fearsome, the amusing, the shameful, etc. But what about other kinds of evaluative facts, such as the fact that something is good or bad? What about facts involving properties that correspond to so-called thick concepts, such as the fact that some action is courageous or some person generous? It is quite easy to see how the proposed account can be extended to the more general evaluative properties and their concepts. The basic idea is that one can use the general notions of positive and negative emotions to handle such cases. Let us say that to favor something is to have a positive emotion towards it, while to disfavor it is to have a negative emotion towards it. Given this, it can be claimed that something is good (*pro tanto*) just if it is appropriate to favor it, and something is bad (*pro tanto*) just if it is appropriate to disfavor it. Similarly, it seems plausible to tie thick concepts, such as *courageous* or *generous*, to affective responses. For instance, it seems plausible to claim that an action is courageous just if it is done in spite of danger to oneself, and in virtue of this, it is admirable, so that admiring it would be appropriate.

Let us accept, then, that Sentimental Realism is a plausible account of the evaluative. One virtue of Sentimental Realism is that it allows for straightforward answers to two traditional objections to Value Realism, both of which can be traced back to John Mackie (1977, ch. 1). Mackie's so-called Argument from Queerness has two parts: one metaphysical (or ontological) and one epistemological. He claims that "[i]f there were objective values, they would be entities or qualities or relations of a very strange sort, utterly different from anything else in the universe. Correspondingly, if we were aware of them, it would have to be by some special faculty of moral perception or intuition, utterly different from our ordinary ways of knowing everything else" (Mackie 1977, 38). According to Mackie, objective evaluative facts provide "the knower with both a direction and an overriding motive; something's being good both tells the person who knows this to pursue it and makes him pursue it" (Mackie 1977, 40). Since ordinary facts appear to lack the same action-guidingness and motivational force, objective values look "queer." So, we do best without objective evaluative facts in our ontology.[10]

The good news is that Sentimental Realism is well-placed to answer both parts of the argument from queerness. In reply to the epistemological challenge, Sentimental Realism can rely on the claim that emotions are perceptual experiences of values in order to develop an account of the justification of evaluative beliefs, and more generally of evaluative knowledge, which grounds such beliefs on our emotional responses.

There are different ways to go here, but what seems plausible is that our emotions be considered to constitute defeasible reasons for our evaluative beliefs, so that when we form a belief on the basis of an emotional reaction, our belief is *prima facie* justified (see, *inter alia*, Tappolet 2000, 2016, ch. 5; and Kauppinen 2013). In our view, the justification of evaluative beliefs turns on the absence of any reason to believe that the emotion on which the belief is based be inappropriate. So, your belief that someone is admirable would be justified on condition that it is based on the admiration you feel for that person, and you have no reason to believe that what you feel is biased or more generally has to be distrusted. One question that arises is how we go about determining what conditions count as unfavorable. As David Hume stressed, finding out what conditions are favorable or unfavorable is something that we undertake with others. What Hume argues is that the very possibility of understanding each other presupposes that we adopt what he calls a "common" or "general point of view," that corrects for perspectival effects, such as biases due to spatial or temporal distance (1739–1740, III, 3, i). The idea is that in order to determine what counts as an unfavorable condition—a defeater, in epistemological terms—we need to take into account not only our own experiences at different times and in different conditions, but also the experiences of other persons.[11] In Philip Pettit's terms (1991, 600–1), when we want to determine whether something should count as a defeater, we have to look to our shared practice of discounting certain conditions as likely to interfere with our responses, a practice that aims at making sense of intra-personal, but also of interpersonal, discrepancies. In any case, evaluative knowledge can then be conceived of in terms of justified evaluative beliefs that are based on emotions that are not defeated. Suppose you form the belief that a friend is admirable on the basis of your feeling of admiration. Your belief will be justified in case you have no reason to believe that your admiration is interfered with. But of course, your admiration might be nonetheless misguided. So, to know that the action is admirable, your belief need not only be justified, it also needs to be such that *de facto* no defeater interferes with your admiration.

Let us now address the metaphysical part of Mackie's argument. Given that emotions normally come with related motivations, Sentimentalist Realism is also in a position to handle the challenges regarding the action-guidingness and motivational force of values. Value Realism is generally associated with a defense of the externalist thesis according to which evaluative judgments are not essentially, or necessarily, tied to motivation. Now, even though it can be agreed that making an evaluative judgment does not necessarily involve a motivation to act, Sentimental Realism can hold that evaluative judgments will normally be accompanied by such a motivation, given that such judgments are grounded in the corresponding emotional reactions. This simply follows from the fact that emotions, or at least most of them, at least normally involve motivations. So, insofar as evaluative judgments are concerned with facts that are relevant to what we ought to do, so that they are in this sense action-guiding, they will normally involve a motivation to act accordingly. Given this, it is not necessary to postulate objective entities possessing magical motivational properties, in order to account for the motivational force of evaluative judgments.

If these arguments are on the right track, Sentimental Realism is an attractive account, which might well be better placed than other Value Realist accounts. But this does not entail that Sentimental Realism is immune to all possible objections against Value Realism.

3 The Evolutionary Debunking Argument against Value Realism

In this part, we want to discuss one important objection, in particular, which has attracted considerable interest in recent years and which is known as the so-called Evolutionary Debunking Argument against Value Realism.[12] We shall argue that Sentimental Realism can successfully address this objection. We will proceed as follows. We will start by presenting the general version of the Evolutionary Debunking Argument. We will then show how the argument applies to the sentimental realist account proposed in the first part of the paper. After presenting some replies offered in the literature, we will put forward a new version of the "third factor argument" against the Evolutionary Debunking Argument.[13]

The debunking argument starts from the empirical claim that evolutionary forces have greatly shaped our evaluative faculties and that they have thereby disposed us to form specific evaluative beliefs. The argument proceeds by noticing that evolutionary forces aim at fitness, not at detecting mind-independent evaluative facts. If so, we have reason to think that evolutionary forces have selected for evaluative faculties that yield fitness-enhancing beliefs, not beliefs tracking mind-independent evaluative facts. Indeed, it would be a massive coincidence—something akin to a miracle—if faculties shaped by forces that aim at survival and reproductive success also managed to deliver beliefs that track a stance-independent evaluative reality. Thus, absent some other reason to think the contrary, we have decisive reason to conclude that our evaluative faculties are unreliable, i.e., that they tend to generate mistaken evaluative beliefs. Insofar as Value Realism is an ontological and not an epistemological stance, this is not yet a conclusion that shows Value Realism to be wrong. However, the stark skepticism that would follow would put Value Realism in an uncomfortable position.

It is clear that Sentimental Realism is a target of the Evolutionary Debunking Argument. As we have seen, Sentimental Realism is committed both to the existence of objective evaluative facts and to the possibility of having knowledge of such facts. More specifically, Sentimental Realism emphasizes the role of emotions, which are conceived of as perceptual experiences of evaluative properties, for the acquisition of evaluative knowledge. This is where the Evolutionary Debunking Argument kicks in. According to it, sentimental realists must recognize that evolutionary pressures have exercised a large influence on our emotions. Indeed, from an evolutionary perspective, emotions (at least basic emotions, that is, innate and pan-culturally shared emotions[14]) can be seen as adaptations which gave our ancestors an evolutionary edge by allowing them to deal more effectively with the specific challenges that they typically encountered. If this is true, then it is hard to maintain that emotions reliably track genuine evaluative facts. For this to be true, it would have to be the case that reliably detecting objective evaluative facts somehow increased our ancestors' fitness. According to the debunker, however, this is scientifically questionable. There are two more plausible options. The first consists in maintaining that emotions have nothing to do with evaluative properties and that, insofar as they are perceptual experiences of some sort, they are perceptual experiences of non-evaluative properties whose detection was significant for our ancestors' survival, but which are uncorrelated to evaluative properties. The second consists in maintaining that, while emotions are indeed perceptual experiences of evaluative properties, they are unreliable ones. Accordingly, when we experience an emotion towards a particular object, we perceive such an object as possessing an evaluative property that it does not really possess. Either way, the debunker's line of thought implies that the evaluative beliefs that we regularly form on the basis of our emotions are likely to be systematically mistaken.

In what follows, we will understand the debunker's challenge in the sense of the second option just mentioned. The reason is the following. For the sake of the argument, the debunker grants the assumption that there exists an objective evaluative reality. If evaluative properties exist, then, according to most emotion theorists,[15] they constitute the formal objects of emotions, where the formal object of an emotion type is what determines the correctness conditions of the tokens belonging to that particular type. For instance, the formal object of fear is the fearsome, while the formal object of admiration is the admirable. Moreover, if emotions are conceived of as perceptual experiences, then, insofar as their formal object determines their correctness conditions, it does so by determining whether the content of such emotions is correct or not. But if the formal object of a given emotion is constituted by an evaluative property, then it seems that the formal object determines whether the content of a given emotion is correct in the sense that such an emotion is assessed as correct if and only if, and because, it represents its object as possessing the relevant evaluative property and its object does indeed possess such a property. From this account, it follows that emotions are unreliable only if they systematically misrepresent their objects as possessing evaluative properties that they do not really possess. Framed in these terms, the Evolutionary Debunking Argument claims that the objects of emotions possess properties that are of evolutionary significance, but it adds that these properties do not correlate with the perceived evaluative properties.

Suppose we grant the evolutionary story about the origin of emotions. One may be tempted to insist that this does not necessarily spell trouble for Sentimental Realism. In fact, emotions admit some degree of plasticity. So, even if we assume that emotions were selected for to track fitness-enhancing properties and that such properties are not correlated with evaluative properties, emotions can nevertheless be "corrected," e.g., by submitting them to rational scrutiny, in such a way to ensure that they reliably track genuine evaluative properties. Not only that, but if we think that emotions provide us only with the basis for particular evaluative judgments, which are then used as raw data to construct more general evaluative principles through a method of interpersonal reflective equilibrium, then, to the extent that such a method leads us to abandon the particular evaluative judgments that are clearly mistaken, then we may also think that some degree of error in the initial data is not necessarily an obstacle to the obtainment of evaluative knowledge.

However, things are not that easy. The debunker may reply, first, that we have good reason to think that most of our emotions are incorrect and, second, that we have no independent way of identifying which of them are incorrect. Together, these points can be used to reject the sentimental realist's initial reply. For a start, it follows from the debunker's claims that we cannot simply assume that the standard defeaters (e.g., drugs, fatigue, cultural biases, etc.) constitute the only distorting influences on otherwise correct emotions, since this is precisely what the debunker disputes. In fact, the debunker argues that we cannot make any substantive assumptions as to which emotions are correct, which we could then use to assess and "reprogram" our incorrect emotions, since such assumptions may simply be the product of our evolutionary history. If this is the case, and if we do have good reason to think that most of our emotions are unreliable, then no method that relies on our emotional responses as a starting point can lead us to evaluative knowledge. The method is as good as the data on which it operates and if the data are inadequate, then the method will deliver inadequate results. As it is often put: "Garbage in, garbage out."

The situation may appear bleak for Sentimental Realism. However, appearances are often deceitful. Value realists of different brands have proposed various replies to the Evolutionary Debunking Argument. The sentimental realist can co-opt at least some of them for her own purpose, as we will now show.[16] The starting point of the sentimental realist's counterattack is the observation that the debunker cannot merely assert that, if emotions have been shaped by evolutionary forces, then most of them are incorrect. In other words, the debunker must give us some reason to accept the claim that it would be a massive coincidence if mental states selected for fitness were also reliable detectors of objective evaluative facts. What does she have to say in this regard?

The natural option for the debunker consists in showing that, more often than not, our emotions lead us to form evaluative beliefs that we know, or can confidently hold, to be incorrect. This would provide compelling evidence that evolutionary forces aimed at survival do not track objective evaluative facts. However, as it should be clear by now, this strategy is not open to the debunker, for it would contradict her claim that we have no independent way of identifying which emotions are correct and which are incorrect and, most importantly, it would provide an easy way out to the value realist.[17]

An alternative for the debunker consists in formulating her argument in terms of probabilities. If we cannot make any substantive assumptions about evaluative facts, then we are left with an essentially unlimited set of *possible* evaluative facts. But then, it is extremely unlikely that evolution has endowed us with mental states that track precisely the *actual* evaluative facts. This gives us reason to endorse the massive coincidence thesis against Value Realism. However, as several authors have noticed,[18] this argument can be challenged. For one, we can resist its starting point and deny that, if we forbid all substantive assumptions, then we must take the space of possible evaluative facts to be infinite. The reason is that there may be conceptual constraints on what counts as an evaluative truth, which significantly narrow down the set of possible evaluative facts.[19] Furthermore, the probabilistic argument seems to overgeneralize. Indeed, if we assume that evolutionary forces have exercised a great influence on the development of our sensory perceptual capacities and if we avoid making any substantive assumptions about perceptual facts, then we must accept that the space of possible perceptual facts is infinite. But then, we should conclude that it is extremely unlikely that capacities selected for survival ended up tracking precisely the actual perceptual facts. However, this does not seem to be a conclusion that debunkers would readily accept.[20]

The debunker may explore other possibilities. One of these consists in claiming that, insofar as evolutionary forces have shaped our emotions, the latter are *insensitive* to the evaluative facts, in the sense that our emotions would not have been different, had we lived in a different possible world.[21] Indeed, had the evaluative facts been different, evolutionary forces would have still pushed us to experience the same fitness-enhancing emotions (at least the same basic emotions), thereby getting us to form the same fitness-enhancing evaluative beliefs. Given this, we should conclude that any match between our actual emotions and the evaluative facts would be massively coincidental. A similar option consists in claiming that, insofar as evolutionary forces have shaped our emotions and insofar as our evaluative beliefs are formed on the basis of such emotions, then our evaluative beliefs are not *safely* formed.[22] Indeed, had we evolved differently, we would have experienced different emotions, which, in turn, would have led us to forming different evaluative beliefs. By our current light, however, such beliefs would have been false. This shows that our evaluative beliefs are simply the result of an accident of our evolutionary history. If so, their correspondence to a mind-independent evaluative reality would be massively coincidental.

However, both arguments can be challenged on general grounds. In fact, as Tomas Bogardus (2016, 644) has stressed, recent work in epistemology gives reason to believe that neither sensitivity nor safety is required for justified belief and knowledge. So, even if we grant that our emotions might be insensitive or unsafe in the way suggested above, this has little implications on their epistemic status. In fact, both the argument from sensitivity and the argument from safety are part of a broader argument that aims at showing that emotions cannot lead us to evaluative knowledge. However, if it is shown that neither sensitivity nor safety, in the technical sense at stake, matters for knowledge, then, the alleged failure of emotions to satisfy these conditions does nothing to support the claim that emotions are not reliable grounds for the acquisition of evaluative knowledge.

One last possibility that we wish to consider is for the debunker to argue that, if evolutionary forces have shaped our emotions, then the best explanation of how we typically form our emotions does not presuppose their reliability.[23] For instance, one may argue that, to the extent that mind-independent evaluative facts are causally inert, then the best explanation of how we form our emotions is simply that the latter are caused by those non-evaluative properties of their objects that were evolutionarily salient for our ancestors' survival. Thus, even if we grant that our emotions are perceptual experiences of evaluative properties, we have no explanatory good reason to hold that they are reliable perceptual experiences. The upshot is that it would be a massive coincidence, if they were actually reliable.

The most natural reply against this argument is for the sentimental realist simply to deny its starting point, namely, that evaluative properties are causally inert. After all, Sentimental Realism holds that evaluative knowledge is largely a kind of synthetic *a posteriori* knowledge. This fits comfortably with the thesis that the evaluative is causally efficacious. But the sentimental realist may also consistently opt for a weaker thesis, according to which, if evaluative properties supervene on causally efficacious non-evaluative properties, they possess causal power in a derivative way. Either way, the sentimental realist has an explanation of how we can have perceptual experiences of genuine evaluative properties that is at least as good as that proposed by her opponent.

Let us take stock. So far, we have examined various arguments that the debunker may give in defense of her claim that the correlation between our actual emotions and mind-independent evaluative facts would be massively coincidental. We have seen that the sentimental realist has a reply to all of these arguments. However, a die-hard skeptic may not be satisfied by this essentially defensive strategy. She may insist that the alleged correlation between fitness-relevant properties and evaluative properties remains fishy and ask the sentimental realist for a more positive defense. It is unclear whether the debunker is entitled to such a demand, without having herself provided more convincing evidence about the unreliability of our emotions.[24] Be that as it may, in what follows we want to put forward precisely an argument of this sort. Our argument is a version of an argument that has been variously explored by value realists of different sorts: the "third factor" argument. In its general form, third factor arguments consist in showing that there is (at least) an indirect correlation between fitness-enhancing facts and objective evaluative facts, in the sense that there exists a "third factor" that is responsible for both and explains their correlation. Our goal in the rest of this section is to see how an argument of this sort can be constructed in defense of Sentimental Realism.

Following David Enoch (2010, 428), our starting point is the observation that, in order to reject the debunker's challenge, it is not necessary to show that there exists a perfect correlation between our emotions and the objective evaluative facts. In other words, it is not necessary to show that all emotions are reliable. Rather, it is sufficient to show that our emotions are reliable more often than not. As we have seen, emotions provide the starting points for the acquisition of evaluative knowledge. However, reasoning mechanisms play an important role as well, and provided that they operate on mostly correct data, they can help us getting closer and closer to the evaluative truth, e.g., by eliminating inconsistencies and logical mistakes, and by taking into account defeaters. Notice that this may be possible even if we do not know exactly which particular emotions are correct, provided that we know that they are *mostly* correct.

How then can we proceed in order to show that our emotions are reliable more often than not? As we have seen, the debunker holds that evolutionary forces have selected for fitness-enhancing emotions, that is, emotions providing evolutionarily relevant information for our ancestors and leading to behaviors that increased their chances of survival and reproductive success. The first substantive step of our argument consists in claiming that, more often than not, whatever contributes to an individual's fitness contributes also to the individual's well-being. If this is true, then, by selecting for fitness-enhancing emotions, evolution may have also selected for emotions that are, more often than not, well-being-enhancing, i.e., emotions that either (a) favor, or lead to the promotion of, things that are good for the individual, or (b) disfavor, or lead to the avoidance of, things that are bad for the individual.[25] Consider, as an example, a fairly universal emotion like the fear of snakes. Such an emotion was obviously adaptive for our ancestors, in the sense that those who experienced such an emotion had more chances to survive that those who did not. By helping them survive, such an emotion had an impact also on their well-being, at the very least by ensuring that a preliminary condition for well-being be satisfied, namely, remaining alive.

The debunker may immediately raise an objection against this first step. Isn't the claim that, more often than not, whatever contributes to an individual's fitness contributes also to the individual's well-being a *substantive* claim? And isn't it precisely the sort of claim that is presumptively infected by evolutionary influences, so that it cannot be presupposed without begging the question against the debunker? In response, the sentimental realist may appeal to the idea, mentioned above, that there are some conceptual constraints on what counts as well-being and that denying the existence of even a weak correlation between individual fitness and individual well-being would violate such constraints. In fact, in order to deny the first step of our argument, the debunker would have to deny that the adoption of effective survival strategies contributes, more often than not, to the individual's well-being, in the sense that it is, more often than not, beneficial for the individual or in the individual's interest. However, it is not clear whether this move can be defended without threatening to make the resulting view unintelligible as a view of well-being.

Suppose, then, that the debunker's objection can be resisted. The argument might be still thought to be very limited, since it establishes only a correlation between emotions and goodness for. In fact, if emotions are conceived of as perceptual experiences of non-relational evaluative facts, what we need in order to reject the Evolutionary Debunking Argument is a correlation between emotions and goodness *tout court*. This is the goal of the second step of our argument. Once again, the debunker will be skeptical. After all, it seems that this requires defending another of those substantive claims that the debunker will take to be question-begging. In order to defuse the debunker's worry, the sentimental realist can pursue a similar strategy as the one pursued with respect to the first step.

As we have seen, emotions are conceived of as perceptual experiences of evaluative properties. More specifically, positive emotions are directed at things that supposedly possess positive evaluative properties, e.g., the property of being pleasurable, joyful, etc.; while negative emotions are directed at things that supposedly possess negative evaluative properties, e.g., the property of being fearsome, disgusting, etc. Our claim is that things that *do*, in fact, possess properties of the former kind are also things that *tend* to be good for the individual. Likewise, things that *do* possess properties of the latter kind are also things that *tend* to be bad for the individual. In order to deny these claims, one would have to maintain that pleasurable, joyful, etc., things do not tend to be well-being-enhancing, and that fearsome, disgusting things do not tend to be well-being-diminishing. Arguably, this would once again render the resulting view unintelligible as a view of well-being.

These considerations have some important implications. They imply that, *if* emotions were *reliable* perceptual experiences of evaluative properties, *then* they would typically be well-being-enhancing, since they would favor (or disfavor) things that tend to be good (or bad) for the individual. In fact, as we have seen above, fitness-enhancing emotions *are*, more often than not, well-being-enhancing. Of course, it would be fallacious to infer from this that fitness-enhancing emotions are, more often than not, reliable perceptual experiences of evaluative properties. However, if we combine the previous considerations with the claim that evaluative properties are causally efficacious, then the thesis that fitness-enhancing emotions are, more often than not, reliable perceptual experiences of evaluative properties appears to be *a plausible explanation* (perhaps, *the* most plausible explanation) of why fitness-enhancing emotions are also, more often than not, well-being-enhancing. If this is the case, then the second step of our argument can be vindicated.

We want to conclude by considering one possible objection that the debunker may raise at this stage. According to it, our argument simply presupposes that our evaluative concepts are not subject to distorting evolutionary influences. However, this presupposition is illegitimate in the context of a discussion of the Evolutionary Debunking Argument. We wish to make two points in response. The first is that conceptual assumptions about evaluative concepts seem to stand on firmer grounds than substantive assumptions about evaluative facts. Indeed, while the claims about the latter may reflect the distorting effects of evolution on our emotions, the former seem to be relatively immune to such effects, since conceptual assumptions are more the product of rational reflection than of our emotions, and we have no evolutionary grounds for thinking that rational reflection is unreliable. The second point is that, as our previous discussion shows, the debunker has not given us any reason to think that our substantive assumptions are systematically mistaken. In fact, we have rejected all her arguments aimed at showing that our emotions are unreliable. All in all, the sentimental realist seems thus to have the edge over her opponent.

4 Conclusion

Let us draw together the treads. We started out by asking what evaluative facts, as compared to natural facts, would be if Value Realism were true. In reply, we argued that what is specific of the evaluative domain lies in its ties to the affective. Thus, we proposed a version of Value Realism, Sentimental Realism, which combines the claim that evaluative facts are fully objective, or stance-independent, with the thesis that evaluative concepts are characterized by response-dependence. We argued that Sentimental Realism is in a good position with respect to the Argument from Queerness as well as the Evolutionary Debunking Argument. In both cases, the emotions we feel did the heavy lifting in the counter-arguments.

Notes

1 We gratefully acknowledge the FRQ-SC and the SSHRC for their support, and wish to thank Samuel Dishaw and Matthew Scarfone for his useful comments.

2 Indeed, moral realism and anti-realism are often considered to involve more than an ontological stance. See Sturgeon 1984; Sayre-McCord 1988; Railton 1996; and Tappolet and Rossi 2015a for a general discussion.

3 This section draws partly on Tappolet and Rossi 2015a and Tappolet 2016.

4 See, *inter alia*, Brentano 1889: 11; Ewing 1947, 1959; Wiggins 1976, 1987; Blackburn 1984, 1998; McDowell 1985; Chisholm1986; Gibbard 1990; Anderson 1993; Mulligan 1998; Scanlon 1998; D'Arms and Jacobson 2000a, 2000b; Oddie 2009.

5 The terminology is inspired by Brentano, who claimed that something is good if and only if loving it is correct (*richtig*) (1889: 11). See also Oddie 2014: 66.

6 See Moore 1922; Enoch 2011: ch. 2.

7 See Meinong 1917; de Sousa 1987; Tappolet 1995, 2000, 2016; Prinz 2004; Deonna 2006; Döring 2007; Tye 2008.

8 See Deonna and Teroni 2012 for a useful overview of emotion theories.

9 See Tappolet 2016: ch. 1 for objections, see Deonna and Teroni 2012; Brady 2013.

10 As this formulation makes manifest, Mackie's argument cannot concern normativity in general, including epistemological normativity (see Putnam 2002; Cuneo 2007).

11 This is an insight that has often been highlighted in recent discussions. See Wiggins 1987: 196; Davidson 1991; Pettit 1991, 600–601.

12 Amongst the authors that have put forward versions of the evolutionary debunking argument, see Joyce 2006; Kitcher 2006, 2011; Ruse and Wilson 1986; Ruse 1998; and Street 2006.

13 Amongst the authors that have defended this kind of argument, see Brosnan 2011; Enoch 2010, 2011; and Skarsaune 2010.

14 See Ekman 1972; Griffiths 1997; and Cosmides and Tooby 2000.

15 See for instance, Kenny 1963; de Sousa 1987; Teroni 2007; and Deonna and Teroni 2012.

16 Amongst the value realists that have offered some replies to the evolutionary debunking argument, see Berker 2014; Bogardus 2016; Brosnan 2011; Copp 2008; Enoch 2010 and 2011; FitzPatrick 2015; Kahane 2011; Huemer 2016; Shafer-Landau 2012; Skarsaune 2010; Vavova 2014 and 2015; Wielenberg 2010.

17 On this point, see also Shafer-Landau 2012; Vavova 2014.

18 See, *inter alia*, Shafer-Landau 2012; Vavova 2014, 2015.

19 We will come back to this line of thought below, when developing a "third factor" argument against the evolutionary debunking argument.

20 On this reply to the evolutionary debunking argument, see Shafer-Landau 2012, 12–13.

21 For a discussion of the argument from sensitivity, see also Wielenberg 2010; Clarke-Doane 2012; Shafer-Landau 2012; Bogardus 2016.
22 For a discussion of the argument from safety, see Bogardus 2016.
23 See Shafer-Landau 2012, especially 24–6.
24 For a similar point, see Vavova 2015, 105–6.
25 In Tappolet and Rossi (2015b), we argued that the claim that emotions have been selected for as fitness-enhancing mechanisms does not entail the claim that emotions have been selected for as well-being-enhancing mechanisms. Furthermore, we have also argued that there exists no necessary relation between emotions and well-being. Both of these claims are compatible with the thesis defended here, according to which *some* fitness-enhancing emotions are also well-being-enhancing emotions.

References

Anderson, Elizabeth S. 1993. *Values in Ethics and Economics*. Cambridge, MA: Harvard University Press.

Berker, Selim. 2014. "Does Evolutionary Psychology Show that Normativity Is Mind-Dependent?" In *Moral Psychology and Human Agency: Philosophical Essays on the Science of Ethics*, edited by Justin D'Arms, and Daniel Jacobson, 215–52. Oxford: Oxford University Press.

Blackburn, Simon. 1984. *Spreading the Word*. Oxford: Oxford University Press.

Blackburn, Simon. 1998. *Ruling Passions*. Oxford: Clarendon Press.

Bogardus, Tomas. 2016. "Only All Naturalists Should Worry about Only One Evolutionary Debunking Argument", *Ethics*, 126: 636–61.

Brady, Michael. 2013. *Emotional Insight*. Oxford: Oxford University Press.

Brentano, Franz. 1889. *The Origin of Our Knowledge of Right and Wrong*. Translated by Roderick M. Chisholm, and Elizabeth H. Schneewind. London: Routledge and Kegan Paul, 2009.

Brink, David Owen. 1989. *Moral Realism and the Foundations of Ethics*. Cambridge: Cambridge University Press.

Brosnan, Kevin. 2011. "Do the Evolutionary Origins of Our Moral Beliefs Undermine Moral Knowledge?" *Biology and Philosophy*, 26: 51–64.

Chisholm, Roderick M. 1986. *Brentano and Intrinsic Value*. Cambridge: Cambridge University Press.

Clarke-Doane, Justin. 2012. "Morality and Mathematics: The Evolutionary Challenge", *Ethics*, 122: 313–40.

Copp, David. 2008. "Darwinian Skepticism", *Philosophical Issues*, 18: 186–206.

Cosmides Leda, and Tooby, John. 2000. "Evolutionary Psychology and the Emotions." In *Handbook of Emotions*, edited by Michael Lewis and Jeannette M. Haviland-Jones, 91–115. New York: The Guilford Press.

Cuneo, Terence. 2007. *The Normative Web: An Argument for Moral Realism*. Oxford: Oxford University Press.

D'Arms, Justin, and Jacobson, Daniel. 2000a. "Sentiment and Value", *Ethics*, 110: 722–48.

D'Arms, Justin, and Jacobson, Daniel. 2000b. "The Moralistic Fallacy: On the 'Appropriateness' of Emotions", *Philosophy and Phenomenological Research* 61: 65–90.

Davidson, Donald. 1991. "Three Varieties of Knowledge." Reprinted in Davidson, D. 2001, *Subjective, Intersubjective, Objective*, 205–20. New York: Oxford University Press.

de Sousa, Ronald. 1987. *The Rationality of Emotion*. Cambridge, MA: MIT Press.

Deonna, Julien. 2006. "Emotion, Perception and Perspective", *Dialectica*, 60: 29–46.

Deonna, Julien, and Teroni, Fabrice. 2012. *The Emotions: A Philosophical Introduction*. New York: Routledge.

Döring, Sabine. 2007. "Seeing What to Do: Affective Perception and Rational Motivation", *Dialectica*, 61: 361–94.

Ekman, Paul. 1972. *Emotions in the Human Face*. New York: Pergamon Press.

Enoch, David. 2010. "The Epistemological Challenge to Meta-Normative Realism: How Best to Understand It, and How Best to Cope with It", *Philosophical Studies*, 148: 413–38.

Enoch, David. 2011. *Taking Morality Seriously: A Defense of Robust Realism*. Oxford: Oxford University.

Ewing, Alfred C. 1947. *The Definition of Good*. London: Hyperion Press.

Ewing, Alfred C. 1959. *Second Thoughts in Moral Philosophy*. London: Routledge and Kegan Paul.

Fitzpatrick, William. 2015. "Debunking Evolutionary Debunking of Ethical Realism", *Philosophical Studies*, 172: 883–904.

Frijda, Nico. 1986. *The Emotions*. Cambridge: Cambridge University Press.

Gibbard, Allan. 1990. *Wise Choices, Apt Feelings*. Cambridge, MA: Harvard University Press.

Griffiths, Paul E. 1997. *What Emotions Really Are: The Problem of Psychological Categories*. Chicago: University of Chicago Press.

Huemer, Michael. 2016. "A Liberal Realist Answer to Debunking Skepticism: The Empirical Case for Realism", *Philosophical Studies*, 173: 1983–2010.

Hume, David. 1739–1741. *A Treatise of Human Nature*, edited by David F. Norton and Mary J. Norton. Oxford: Oxford University Press, 2000.

James, William. 1884. "What is an Emotion?" *Mind*, 9: 188–204.

Joyce, Richard. 2006. *The Evolution of Morality*. Cambridge, MA: MIT Press.

Kahane, Guy. 2011. "Evolutionary Debunking Arguments", *Noûs*, 45: 103–25.

Kauppinen, Antti. 2013. "A Humean Theory of Moral Intuition", *Canadian Journal of Philosophy*, 43: 360–81.

Kenny, Anthony. 1963. *Action, Emotion and Will*. London: Routledge and Kegan Paul.

Kitcher, Philip. 2006. "Biology and Ethics." In *The Oxford Handbook of Ethical Theory*, edited by David Copp, 163–85. Oxford: Oxford University Press.

Kitcher, Philip. 2011. *The Ethical Project*. Cambridge, MA: Harvard University Press.

Lange, Carl G. 1885. "The Emotions: A Psychological Study." In *The Emotions*, edited by Carl G. Lange, and William James, 33–90. Baltimore: Williams and Wilkins, 1922.

Mackie, John L. 1977. *Ethics: Inventing Right and Wrong*. London: Penguin.

McDowell, John. 1985. "Values and Secondary Qualities." In *Morality and Objectivity: A Tribute to John Mackie*, edited by Ted Honderich, 110–29. London: Routledge and Kegan Paul.

Meinong, Alexius. 1917. *On Emotional Presentation*. Translated by Marie-Luise Schubert Kalsi. Evanston Ill: Northwestern University Press.

Moore, George E. 1903. *Principia Ethica*. Cambridge: Cambridge University Press, revised edition 1993.

Moore, George E. 1922. "The Conception of Intrinsic Value." In *Philosophical Studies*, edited by Moore, George E. (253–275). London: Routledge and Kegan Paul, 1965.
Mulligan, Kevin. 1998. "From Appropriate Emotions to Values", *The Monist*, 81: 161–88.
Nichols, Shaun. 2004. *Sentimental Rules: On the Natural Foundations of Moral Judgment*. New York: Oxford University Press.
Nussbaum, Martha C. 2001. *Upheavals of Thought*. Cambridge: Cambridge University Press.
Oddie, Graham. 2009. *Value, Reality, and Desire*. New York: Oxford University Press.
Oddie, Graham. 2014. "Thinking Globally, Acting Locally: Partiality, Preferences and Perspective", *Les Ateliers de l'éthique/The Ethics Forum*, 9: 57–81.
Pettit, Philip. 1991. "Realism and Response-Dependence", *Mind*, 100: 587–626.
Prinz, Jesse. 2004. *Gut Reactions: A Perceptual Theory of Emotion*. New York: Oxford University Press.
Prinz, Jesse. 2007. *The Emotional Construction of Morals*. New York: Oxford University Press.
Putnam, Hilary. 2002. *The Collapse of the Fact/Value Dichotomy and Other Essays*. Cambridge, MA: Harvard University Press.
Railton, Peter. 1996. "Moral Realism: Problems and Prospects." In *Moral Knowledge? New Readings in Moral Epistemology*, edited by Walter Sinnott-Armstrong and Mark Timmons, 49–81. New York: Oxford University Press.
Ruse, Michael. 1998. *Taking Darwin Seriously*. New York: Blackwell.
Ruse, Michael, and Wilson, Edward O. 1986. "Moral Philosophy as Applied Science", *Philosophy*, 61: 173–92.
Sayre-McCord, Geoffrey. 1988. "Moral Theory and Explanatory Power." In *Midwest Studies in Philosophy*, edited by P. A. French, T. E. Uehling, and H. Wettstein, 433–56. Minneapolis: Minneapolis University Press.
Scanlon, Thomas M. 1998. *What We Owe to Each Other*. Cambridge, MA: The Belknap Press.
Scarantino, Andrea. 2014. "The Motivational Theory of Emotions." In *Moral Psychology and Human Agency*, edited by Daniel Jacobson, and Justin D'Arms, 156–85. New York: Oxford University Press.
Shafer-Landau, Russ. 2012. "Evolutionary Debunking: Moral Realism and Moral Knowledge", *Journal of Ethics and Social Philosophy*, 7: 1–37.
Skarsaune, Knut. 2010. "Darwin and Moral Realism: Survival of the Iffiest", *Philosophical Studies*, 152: 229–43.
Solomon, Robert C. 1976. *The Passions*. Indianapolis, Ill: Hackett Publishing Company.
Street, Sharon. 2006. "A Darwinian Dilemma for Realist Theories of Value", *Philosophical Studies*, 127: 109–66.
Sturgeon, Nicholas. 1984. "Moral Explanation." In *Morality, Reason, and Truth*, edited by David Copp, and David Zimmerman, 49–78.Totowa, NJ: Rowman and Allenheld.
Tappolet, Christine. 1995. "Les émotions et les concepts axiologiques." In *La Couleur des Pensées*, edited by Patricia Paperman and Ruwen Ogien, *Raisons Pratiques*, 6: 237–257.

Tappolet, Christine. 2000. *Emotions et valeurs*. Paris: Presses Universitaires de France.

Tappolet, Christine. 2016. *Emotions, Values, and Agency*. Oxford: Oxford University Press.

Tappolet, Christine, and Rossi, Mauro. 2015a. "What Is Value? Where Does It Come From? A Philosophical Perspective." In *The Handbook of Value: The Affective Sciences of Values and Valuation*, edited by Tobias Brosch and David Sander, 3–22. Oxford: Oxford University Press.

Tappolet, Christine, and Rossi, Mauro. 2015b. "Emotions and Wellbeing", *Topoi*, 34: 461–74.

Teroni, Fabrice. 2007. "Emotions and Formal Objects", *Dialectica*, 61: 395–415.

Tye, Michael. 2008. "The Experience of Emotion: An Intentionalist Theory", *Revue Internationale de Philosophie*, 243: 25–50.

Vavova, Katia. 2014. "Debunking Evolutionary Debunking." In *Oxford Studies in Metaethics*, Vol. 9, edited by Russ Shafer-Landau, 76–101. Oxford: Oxford University Press.

Vavova, Katia. 2015. "Evolutionary Debunking of Moral Realism", *Philosophy Compass*, 10: 104–16.

Whiting, Demian. 2011. "The Feeling Theory of Emotion and the Object-Directed Emotions", *European Journal of Philosophy*, 19: 281–303.

Wielenberg, Erik J. 2010. "On the Evolutionary Debunking of Morality", *Ethics*, 120: 441–64.

Wiggins, David. 1976. "Truth, Invention and the Meaning of Life." In his 1998, *Needs, Values, Truth: Essays in the Philosophy of Value*, 87–127. Oxford: Basil Blackwell.

Wiggins, David. 1987. "A Sensible Subjectivism." In his *Needs, Values, Truth: Essays in the Philosophy of Value*, 185–214. Oxford: Basil Blackwell Press.

10 How to Be a Relativist

Kenneth A. Taylor

1 Preliminaries

Rampant moral relativism is widely decried as the leading source of the degeneracy of modern life (Ratzinger and Pera 2006). Though I proudly count myself a relativist, I rather doubt that relativism has anything like the cultural influence that its most ardent critics fearfully attribute to it. Much of what gets criticized under the rubric of relativism is often really no such thing. Relativists need not be hedonists, egoists, nihilists, or even moral skeptics. Moreover, when it comes to the upper reaches of our intellectual culture, relativism is more often dismissed than defended.[1] I don't deny that in certain literary corners of academe, relativism retains a fashionable post-modern cachet.[2] But in more sober philosophical circles, the catalog of ills from which relativism is widely thought to suffer is impressive.[3] When taken as a characterization of the nature of moral discourse and moral argument, relativism is often thought to be descriptively inadequate. Contra the relativist, we do not treat moral disputes as rationally irresolvable. We do not tolerate all alternative moral "codes" as equally valid. Relativism may be true of merely cultural norms or practices. But morality has a felt universality that makes it quite different in character from a system of merely cultural norms or practices. In the face of morally abhorrent practices, we don't simply shrug our shoulders and say that while the relevant practices may be wrong for us, they are alright for them. Relativism is sometimes even said to be self-undermining. It makes the very thing it purports to explain—the possibility of rationally intractable disagreements—impossible in the first place. Partly because of its supposedly self-undermining character, relativism is sometimes accused of being a strictly incredible doctrine. Those who profess to be relativists must, if this is true, either be insincere, confused, or self-deceived. Though someone might well sincerely hold the mistaken second-order belief that she believes that she believes that relativism is true, no one, in his or her deepest heart of hearts, sincerely, non-self-deceptively and informedly believes that relativism is true.

In this essay, I swim against the predominant anti-relativistic philosophical tide. My minimal aim is to show that relativism is neither descriptively inadequate nor self-defeating. My maximal aim is to outline the beginnings of an argument that relativism is a truth resting on deep facts about the human normative predicament. And I shall suggest that far from being a source of cultural degeneracy, the fact of relativism has the potential to ground a culture that is deeply life-affirming. My argument against the twin charges of descriptive inadequacy and self-defeat turns on a distinction between *tolerant* and *intolerant* relativism. I concede that many of the standard arguments against relativism do have force against tolerant relativism. But against intolerant relativism, those arguments are entirely unavailing. The crucial difference between the tolerant and intolerant relativist is that although the intolerant relativist agrees with the tolerant relativist that norms are relative, she insists that agents are sometimes entitled to hold others to norms by which they are not bound. I shall argue that just because the intolerant relativist allows that we are sometimes entitled to hold others to norms by which we are bound but they are not, she is able to escape both the charge of descriptive inadequacy and the charge of self-defeat. In particular, I shall show that the intolerant relativist has a coherent and satisfying account of the nature of moral disagreement and moral argument. Establishing the ultimate truth of relativism, however, would take more than showing that one form of relativism escapes certain standard arguments against relativism. Though I do not pretend to conclusively discharge the burden of showing that relativism is true in the space of this essay, I do sketch the beginnings of an account of what I call the bindingness of norms that has intolerant relativism as a more or less straightforward downstream consequence. If there are independent grounds for accepting that account of bindingness, then there are independent grounds for accepting intolerant moral relativism.

2 A Metaphilosophical Prelude

The account of the bindingness of norms on offer in this essay is psychologistic and naturalistic. In order to forestall certain objections to my account that may arise just because of its psychologistic and naturalistic character, let me be clear from the outset what I do and do not claim to show. The pretensions of the theory on offer here are descriptive and explanatory rather than normative and justificatory. I do not seek to justify any particular set of norms. Rather, I seek merely to describe what the bindingness of norms might plausibly consist in. My guiding question is a how possibly question. I want to know what in the natural order norms of rational self-management might be such that an agent might be bound by such norms in virtue of merely natural and psychological facts about that agent. What makes this question at all gripping and challenging is the evident fact that there exists a certain conceptual distance between our ordinary, intuitive conception of the normative and our ordinary, intuitive conception of the merely natural.[4] Because of this conceptual distance, we don't know in advance how to rationally coordinate the explicitly naturalistic concepts by which we cognize the denizens of the natural order and the explicitly normative concepts by which we cognize the denizens of the normative order. We have no antecedently available means of re-identifying that which we proto-typically re-identify via the deployment of normative concepts as merely further aspects of the natural order. If we are to achieve rational coordination between the natural and the normative, we need more concepts than are currently dreamt of in either our commonsense intuitive conceptions of the natural or our commonsense intuitive conceptions of the normative. And those new concepts must bridge the conceptual distance between the natural and the normative as we currently conceive of them.

My aim in this essay is to offer up just such a set of intermediate or bridging concepts. Consequently, the central claims on offer here should not be understood as conceptual-analytic claims about our intuitive understanding of normativity and its relationship to the natural order. I am prepared, if need be, to adopt a quite revisionary attitude toward our ordinary understanding of our ordinary normative practices. Though there is a budget of folk concepts and notions that we typically use to understand our own normative thought and talk, I am prepared to find that those concepts give us a poor cognitive hold on a certain real phenomenon in the world. I do not take it as a condition on the adequacy of the theory of norm bindingness on offer here that it should preserve intact our ordinary conceptions, intuitions, and notions. My account fairly bristles with theoretical notions and distinctions neither directly nor explicitly countenanced by our ordinary commonsense understanding of normativity. This is not to say that I remain entirely indifferent to the deliverances of commonsense. I do take it to be a condition on the adequacy of my account that where it has consequences that appear to conflict with certain ordinary intuitions and notions, that I should, ultimately, be able to either explain or explain away those intuitions, but in my own privileged theoretical vocabulary. This I will do, for example, with the widely shared intuition that morality has a felt universality that renders it incompatible with relativism. Morality does have a kind of universal purport, it will turn out, but of a kind that is entirely consistent with intolerant relativism.

One way to think of this essay is as an exercise in Martian Anthropology. It is as if I am a Martian Anthropologist, on a scientific expedition to planet Earth. My aim is to understand what in the natural order of things the alien human practice of guiding their lives by norms of rational self-management comes to. Qua episode in Martian Anthropology, my investigations are not normative inquiries into the question by which norms ought humans to live. For the purposes of my merely anthropological investigations it is as if I stand outside and apart from all human normative communities and all human normative disputes. Qua outsider, my aim is merely to describe and explain what humans are doing when they undertake to manage their cognition and conation in accordance with norms of rational self-management and to show that those doings are not, in the end, something outside of the natural order, but something that subsist wholly within and as a part of that order.

Now, since this exercise in Martian Anthropology is intended as an exercise in philosophical rather than scientific anthropology, I will count myself successful if I can show that there are plausibly nearby possible worlds of which my naturalistic and psychologistic account of the bindingness of norms is plausibly true. For then I will have shown that the normative really *could* have a place in the natural order. Admittedly, I will not thereby have shown that normativity *actually does* have a place in the natural order. But it is, I hope, not unreasonable to expect that the stock of concepts and distinctions I develop in brief compass in this essay and more fully elsewhere will ultimately prove to have application not just to nearby possible worlds, but to our very own as well.[5] Establishing that, however, is a task for another day. For the nonce, I will be satisfied if you gain fuller imaginative acquaintance with a possibility—the possibility that norms and their binding force are a real part of the natural order. If we are able to gain fuller imaginative acquaintance with that possibility, we should be left with less lingering temptation to see normativity as *sui generis* and irreducible. And given that my defense of relativism flows directly from my account of the metaphysics of normativity, our exercise in imagining should also lower any antecedent resistance to and fear of relativism.

3 Norms vs. Normative Statuses

Naturalistically minded philosophers have thought about the subsistence of norms in a number of different ways. Some believe that the biological world is replete with normativity. They see normativity in the "proper functioning" of the parts of animals and plants and in the way the coordinated functioning of those parts enable living things to thrive and reproduce.[6] I am not, in the first instance, concerned with such putative norms of proper functioning. Indeed, I take no stand on whether norms of proper function are normative in any robust sense—though I rather doubt that they are. My concern is rather with what I call norms of rational self-management. Norms of rational self-management are a very special kind of thing, addressed to very special kinds of creatures. They are addressed, in the first instance, to cognizing agents who enjoy the capacity for a kind of self-mastery over their own cognition and conation. Norms of rational self-management direct cognizing agents to govern their cognition and conation in one way rather than another. When agents are bound by such directives they are often thereby "committed" to manage their cognition and conation in accordance with those directives. And others may thereby be entitled to hold them to such commitments. One central subsidiary aim of this essay is to sketch a naturalistic, psychologistic account of how possibly norms of rational self-management manage to bind us and to explain how possibly commitments and entitlements are generated by the norms by which we are sometimes bound.

What exactly is a norm? One way to think about norms is as "ought-to's," where an ought-to is a directive articulating what (putatively) ought to be, be done, or be believed. Such directives can be more or less general. They can articulate what a given agent ought to do or believe at a given time or in a given set of circumstances. Or they can articulate general constraints on action or belief. If you are prone to reify norms, you may, for the nonce, think of the totality of norms as subsisting in a sort of abstract norm space, roughly on a par with the space of propositions. You may think of this abstract space as a plenum, containing every possible ought-to, from the most specific to the most general. If one were to think of norms as abstract real existents of this sort, one might believe it worthwhile to investigate the, as it were, fine structure of this plenum. For two reasons, that is not a task I shall undertake here. First, our current problem is not to determine which norms *subsist* in the plenum of all possible norms, but to say which norms *bind* self-managing cognizing agents and to say in virtue of what they do so. Separating questions about which norms are subsistent from questions about which norms are binding is crucial for our anthropological inquiry. Once we recognize that norms may subsist even when they bind no one, we can view the totality of norms as constituting a kind of possibility space. We want to know in virtue of what natural and psychological facts merely subsistent norms actually bind cognizing agents.

In this quasi-Platonistic mode of thinking of norms as abstract real existents, it may also seem natural to think of the plenum of norms as being metaphysically on a par with the plenum of propositions. Thinking that way about norms may lead one to believe that norms are the kinds of thing that can be true or false. But even in our quasi-Platonistic mode, we should not give in to that temptation. A plenum of norms would not be a plenum of propositions. It would be a plenum of directives. [7] As such, norms would not themselves be directly in the business of being true or false. This is not to deny that the plenum of norms would, if such a thing really did subsist, be in a related business—the business of binding or failing to bind cognizing agents. To deny that norms are propositions is not to deny that there subsist, or might subsist, normative propositions about what ought to be, be done, or be believed. Normative propositions would indeed be the sorts of things that could plausibly be said to be true or false. My point is about the relative priority of norms and normative proposition. If there are such things as normative propositions, they are made true, if they are true, by facts about norm bindingness. If Smith is bound by a norm of rational self-management that directs the prompt completion of her relativism paper, that makes it true, at least in one sense, that Smith *ought* to finish her paper soon. So in order to know which normative propositions are true or false it would behoove us to say just what it takes for an agent to be bound by a norm.

I am not entirely comfortable with talk of an abstract plenum of subsistent norms. But I am content to leave that talk stand for the nonce, as long as one is willing to take such talk as a mere *façon de parler*. Ultimately, I seek to replace talk of norms with talk of *normative statuses*. A normative status is defined by a pairing of upstream entry conditions and downstream consequences.[8] To specify a normative status S, we specify: (a) a set upstream entry conditions, $\phi_1 \ldots \phi_n$ such that if x satisfies $\phi_1 \ldots \phi_n$ then x has the status S and (b) a set of downstream consequences, $c_1 \ldots c_n$, such that if x has status S, then $c_1 \ldots c_n$ obtain. The entry conditions for a normative status may be either normative or non-normative. The downstream consequences that define a normative status will typically be characterized in terms of a set of entitlements and commitments. Consider bankruptcy. There is a set of conditions that one has to satisfy in order to count as being bankrupt. There is also a set of entitlements enjoyed by relevant creditors and commitments undertaken by the relevant debtor that are consequences of the debtor's status as bankrupt. The pairing of the particular entry conditions with the particular downstream entitlements and commitments that define a given normative status will often be a consequence of the social-dialectical role of the relevant status in some collectivity. It is because the status of being bankrupt is a social-dialectical instrument for coordinating commitments and entitlements among creditors and debtors that it consists in just this rather than that pairing of entry conditions and downstream consequences.

Bankruptcy is just one normative status among others. There are a plethora of such statuses, including being innocent or guilty in the eyes of the law, having a failing or passing grade, being called out on strikes in baseball, being the President of the United States, being married, being divorced, being rational, being irrational, being virtuous or vicious and, according to some, believing that snow is white. Many of the normative statuses just mentioned are what I call *explicitly conferred* statuses. Others—like being rational or being virtuous—may seem *automatic* rather than conferred. Qua automatic, a normative status enjoys its standing as normative independently of anything that we do or are. Apparently automatic normative statuses may be thought to be constitutively tied to certain bedrock normative domains. Does it not come with the bedrock normative turf of morality, for example, that one who has killed an innocent child merely for the pleasure of it has the normative status of being vicious or evil? Does it not come with the bedrock normative domain of rational belief-fixation, that one who affirms the consequent is illogical or irrational?

The tacit belief that certain normative statuses are constitutively tied to certain "bedrock" normative domains may appear to militate against both relativism and naturalism. In virtue of their presumed fixity and independence from what we do or are, automatic statuses may seem to undermine the relativistic claim that morality is, in some sense, entirely up to us. Because of their presumed constitutive ties to putatively bedrock normative domains, automatic statuses may be thought to be normative "all the way down" and thus never to bottom out in anything merely natural and non-normative. Ultimately, I shall reject the very idea of automatic normative statuses that enjoy their standing as normative independently of anything that we do or are. Normative status rests always and only on the merely natural and evolved psychological power of the human mind to *confer* normative status. No normative status achieves standing as a status for us except that we *take it up* as a status and thereby make it a status for us. Seeing that normative statuses one and all bottom out in merely conferred statuses is the ultimate key to appreciating both the truth of relativism and the truth of naturalism.

I said earlier on that the entry conditions for a given normative status may well involve conditions that are already normative. One doesn't count as bankrupt, for example, unless one has legally enforceable debts that one is unable to pay. Having a legally enforceable debt is already a normative status. Moreover, the downstream consequences that partially define a given normative status will typically be a set of entitlements and commitments. Entitlements and commitments may themselves seem inherently normative. Consequently, if we are to give a fully naturalistic account of normativity and normative statuses, we must meet two conditions. First, the hierarchy of entry conditions must ultimately bottom out in a set of ground level conditions that can be specified wholly naturalistically. Second, we must be able to tell a naturalistic story about the generating of entitlements and commitments. In particular, we must be able to characterize the issuing of entitlements and the undertaking of commitments in wholly naturalistic terms. We must show, in effect, that status-conferring power of the human mind—the power by which it issues entitlements and undertakes commitments—is a wholly natural power that can be fully described and explained in a psychologistic and naturalistic vocabulary.

4 Binding and the Conferral of Status

I conjecture that a cognizing agent is "bound" by a norm *N* just in case she does or would "endorse" *N* upon what I call culminated competent reflection. Through such an endorsement, an agent confers a certain status upon herself. That binding involves what we might call the self-conferral of a status is the crucial initial point. Others may confer normative status upon me. And I may either accept or reject the status conferred on me by others. Indeed, we shall have a great deal to say in what follows about, as it were, the give and take of normative status among status-conferring creatures like ourselves. But we begin by singling out a special kind of status-conferral—what we might call self-conferral. Our main conjecture is that norm bindingness ultimately amounts to the self-conferral of normative status.

My conjecture should be understood as a conjecture about just what in the natural order being "bound" by a norm consists in. The conjecture is not itself a normative claim, but a substantive explanatory hypothesis in the naturalistic metaphysics cum psychology of normativity. Like all substantive hypotheses, it is to be tested by its consequences for our understanding of the real world phenomenon it seeks to describe, interpret, and explain. In that spirit, it is crucial to understand phrases like "culminated competent reflection," and "endorsement" as purely psychofunctional role concepts, systematically interdefinable in terms of one another. As deployed here, none of these notions is intrinsically or irreducibly normative. They are intended to function as explanatory notions, posited for the sole purpose of locating within the natural order what human beings are doing when they self-confer normative status and thereby bind themselves to norms. They are defined theoretical terms with no antecedent meanings. Their meanings are constituted by their roles in a privileged theoretical framework that serves as their home turf. Our privileged theoretical vocabulary is not intended to bear any tight connection to our ordinary, commonsensical notions of endorsement or competence or the like. We are in the business of speculative theory construction, not the business of analyzing commonsense. Nor are we in the business of justifying any particular normative practices. The measure of the adequacy for our framework has to do solely with its power to enhance our ability to explain and systematize the phenomena under investigation—even if that explanation and systematization sometimes runs counter to our commonsense pre-theoretical understanding of those phenomena.

We begin by considering in more detail the competence condition appealed to in our guiding conjecture. A form of reflection counts as competent, for a given dialectical cohort, if exercises of that form of reflection historically played, or currently plays, a decisive causal role in spreading and sustaining normative community among the members of that dialectical cohort. Within a cohort, *current* exercises of the historically decisive or currently effective form or forms of reflection count as episodes of competent reflection. A cognizer reflects competently, in other words, if she is disposed to reflect in ways that have historically sustained or currently function to sustain normative community among a dialectical cohort of which she is a member.

If, and only if, you exercise considerable caution in so doing, you may think of competent reflection as a kind of "ideal" reflection. But the perils of this way of thinking are manifold. Some philosophers think of ideal reflection as reflection that tracks the "objectively good" whatever exactly that is. Others believe that under "ideal" reflection, rational agents are guaranteed to converge on endorsements of the same standards or norms. As used here, "competent" carries no such connotations at all. The question my competence condition is intended to enable us to answer in the course of our anthropological investigations is not the question which form of reflection objectively "deserves" to play a causal role in the sustaining and spreading of normative community. The question is, rather, what kind of reflection has in fact played the decisive causal role. Recall our perspective as Martian anthropologists. Our goal is merely to locate in the natural order, the possibly diverse forms of reflection, whatever they are, that have historically played, or are currently playing, a decisive causal role among the extant dialectical cohorts into which we find the human species arrayed. From our anthropological perspective, we need not ipso facto take a critical stance toward any or all of these dialectical cohorts and their community sustaining forms of reflection. That is, we need not ourselves confer any normative status on the extant forms of reflection. Indeed, though we are prepared to find that our subjects have themselves conferred normative status on one or another form of reflection, we are equally prepared to find that that they have not done so. We do not presume that our subjects are either highly self-aware or highly self-endorsing. There may indeed be highly reflective agents who have reflected upon their own forms of reflection and have developed a theory of such reflection. Moreover, as a consequence of such theoretical self-awareness, they may even confer normative status upon the relevant form of reflection. But neither reflective self-awareness nor reflective self-endorsement is required for us, from our anthropological perspective, to count a form of reflection as "competent." We are after those extant forms of reflection, whatever they turn out to be, that have done a certain job among extant and/or past dialectical cohorts—the job of spreading and sustaining normative community. It matters not for our purposes whether the members of the relevant cohorts are either self-aware or self-endorsing.

A further word about dialectical cohorts is required. Characterized at the highest level of abstraction, a dialectal cohort is a collection of cognizing agents who engage in modes of reasoning to which the members of that cohort mutually "resonate." If I offer you arguments that move you in ways that also move me and if there is some causal/historical explanation of how we came to be so related, then we count as members of a common dialectical cohort. The relativization of competence is meant to mark the possibility that different modes of reflection may play the cohort-sustaining role in different cohorts. Forms of reflection that spread and sustain normative community among pre-scientific, pre-literate, or pre-philosophical cohorts may differ radically from the forms of reflection that are extant among more scientific, literate, and philosophical cohorts. Even within a dialectical cohort, intellectual progress may give rise to progressively more refined forms of reflection. When intellectual progress does happen, the competence condition for a dialectical cohort will specify the form of reflection that currently plays the decisive role in spreading and sustaining normative community. Dialectical cohorts may also fragment and divide. Out of this fragmentation, a new array of dialectical cohorts may constitute themselves. At the very extreme, a given cognizing agent may even come to form a dialectical cohort of one. A cognizer may count simultaneously as a member of multiple dialectical cohorts subject to different competence conditions. But such an agent is likely to suffer from a kind of internal fragmentation.

I said that at the highest level of abstraction, a dialectical cohort may be characterized as a collection of cognizing agents who mutually resonate to shared forms of reasoning and reflection. Closer in, one dialectical cohort is distinguished from another by what I call epistemic fine structure. The epistemic fine structure of a dialectical cohort is determined by the set of background theories, principles, and cognitive dispositions that jointly function as warrant spreading machinery within the relevant local community. But by warrant I do not mean "objective" warrant—whatever that might be—but warrant by the shared lights of the members of the relevant dialectical cohort. [9] We might call this sort of warrant internal warrant. To explain the epistemic fine structure of a dialectical cohort is to characterize the mediating structures that spread internal warrant within and across agents. For example, there may be shared standards that determine what counts as evidence for what and with what weight. These may vary from cohort to cohort. Similarly, the distribution of epistemic authority may vary. The members of a shamanistic cohort may invest epistemic authority in the deliverances of the shaman, where the members of a scientific cohort invest such authority in the deliverances of science.

Think of the diverse dialectical cohorts, with their varying epistemic fine structures, as so many local configurations of reason. There is a rich and complex story to tell about the formation, deformation, and reformation of dialectical cohorts over the long sweep of human history. The history of such formation, deformation, and reformation represents reason's actual historical walk through the space of all possible local configurations of reason. There can be no *a priori* anticipation of reason's trajectory through the space of possible local configurations.[10] Reason's walk through history is determined by no simple principle. It is a walk fraught with contingency, with dead ends and wrong turns, but also with decisive and clarifying ruptures. What bears stressing here, however, is that the local configurations of reason are one and all configuration *of reason*. From our perspective as Martian anthropologists, our task is not to choose among the ways that human reason has configured itself locally through history. Our task is merely descriptive and explanatory.

The diversity of local configurations of reason raises some deeply challenging and important issues. From our Martian perch, we may regard that diversity with equanimity. But no such equanimity is demanded of those who stand within any one local configuration. The members of a given dialectical cohort need not, and likely will not, take alternative configurations of reason as rationally on a par with their own. That is because one's own normative lights may illuminate the entire history and present dynamics of reason. When illuminated in this one-sided way, from a peculiar normative perch, some of reason's alternative local configurations may be presented as engines of intellectual progress, while others may be presented as sources of darkness and error. It goes without saying that what one set of normative lights presents as engines of intellectual progress, other lights may present as instruments of intellectual decline.

One might worry that the members of a given dialectical cohort must altogether lack the capacity to recognize alternative configurations of reason as configurations of reason at all. It was something like this worry that lay at the heart of Davidson's rejection of the very idea of a conceptual scheme and to insist that rationality must always and only be rationality by our own lights. I have argued against this view elsewhere and will not rehearse those arguments here. But consider a multi-part distinction in terms of which we may measure our distance from the rational other. Closest to us will be the rational other with whom we stand in what I call full rational solidarity. When we stand in full rational solidarity with another rational being, we enjoy with that other a community of reasons, of mutually conferred and endorsed normative statuses. Further away, are rational others whom we recognize as reasoning but whom we may condemn as unreasonable. In the reasoning, but unreasonable rational other, we recognize the clear traces of reason at work. But our recognition does not take place within a fully shared normative framework. Further away still, will be others whom we take to be not merely unreasonable, but "irrational." Here we may begin to doubt that we firmly recognize reason at work in the other. Nonetheless, in condemning such another as irrational, we do not imply that she has no place in the realm of reason. It is as if we recognize in the other mere remnants of reason rather than reason fully formed. I say the remnants of reason because the irrational other is not as distant from the realm of reason as the a-rational other is. The a-rational is the most distant from us. It includes the entire unreasoning order—including rocks, trees, and many animals. But the a-rational is also a divided realm. For just as we recognize the mere remnants of reason in certain rational others that sit near the outside edges of the rational order, so we recognize precursors of reason in certain a-rational others that sit near the edges of the a-rational order. But the point that bears stressing is that between the extremes of full rational solidarity, on the one hand, and the a-rational order, on the other, fall both those with whom, though we confidently take them to be part of the "rational" order, we do not enjoy full rational solidarity and those in whom we recognize only what we take to be mere remnants and/or fragments of reason.

From our perspective as Martian anthropologists, the illuminations cast back and forth by the diversity of normative lights are just more facts on the ground. We are, to be sure, interested in charting the growth and decay of dialectical cohorts over historical time. Consequently, we must ultimately explore the space of all possible local configurations of reason and explain the dynamical principles governing reason's actual historical walk through that possibility space. As a stage of that inquiry, we will want to know which, if any, local configurations enjoy large basins of attraction, and which, if any, represent stable equilibrium points. Such an inquiry has the potential to discover that some local configurations are "dynamically favored" over others. But such a finding should not be taken as evidence that the favored configurations rest on some privileged transcendental ground, fixed once and for all, outside of history and culture, a ground from which we may determine by whose lights the ultimate "truth" is to be measured. Dynamically favored dialectical cohorts are likely to narrate the history of reason up to the time of their own emergence and consolidation in their own normative terms. But qua Martian anthropologists, we should refrain from endorsing the self-told narratives of the dynamically favored merely because they are dynamically favored.

Consider next what I call conceptions of the good. A conception of the good is a set of convictions and commitments about what is to be, be done, or be believed. That is, a conception of the good concerns what is good in the way of action, good in the way of being, and good in the way of believing. The set of convictions and commitments that constitute a conception of the good may be of varying strength and intensity. They may be more or less articulate, more or less determinate. A conception of the good may be either initial or considered. A conception of the good is initial when, although it is in some sense there, inside the agent, it does not yet enjoy the agent's full rational backing. A conception of the good is considered when an agent has decisively owned, through culminated competent reflection, that conception of the good as her own. She has thereby decisively undertaken to govern her conation and/or cognition in accordance with norms that license that conception of the good. She has thereby conferred a certain normative status upon herself.

Now initial conceptions of the good are shaped and conditioned in a variety of ways. Mechanisms of socialization, for example, play an important and powerful role in determining one's initial conception of what is good in the way of acting or of being or of believing. Before even the first dawning of reflective self-awareness, human beings are typically thrown into various collectivities in which our still developing normative lights are assaulted from without by the relentlessly droning other. Others attempt to mold and shape us, from the ground floor of our selves, into beings fit for a life within the local collectivities into which we find ourselves thrown. At the eventual dawning of reflective self-awareness, we may find ourselves already furnished with an initial conception of the good, one that may be deeply psychological entrenched and thus, in one sense, firmly held. But as long as that already given conception of the good lacks our own full rational backing, however firmly psychologically entrenched it may be, it remains still a merely initial conception, rather than a considered conception. A conception becomes considered only when one makes it fully one's own through "culminated" and "competent" reflective endorsement. When one does make a conception of the good fully one's own, one thereby undertakes, with all one's rational powers, to govern one's life as one's own.[11]

If initial conceptions of the good are the (initial) inputs to reflection, endorsements are the outputs of reflection. Again, I use the term 'endorsement' as a purely non-normative, psychofunctional term of theoretical art.[12] To a first approximation, a state x is an endorsement if it is a state of a kind K such that (a) culminated courses of reflection typically culminate in states of kind K and (b) states of kind K typically cause pro-attitudes toward actions, attitudes, and states of affairs appropriate to states of kind K. If I endorse Barack Obama for President that will typically cause me to have a pro-attitude toward any or all of the following: (a) the state of affairs of Obama's being or becoming president; (b) my own or another's desire to see Obama become President; and (c) actions taken by me or others that are intended to bring about or sustain an Obama presidency.

All manner of states and properties will present themselves to our subjects as *candidates* for their reflective endorsement, including emotions, desires, and beliefs. This is a deep fact about the psychological architecture of self-governing rational intellects and wills. For a self-managing rational being, having a belief, desire, emotion, or urge merely occur within the psychic economy is not yet for that state to be "owned" by that cognizing agent. But it would be a mistake to conclude that a state that merely occurs within the psychic economy of a self-managing rational agent is, therefore, merely an alien interloper until it has been reflectively owned. Through the mere occurrence of a state within the psychic economy a question is indeed put—even if not yet explicitly and self-consciously so—viz., the question whether what she merely finds her believing, feeling, or desiring is to be taken up as *her own* believing, feeling, or desiring. Through culminated competent reflection, self-managing cognizing agents decisively answer such questions for themselves. When an agent answers such questions in the affirmative, we will say that she has *ratified* the relevant beliefs, desires, or emotions. When one ratifies one's inner states, one thereby undertakes to stand behind those states in what I call the contest of reasons. One undertakes, thereby, to be responsive to rational pressures of various sorts, including possibly self-generated rational pressures for coherence and consistency, rational pressures directed toward the self from other rational beings, and worldly rational pressures from below.

I digress briefly to reflect on the nature of my conjecture that ratification involves a kind of undertaking, a concrete cognitive doing, and thus, in one sense, merely a further ingredient in the churning of an inner psychic stew. It is often objected that precisely because naturalistic conjectures purport to be merely descriptive and explanatory, such conjectures are bound to miss the mark when it comes to normativity. It is sometimes said, for example, that no merely third-personal characterization of a set of mere facts or mere happenings could be adequate to answer what Korsgaard has called the normative question of what is to be, be done, or be believed. And whatever else a merely descriptive, merely psychologistic account achieves, the argument sometimes goes, if it does not provide us with a hold on that which answers the normative question, it has not succeeded in locating normativity in the order of things. [13]

But this line of reasoning rests on a persistent confusion that is surprisingly resistant to correction. It is no part of the naturalistic enterprise to provide direct answers to the normative question. Asking and answering the normative question is something that deliberating, reasoning agents do—not something that we Martian philosophical anthropologists seek to do on behalf of those very agents. It may be helpful to compare and contrast our Martian philosophical anthropologist with another philosophical trope—the Socratic midwife. The Socratic midwife attempts to ferret out and render explicit what is merely implicit in our practices. Socratic midwifery is premised on the thought that we already tacitly if confusedly grasp certain truths about our practices and need the help of the Socratic midwife to render that implicit understanding explicit. To be sure, once that understanding is made explicit, our practices will be held up for further explicit critical scrutiny. The enterprise of Socratic midwifery is not unrelated to Martian philosophical anthropology. But there are important differences between the two. Martian philosophical anthropology does not presume that we already possess a tacit, if confused understanding of the true nature of our practices. The Martian philosophical anthropologist is not attempting to describe those practices back to us in a vocabulary we already tacitly possess. The Martian anthropologist finds herself faced with a metaphysically puzzling phenomenon in the world—the phenomenon of deliberating, reasoning agents putting to themselves and to others questions about what is to be, be done, or be believed. Her primary task is to understand what in the natural order of things this puzzling phenomenon amounts to. If she can show what in the natural order that phenomenon amounts to she is done—at least with that stage of her inquiry. To be sure, like the Socratic midwife, she may subsequently offer up her finished theory of the normative in nature to us as an additional instrument for our critical self-understanding. But our decision whether to take up her findings in our reasoning and deliberation about what is to be, be done, or be believed, is entirely irrelevant to judging the adequacy of her final theory. The theory may successfully represent back to us our own metaphysical natures as norm-mongering, status-conferring creatures, without *ipso facto* becoming a *deliberative* instrument, without, that is, guiding us in our deliberations about what is to be, be done, or be believed.

This is not to say that the finished theory of our own nature *must* leave our practices as they stand. Our practices may come in for thorough revision once the true metaphysical details of those practices are in. But here it is important to distinguish between the practices themselves and the possibly false philosophical and theological lore that has grown up around those practices. Our practices are as old as the species itself. The capacity for rational self-management, and with it the ability to confer normative status and form normative communities, emerged during the evolutionary prehistory of our species. The explicit philosophical and theological lore surrounding those capacities emerged much later. And it is an unfortunate fact that when philosophy and theology first began to theorize about normativity, the two of them together misunderstood the entire natural order and possessed only the shallowest understanding of what the human being in nature really amounts to. It would not be at all surprising if centuries of philosophy and theology together imported false understandings of the human being in nature first into explicit intellectual cultures and through them into our commonsense understandings of ourselves. Indeed our Martian philosophical anthropologist, who has no deep allegiance to what we may be pleased to call commonsense, is perfectly prepared to find that centuries of misguided philosophy and theology have led us to have false conceptions of what we are doing when we do perfectly ordinary things like binding ourselves to norms. Indeed, she is even prepared to find that entire cultural formations were built on these fictions and falsehoods. Consequently, when its results are fully taken up into our self-understanding, our exercise in Martian philosophical anthropology may lead us to throw off the anti-naturalistic blinders that archaic philosophy and theology, sometimes together sometimes separately, have introduced into our self-understanding. But such reformation in our self-understanding may still leave our ancient practices pretty much as they already were. Reformation of the mistaken philosophical lore surrounding normativity is one thing. Reformation of our practices is an entirely different thing.

But let us end our digression and return now to the nature of endorsement. There is no *a priori* guarantee that a cognizer will in fact desire or believe that which she would endorse, upon culminated competent reflection, as "worthy" of belief or "worthy" of desire. Nor is there an *a priori* guarantee that a cognizer would endorse, upon culminated competent reflection, that which she in fact desires or believes.[14] A psychologically well-ordered cognizer may strive to bring it about that she believes only what she deems worthy of belief and desires only what she deems worthy of desiring, but she is not guaranteed of success in that endeavor. In one sense, our beliefs, desires, and commitments may not be entirely up to us. Even a psychologically well-ordered cognizer may be causally determined to believe or desire that which, upon culminated competent reflection, she would deem unworthy of believing or desiring. Imagine a cognizer who believes that *p* as the result of hypnotic suggestion and lacks any further grounds for believing that *p*. Imagine that if she were to competently reflect upon hypnosis as a method of belief-fixation, she would not endorse it. Even if our cognizer were to reflectively conclude that her belief is not worthy of belief, she might still be unable to rid herself of that belief. This points to the possible real world limits to our capacity for self-management.

Endorsements are not all created equal. Some actual endorsements are not the outcome of a culminated course of competent reflection—either because though there was a culminated course of reflection, it was incompetent or because the endorsement was not the product of reflection at all. Only endorsements that are outcomes of culminated courses of competent reflection suffice to bind an agent to a norm. We might call those endorsements that are the outcome of culminated competent reflection deep endorsements. Deep endorsements matter because they amount to the undertaking of decisive rational commitments to norms of rational self-management. One who deeply endorses a norm thereby undertakes responsibility for rational self-management in accordance with that norm. She is thereby responsible both to herself and to any with whom she stands in rational community in the mutual endorsement of the relevant norm. She has thereby endowed herself with a certain normative status that commits her, both to herself and to those with who she stands in rational community, to manage her claims, intentions, beliefs, desires, and actions in accordance with the relevant norm. She may even thereby entitle others to *hold* her to the relevant norm. Now when an agent is in fact causally regulated only by attitudes and states that enjoy her own deep endorsement, and when the fact of deep endorsement is decisively causally responsible for bringing about such causal regulation, we may say that the agent is a fully self-managing agent.

Out of an abundance of caution, lest the nature of our inquiry be subject to the sort of persistent confusion that often arises in disputes over naturalism, it bears stressing once again that by calling a belief or desire "worthy" or in talking about entitlements and the like, we theorists confer no normative statuses of our own. Only the subject of our Martian philosophical anthropology is in the business of conferring such a status.

What we are offering is a substantive conjecture about the workings of the psychological mechanism through which the conferral of status is affected. Our conjecture is that status is conferred through culminated, competent reflective endorsement. Making good on that conjecture does not involve a normative inquiry into what is to be, be done, or be believed. Rather, it requires us to plumb the deep psychology of status conferral. And that is precisely what we are doing here.

There is a further question that a subject might put to either herself or another. That is the question, "What status *ought* to be conferred upon a given state or attitude?" That question *is* a normative question. But this normative question is a question only for our subjects. It is not a question for us, in our guise as Martian philosophical anthropologists. Qua Martian philosophical anthropologists, our aim is merely to locate human beings and their capacity to monger norms in nature, to show where humans, their capacity, and the "norms" they monger sit in the natural order. We want to know what in nature our subjects are doing when they ask and answer the normative question. In this connection, we may note, for example, that the normative question often arises in the context of what I below call the dialectic of ratification—especially when one rational being *challenges* a status conferral made by another. Understanding the dynamics that govern the dialectic of ratification is indeed part of our anthropological inquiry into what our subjects are doing. But here too *our* task is descriptive and explanatory rather than normative.

Our inquiry into the dialectic of ratification will no doubt carry us beyond matters of deep psychology into the realm of the social. But that is an entirely to be expected outcome. The evolutionary prehistory of the human species is bound up with the gradual emergence of a distinctive form of social life—a form of social life grounded in our nature as norm-mongering creatures. From the lowly ant to the highest of the apes, coalition-forming creatures must solve certain coordination problems. And nature has endowed its coalition-forming creatures with a wondrous variety of solutions to such problems. To humankind alone, however, she has given the gift of rational self-management and normativity. The distinctively human adaptive trick was to solve coordination problems by evolving the capacity to regulate shared cognition and conation by means of mutually conferred and endorsed normative statuses. But if that is so, it is not surprising that an inquiry into the metaphysical nature of normativity would quickly carry us beyond matters of deep psychology into matters social and dialectical. To explore the deep psychology of the capacity for normativity is to explore one of the primary conditions for the possibility of normative community. But it is not yet to explore the social-dialectical dynamics governing the growth and decay of normative community over evolutionary and historical time. To that task, we will come in due course.

Consider next the notion of culmination. At first blush, such talk may appear to have a quasi-normative feel. One is tempted to say that reflection culminates when it reaches an "appropriate" stopping point. One then wants to know what *standards* of appropriateness amount to. But once again our anthropological talk of culmination is intended in an entirely non-normative manner that involves no reference to any antecedent standard of appropriateness. The culmination of reflection is a matter of reflection coming to a stopping point, at least temporarily. Our subjects, we find, may reflect and reflect, but until reflection culminates, they have not pulled off the seemingly mysterious hat trick of binding themselves to any determinate norm. We will say that reflection culminates for a subject when it produces endorsements that are "stable" in light of all currently relevant rational pressures on reflection. Reflection culminates for subject, that is, when further reflection would yield the same endorsement at least given the same rational pressures.[15]

But the stability in which reflection culminates is typically a merely local, merely temporary stability. The inputs to reflection change in a myriad of ways and for a plethora of reasons. They change in response to social and personal upheaval, in response to new voices, demanding recognition and respect, in response to new discoveries about either our individual lives or about our collective places in the order of things. Reflection is practically inexhaustible. Deliberating, reasoning agents are subject to constant moral testing, to constant opportunities for discovery, for growth, for failure, for success. What stability and fixity their reflection achieves, in light of the constant churning of the moral whirlwind, is likely to be but the fixity and stability of the dialectical moment. Still when reflection does culminate in a stable and fixed endorsement, if even only for a dialectical moment, a subject has *decisively* committed to govern her life by the endorsed norm. For at least this moment, she has given that norm what we may call her full rational backing.

There is no *a priori* guarantee that reflection will in fact culminate in stable endorsements. There may be incoherence or indeterminacy in the initial inputs to reflection. Instability, incoherence, and/or indeterminacy in the set of initial commitments and convictions may lead to instability, incoherence, and/or indeterminacy in the set of reflective endorsements. Where reflection does not culminate in a set of coherent, stable, and determinate endorsements, the agent is not stably bound by relevant norms. An agent in such a normative predicament is still an agent, but she is not yet a fully self-owning, self-governing one. She has reached no equilibrium point in rational self-management and has not decisively conferred a determinate normative status upon that upon which she reflects. Consequently, there are no determinate and stable facts of the matter about which of the would-be reasons, as it were, are really and truly *her* reasons.

Now a course of reflection may sometimes bring stability, coherence, and determinacy, even when the initial inputs to reflection originally enjoyed no such coherence, stability, or determinacy. Reflection that creates stability, coherence, or determinacy de novo I call *Sartrean reflection*. Reflection that merely elucidates coherent, stable, and determinate commitments and convictions that were already there, but not yet fully "owned" or acknowledged, I call *elucidative reflection*. Elucidative and Sartrean reflection each plays a significant role in our self-constitution and self-governance. Through elucidative reflection, we invest our full normative authority in commitments that are, in sense, already there, but not yet fully avknowledged and owned as our own. Through elucidative reflection, we decisively commit to owning up to our initial commitments as considered commitments. In so doing, we decisively undertake the rational self-management of our lives in accordance with those now elucidated commitments. Through Sartrean reflection, by contrast, we normatively configure ourselves, as it were, de novo. Through Satrean reflection, one may decisively break with previous commitments after, for example, continually finding oneself incapable of living up to the commitments of old. In an act of Sartrean reconfiguration, one forswears the old commitment and gives oneself a new commitment in accordance with which one henceforth decisively undertakes to manage one's life. Alternatively, Sartrean reflection may lead one to decisively choose one side of a previously intractable conflict of commitments.

Accounts of bindingness that emphasize the role of reflective endorsement are sometimes accused of being overly intellectualized. It is not just those highly reflective few who are in the habit of explicitly reflecting that are, and are regarded as, subject to norms. Indeed, we often take ourselves to be entitled to hold even the most unreflective agents to norms of various sorts. But how could this be, the worry goes, if reflective endorsement were a sine qua non of norm bindingness?

We need to step back a bit. We have been talking a great deal about 'binding' and the conferral status. But there are at least two different ways to be subject to a norm. On the one hand, an agent is "subject to a norm" if she is, in the sense we have so far been considering, bound by that norm through the psychic mechanism of culminated competent reflection. But an agent may also be "subject to a norm" if another is entitled to hold her to the relevant norm, even if she is not herself bound by that norm in my privileged sense. Something like this distinction is tacitly recognized in our commonsense practices. But it is often assumed, without much real argument, that one is subject to a norm in the "entitlement to be held to it" sense if and only if one is subject to that norm in the "being bound by it" sense. But this, I shall argue below, is a mistake, a mistake that lies at the core of many arguments against relativism.

Before delving further into that issue, however, let me more directly answer the potential concern about the overly intellectual character of my account. To do that, we need another distinction. We distinguish between being merely tacitly bound by a norm and being explicitly bound by a norm. An agent is merely tacitly bound by a norm when, although she has not in fact engaged in an episode of culminated competent reflection, nonetheless, the current actual facts of her psychology and situation suffice to make true counterfactuals about what she would endorse upon culminated competent (elucidative) reflection. When an agent is merely tacitly bound by a norm, she has not thereby explicitly committed herself to that norm. She has not yet explicitly and concretely undertaken to manage her life in accordance with that norm. So she has not self-conferred a normative status. That means that she is not yet subject to the relevant norm in the "bound to it" sense. But she may, nonetheless, be subject to the norm in the "entitlement to be held to it" sense. That is because when agent A recognizes that agent B would endorse a certain norm upon culminated competent elucidative reflection, even if B has never, in fact, explicitly reflected on the relevant norm, A may *self-generate* an entitlement to hold B to the relevant norm. A thereby confers a normative status on B. Though A's self-generated entitlement to hold B to the conferred status is not yet *ratified* by B, nonetheless, A's self-generated entitlement to hold B to N is, in a way, responsive to normatively relevant facts about B herself. As such, the status conferred on B by A purports to be more than a mere imposition from without. It is a conferral that is at least responsive to facts about B's normative lights.

For example, parents often hold their immature children to norms which they believe, rightly or wrongly, the still developing child would endorse upon mature reflection. Parents do so even though their children are not yet reflective enough to endorse the relevant norms. Parents presume that their children are at least tacitly bound to the relevant norms. We may say that parents sometimes hold their still developing child to norms on "behalf" of the child's more reflective future self. But this is just one example of the way that tacit binding may play a significant social-dialectical role. With more reflective agents, the occasion of being held by another to a norm that one has not yet explicitly endorsed may occasion reflection that does culminate in one's either decisively taking up or decisively rejecting the relevant norm. If one decisively takes up the norm, one becomes explicitly bound by the norms. If one decisively rejects the norm then any further attempt by another to still hold one to that norm will be an imposition of normative status from without and no longer responsive to facts about one's own normative lights.

It may be doubted that one person may know what counterfactuals are true of another's psyche in the absence of a well-worked out psychological theory of the other person's mind. But if we cannot know what norms another would endorse, then we cannot legitimately hold them to such norms—at least not on behalf of their own more reflective selves. It must be stressed that I am not supposing that the ability to understand another's normative horizon, as it were, requires systematic theoretical knowledge of the other's deep psychology. We have the practical ability to project ourselves into one another's psyches. No doubt the reach of any such ability will be limited. It will be most limited with respect to those whose normative lights are radically unlike our own and/or radically unfamiliar to us. It will be least limited with respect to those who are very like us and/or very familiar to us. In the everyday give and take of reasons, we employ this practical ability all the time, with no guidance by a deep psychological theory. It would be an interesting task to outline the exact psychodynamics by which we recognize and negotiate the varied normative lights with which everyday experience confronts us. I shall not undertake that task here. Here I make only the negative point that we do not need deep theoretical knowledge to project ourselves into the normative lights of others. Still, it is only the current actual facts of our deep psychology that serve as truth-makers for the counterfactuals that articulate the norms to which we are merely tacitly bound. We must separate questions about what makes it true that a given cognizer is (merely) tacitly bound to a given norm from questions about how one cognizer might go about recognizing that another is tacitly bound to a given norm.

5 The One Authority that Binds; The Many Authorities that Entitle

I have stressed throughout that I alone have the power to bind myself to a norm through culminated competent reflective endorsement. Through such endorsements, one self-confers normative status and decisively undertakes to govern one's cognition and conation in accordance with such conferred status. Others may confer normative status upon me without my cooperation. The conferring other may even attempt to coerce or otherwise socially condition me into living in ways called for by such uncooperatively conferred status. Coercion or conditioning may even succeed at causing me to live in the relevant way. But even when they are successful, neither coercion nor conditioning suffices to endow a norm with *my* "authority" and thus to bind me to that norm. My authority, and my authority alone, is the one authority that may bind me. That is, only through *my own* reflective endorsement can a norm be endowed with binding force over me. Only such endorsements constitute my decisively undertaking, with all my rational powers, to live in accordance with the relevant norm. Only through such decisive rational undertaking am I "bound" to a norm.[16]

When we have told only the story of what in nature norm bindingness consists in, we have told only the barest beginnings of the truth about normativity. We have plumbed the deep psychology of our capacity for normativity, but we have not yet touched on the complex social-dialectical dynamics of normativity. Our foray into that domain begins with the observation that despite the fact that nothing but my own authority can suffice to *bind* me to a norm, another may, as I have already said, be *entitled* to *hold* me to a norm, even to a norm by which I am not bound. We must distinguish, that is, the one authority that binds from the many authorities that *entitle*.

What exactly is an entitlement? I will not stop to give a full dress account of the nature of entitlement here. But an analogy may help. Issuing an entitlement is analogous to giving out of a ticket to the theater or to a ballgame. If I issue you a ticket to my theater, I thereby confer normative status upon both you and myself. In virtue of my act, you are "licensed" to enter my theater and I am "committed" to letting you enter the theater upon presentation of the issued ticket. More generally, issuing entitlements and undertaking concomitant commitments is caught up with the conferral of normative status. We have already suggested that the power to confer normative status generally is rooted in nothing but evolved psychological powers of the human mind-brain. And just because entitlements and commitments are, at bottom, just further but distinctive species of conferred normative statuses, adding them to our story should add few, if any additional perplexities.

Now entitlements to hold an agent to a norm can be conferred in at least two different ways. They can be *self-generated* or *granted by the subject*. x confers upon x a *self-generated entitlement* to hold y to N when x entitles herself to hold y to N. y confers on x a *subject-granted entitlement* to hold y to N when y entitles x to hold y to N. When one self-generates an entitlement to hold another to N, one, in effect, endorses N as a norm for the other. Now *my* conferral of a normative status upon you does not suffice to *bind you* to N. Nor need *my* conferral give *you* any original, non-derivative reason to live up to N. Nonetheless, that does not prevent my conferral of status upon you from giving *me* a reason to hold, or try to hold you to N.

To be sure, when I endorse N as a norm governing you and thereby confer a status upon you merely on my own normative authority, any self-generated entitlement to hold you to N need not be *recognized or acknowledged* by you as a legitimate authority over you. Indeed, by way of making explicit your rejection of my self-generated entitlement, you may self-generate an entitlement of your own—an entitlement to resist my attempts to hold you to N. You thereby refuse to take up my normative authority as an authority for you and in so doing refuse to endorse the normative status I have conferred upon you. Now in self-generating an entitlement to hold you to N, I need not, in turn, acknowledge the authority by which you self-generate an entitlement to resist as a legitimate authority with respect to me. When I fail to acknowledge the normative authority in you as an authority for me and you fail to acknowledge the normative authority in me as an authority for you, we stand in the situation of what I call rational enmity. In the situation of rational enmity, there exists a discord of reasons and a contest over status. What counts as a reason for me remains unratified by you as a reason for you. What counts as a reason for you remains unratified by me as a reason for me.

Some will no doubt be tempted to conclude that self-generated entitlements to hold another to a norm by which she is not bound, which she may even abhor, are rooted in normative hubris and an overreach of normative authority. But that conclusion would be hasty. To help see why, we must distinguish two distinct categories of norms—*traveling norms* and *merely local norms*. Roughly, *N* is endorsed as a traveling norm by *x* if and only if for any agent or (normative) community of agents *y*, *x*'s application of *N* to *y* is licensed by *N* to be unconstrained by *y*'s reflective attitude(s), pro or con, toward *N*. On the other hand, *N* is endorsed as a merely local norm by *x* if and only if for any agent or normative community of agents *y*, *x*'s application of *N* to *y* is not licensed by *N* to be unconstrained by *y*'s reflective attitude(s), pro or con, toward *N*. Contrast norms of etiquette with ethical norms. Norms of etiquette are paradigmatically local. Among one normative community, burping after a meal may be a polite expression of satisfaction. In a different normative community, burping after a meal may be regarded as rude and obnoxious. If the members of the burping community endorse burping as a merely local norm, then they will not self-generate entitlements to hold the non-burping community to their own standards of politeness, at least when the non-burpers remain within their own community. To be sure, members of the burping community may themselves travel to non-burping locales and vice versa. When they do travel, they may be held and may even permit themselves to be held to local norms different from their own. When a norm is merely local, there is no guarantee that when *you* travel, it travels with you.

Norms of etiquette stand in apparently sharp contrast to ethical norms. Ethical norms are often thought to enjoy a certain felt universality. There is, I think, something to this idea, but less than many have thought. The felt universality of morality results from nothing but the fact that distinctively ethical norms are endorsed as traveling norms. To endorse a norm as a norm for the entire rational order is to entitle or license oneself to hold every other rational agent "answerable" to that norm, independently of his own reflective attitude toward it. Indeed, even if one recognizes that another would upon culminated competent reflection abhor the relevant norm, still if the norm is endorsed as a traveling norm and is thereby proffered up as at least a candidate for governing the entire rational order, one thereby self-generates an entitlement to hold others to the relevant norm, to evaluate and perhaps even sanction them in light of the norm. Consider, for example, an abolitionist community that endorses the ending of slavery as a traveling norm. Even if a slave-holding community endorses a norm that permits slave-holding as either a traveling or merely local norm, the abolitionist community may self-generate an entitlement to subject the slave-holders to their abolitionist norms. They will license themselves to condemn, to seek to persuade, and perhaps even to coerce the slave-holding community into freeing their slaves. And they need not regard the slave-holding community's abhorrence of their abolitionist norms as legitimately blocking their self-generated entitlement to do so. At the same time, the slave-holding community may well refuse to recognize the normative authority by which the abolitionists entitle themselves to condemn, persuade, or coerce as a legitimate or governing authority for them. That is, they may refuse to *ratify* the abolitionist's self-generated entitlements with subject-granted entitlements. Indeed, the slave-holders may self-generate entitlements to resist and reject all condemnation, argument, and coercion from the abolitionist community.

When two normative communities endorse two incompatible norms, with at least one of the norms being endorsed as a traveling norm, there arises the possibility of intractable moral conflict between them. Moral conflict arises, that is, when we take what is merely our own normative authority as a normative authority for another, often through the endorsement of a norm as a traveling norm. Such norms will very often meet with normative resistance as we try to make them travel.

I do not mean to say that moral conflict is inevitable or that all moral disputes are rationally irresolvable. Moral conflict is often a mere way station on the path toward more encompassing normative community. As a consequence of our evolved nature as norm-mongering creatures, human beings hunger for normative community with others—though not necessarily with *all* others. Often when we do self-generate entitlements to hold others to our traveling norms, we offer those norms up to the other as candidates for their reflective endorsement as well. We ask others to ratify our self-generated entitlement by granting us subject-granted entitlements. When agents do ratify one another's self-generated entitlements with subject-granted entitlements they thereby achieve *mutual ratification* of a system of traveling norms. They thereby make the system of traveling norms mutually and reciprocally binding on one another. They no longer enjoy merely self-generated entitlements. They have granted one another *mutual and reciprocal entitlements* to hold one another to the norms by which they are now mutually and reciprocally bound. They have acknowledged each other as full and equal partners in normative community. To acknowledge one another in this way is for each to say to the other that the normative authority of one is also a normative authority for the other.

None of this is automatic. It grows haltingly and dialectically from an initial tension generated by agents' competing self-generated entitlements. These self-generated entitlements reflect first and foremost our self-recognition and self-valuing. Each fully reflective intact rational being recognizes herself to be an original, non-derivative source of reasons for herself. I take such self-recognition and self-valuing to be an architectural consequence of the deep psychological structure of a rational will. But almost without hesitation, we sometimes take what are merely reasons of our own as reasons for other rational beings. Our tendency to extend our own reasons beyond our own domain is typically brought short by resistance from the rational other. But mere resistance from the rational other does not automatically put the brakes on our attempt to extend our normative authority. To be sure, we may recognize that the rational other values and esteems herself in just the ways that we value and esteem our own dear self. To recognize another as a fellow reflective rational being and a fellow status-conferring creature, is to recognize that other as an original and non-derivative source of reasons for herself. In this mere recognition of the rational other, we have already elevated the other above the whole of non-rational nature. Non-rational beings who lack the power of reflection are nothing at all either to themselves or for themselves. They are at best derivative sources of reasons for any rational being. Non-rational beings can indeed be sources of reasons for us, but only in virtue of the rationally optional interests that we happen to take in them. We may esteem non-rational beings as instruments, as objects of wonder and awe, even as objects of a peculiar kind of sympathy or love. But they are not the kinds of beings for which even the possibility of normative community arises. For though we can and do confer status upon such creatures, they cannot confer status back upon us and cannot take up our offered up statuses as their own.

The mere recognition of another as a fellow rational being—as a fellow norm-mongering, status-conferring creature, and as a being capable of the deepest self-valuing and highest self-estimation—is not yet the achievement of normative community. In the bare recognition of the rational other, one has not thereby reflectively owned the other as a non-derivative rational source for oneself. Nor has one thereby limited the presumed reach of one's own normative authority. Recognition does, however, set the question, "What, if anything, shall we do, be or believe together as fellow rational beings?" This happens when we confront each other with concrete demands for respect and recognition of the normative authority that lies within. I claim here and now a right to what I take to be mine. I demand recognition and respect of my claim from you. Correlatively, you claim rights to what you take to be yours. Our claims may conflict. We are confronted with a question. How, if at all, shall we be reconciled? How, if at all, shall we live together? The struggle to arrive at mutually acceptable answers to such questions, a struggle in which we sometimes succeed and sometimes fail, is what I mean by the dialectic of ratification.

Through the dialectic of ratification, I try to get you to ratify me, my norms, and the statuses I seek to confer. I try thereby to make it the case that me, my norms, and my status-conferring powers govern your life. Simultaneously, you try to get me to ratify you and your norms. You try thereby to make it the case that you, your norms, and your status-conferring powers govern my life. When we each are governed by the other, we constitute a normative community. We have made ourselves into original normative authorities and non-derivative sources of reasons for each other.

Normative communities are among humanity's highest achievements. Through the constitution of normative communities, we extend the reach of our own rational powers. For example, through the mediation of mutually ratified norms of inquiry and communication which direct the truth to be sought and told, my having reasons for believing a certain proposition may give you a non-derivative reason for believing that proposition as well. Through the mediation of mutually ratified norms of conduct calling for mutual aid and cooperation, my having a reason for pursuing some good may give you a non-derivative reason either to refrain from interfering with my attempts to pursue that good or perhaps even a reason for aiding me in my attempts to achieve that good. Mutually ratified norms are thus the rails along which reasons may be transmitted from cognizing agent to cognizing agent. Within a normative community, the rational powers of one become rational resources for all. Normative community thus makes possible the emergence of complex cooperative rational activity, including shared forms of inquiry, deliberation, and argument.

Contrary to the dreams of, say, Kant, however, an all-encompassing community of reasons is not an *a priori*, rationally mandatory imperative categorically binding on all rational beings as such. Rather, they are historically contingent, culturally specific rationally optional achievements. Now, there are myriad ways in which we might fail to achieve thoroughgoing community, despite the full rationality of all who are a party to the failure. The norms by which I would see the world governed, that I most urgently offer up for mutual acceptance to the entire rational order, may simply be rejected. That would make them an insufficient basis for normative community. But it need not make them any less dear to me, nor in any way weaken my rational backing of them. Not out of mere hubris or self-love, but out of deep concern for the entire rational order, one may self-generate an entitlement to shape the unyielding world by one's own normative lights. One may prefer to shape the world by the force of argument, if argument will suffice. But by what imperative must we abandon our deepest convictions about the governance of the world, if argument should fail? Yet, were one to succeed through mere coercion in imposing norms upon a reluctant world, one would not have achieved true normative community, but the mere domination of one over another. With fellow rational beings who succeed through coercion in holding me to norms of their own endorsing, despite my abhorrence of those norms, there can only be rational enmity and a discord of reasons. Even if I appear to endorse their domination over me through incompetent or non-culminating reflection, that amounts to a mere semblance of normative community, not its reality.

6 Conclusion: Relativism Revisited

The account of the capacity for normativity on offer in this essay provides us the resources to defend a version of relativism that is subject to none of the standard arguments against it. The intolerant relativist maintains that agents are sometimes entitled, via merely self-generated entitlements, to hold others to norms by which they are not bound. Precisely because she distinguishes the authority that entitles from the authority that binds, the intolerant relativist may coherently deny that all moralities must be regarded as "equally valid." She can allow that those who owe allegiance to one moral system may entitle themselves to condemn or criticize those who are bound by the norms of a different moral system. Liberal, secular moderns may condemn, on their own normative authority, what they regard as benighted and archaic fundamentalisms. The intolerant relativist does deny, however, that such condemnation enjoys the ultimate backing of some external, transcendent or "impartial" normative authority. There is, she insists, no such normative authority either in heaven or on earth. There is only the normative authority that lies within each of us.

Intolerant relativism does not imply that the normative authority that lies within each rational being glides upon a frictionless plane, never meeting with resistance from the rational other. For when we do hold the rational other to norms grounded in nothing but our own normative authority, the other will often self-generate entitlements to resist. Equally often, we may, in turn, refuse to ratify the resisting other's self-generated entitlement to resist. In such a situation, we find ourselves in rational enmity, rather than rational solidarity, with the rational other. In situations of rational enmity, the question naturally arises: by what authority, with what right do we refuse to ratify the resisting rational other's self-generated entitlement?

Faced with the very possibility of rational enmity among rational cognizers, the transcendental absolutist and the tolerant relativist turn into strange bedfellows. For both will agree that if we refuse ratification of the resisting rational other with an authority that is merely our own, nothing but hubris could ground that refusal. In fear of such hubris, the tolerant relativist retreats into the morass of equal validity, refusing to criticize, sanction, or judge by normative lights that are "merely" her own. In denial of such hubris, the absolutist lays claim to an authority that is more than her own. She claims, that is, to judge not merely on the basis of her own authority, but on behalf of a transcendental or external authority to which all rational beings, qua rational, are automatically subject—the voice of god, the unwavering voice of human reason, the mandate of history or of universal human sentiment.

Both the tolerant relativist and absolutist misunderstand the true reach, source, and nature of our normative authority. To explain where they go wrong, we need to distinguish between that which, by its very nature, goes on parade, in search of external vindication and that which stands in no need of such vindication. Some of our attitudes clearly do parade in search of external vindication. But not all of them do. On the one hand, there are our beliefs. They are semantically answerable to how things stand by a largely mind-independent world. As such, they do parade in search of vindication by the very world to which they are semantically answerable. The belief that snow is white cannot stand as rightly held in splendid indifference to whether snow is, in fact, white. The propriety of believing that snow is white is hostage to how things are by snow. If snow is not white, then, at least to the extent that the world throws up evidence to that effect, the propriety of believing that snow is white is undermined. Our merely *taking* snow to be white does not confer propriety on the belief that snow is white. If our conferrals of normative status were like our beliefs, then it would be part of their very nature too to parade in search of external vindication. The propriety of our conferrals of status would then be hostage to whether that on which we conferred the relevant status "deserved" that status, in and of itself and independently of our conferrals.

But conferring normative status is not like believing in this regard. Though we cannot make snow to be white and cannot make the belief that snow is white to be rightly held merely by taking snow to be white, nothing but our taking a thing to have a normative status makes it to have that status. That is to say, the conferral of normative status is always and only rooted in nothing but our own status-constituting attitudes. And those status-conferring attitudes need not be answerable to anything antecedently present within that upon which a status is conferred. The conferral of status is simply not hostage to how things are in some external, objective, or transcendental normative order.[17]

But I hasten to stress again that to reject the very idea that our status-conferring attitudes are answerable to an objective normative order is decidedly not to say that our status conferrals never meet with resistance. We have already noted that they may meet with resistance from the rational other, that is, from other status-conferring creatures who may either ratify or fail to ratify, on their own normative authority, the statuses we have thrown upon them. The question naturally arises whether resistance from the rational other rationally "trumps" our self-generated entitlements to hold the other to norms of our own endorsing. The tolerant relativist says that resistance from the rational other does trump any merely self-generated entitlement. By contrast, the intolerant relativist holds that we may coherently stand our normative ground and not retreat in the face of resistance from the rational other. She admits that we cannot *bind* the entire world to norms merely of our own endorsing. But she insists that we can nonetheless *entitle* ourselves, on our own rational authority, to hold the world to the norms by which we would most urgently see it governed. Our self-generated entitlements do not, she insists, lose their standing as entitlements just because of the fact of resistance from the rational other.

Now the transcendentalist transcendental absolutist also maintains that we can coherently stand our normative ground in the face of resistance from the rational other. But she insists that doing so requires a mandate that is not merely our own—the mandate of impersonal reason. Against the transcendental absolutist, the intolerant relativist insists that although we may indeed speak with only our own locally generated mandate for the governance of the world, that mandate is not ipso facto defeated by the mere fact of resistance from the rational other. Indeed, the intolerant relativist suspects that the belief in an external normative authority rests on bad faith. In the contest over normative status, there is no external judge who may decisively settle our disputes. We have only our own authority and the counter-authority of the ever-resistant rational other. To silence the resisting other, we posit a final judge, an external authority that trumps all rational resistance. And we declare that the final judge has decided for our side. This declaration serves as club with which we can beat down the resisting other. And if the resisting other will not be silent, we declare that reason has gone silent in the resisting other. But this is all an illusion. The illusion is generated by our unwillingness to recognize and acknowledge that it is only our own authority on which we ultimately stand and by the concomitant desire to silence resistance that will not be silenced merely by our own voices. In a contest with no neutral judge, we invent a judge and declare that he has decided the case in our favor.

By contrast, the intolerant relativist honestly acknowledges that there is no neutral judge that will impartially decide all disputes. She recognizes that reason sometimes speaks in a cacophony of competing voices and that these competing voices may never be reconciled. But she refuses to surrender her own voice in the face of that cacophony. That refusal should not be read as a denial that human beings do hunger for normative community. Indeed, the intolerant relativist sees that hunger as one among the great driving engines of human history. And she happily acknowledges that the hunger for normative community sometimes may function as a great blocking constraint on our self-generated entitlements—especially when the distribution of coercive powers does not enable one to dominate the other. She acknowledges, that is, that unless we do adjust our proffered up traveling norms in the face of resistance from the rational other, we can never in fact achieve normative community with them. Since the dawn of humanity, no doubt, we human beings collectively have been engaged in fraught struggles to constitute ourselves in moral community one with another. That struggle has involved and will no doubt continue to involve an intense contest over conferred normative statuses. The contest over status has sometimes been settled by the force of better argument, but it has also often been settled by the force of the better arms. In face of the multiplicity of ways that reason has configured itself locally throughout history, the intolerant relativist insists only that there can be no *a priori* anticipations of the outcome of the contest over status. Indeed, she allows that if there is such a thing as a final configuration of reason, it may take radically different forms. The march of reason through history may culminate in either thoroughgoing moral fragmentation and rational enmity or thoroughgoing normative community and rational solidarity. It is no doubt difficult to discern the truth of this conjecture from within any given local configuration of reason. From within any such configuration, we typically entitle ourselves to narrate the history of the world by our own normative lights. And that may lead us to deny that the many alternative configurations of reason are equally configurations of reason. But we have adopted the guise of a Martian philosophical anthropological perspective precisely to allow us to gain a purchase on the nature of normativity as if from outside the unending contest over status. From that perspective, the fact that both thoroughgoing fragmentation and enmity, on the one hand, and thoroughgoing community and solidarity, on the other, are both really possible as "final" configurations of reason, is among the deepest truths about the true nature, reach, and limits of the human capacity for normativity.

Notes

1 For some by now classical philosophical defenses of relativism see Wong 1984; Harman 1996 and 2000. A fair number of other philosophers defend views with strong relativistic tendencies, even if they don't flat-out embrace relativism. Two prime examples are Blackburn 1993, 1998 and Gibbard 1990, 2003. Blackburn labors quite explicitly and mightily to keep various forms of relativism at bay, but he does not anticipate my distinction between tolerant and intolerant relativism.

2 Many literary theorists appear to draw relativist inspiration from some of the works of Richard Rorty. See, for example, his "Justice as a larger Loyalty" and "Kant vs. Dewey: the current situation of moral philosophy," both in Rorty 2007 and also his essays on solidarity in Rorty 1991.

3 For a recent sustained attack on relativism, especially epistemic relativism, see Boghossian 2006.

4 I take Moore's (1903) justly famous open-question argument to be a pretty decisive demonstration that there is conceptual distance between our ordinary concepts of the normative and our ordinary concepts of the natural. But I take the open-question argument to be of no further philosophical importance. In particular, it shows nothing at all about the metaphysics of normativity and in no way constrains the future co-evolution of our concepts of the normative and the natural.

5 See Taylor 2003, especially essays 13 and 14. These concepts receive their fullest development in Taylor (in progress).

6 The exemplar is, of course, Millikan 1984.

7 As such, norms are akin to what Castañeda (1975) calls practitions. Ultimately, however, in the longer work from which this essay is drawn, I reject talk of norms, understood as abstract existents, in favor of what I call normative statuses. A normative status is defined by a set of (normative or non-normative) entry conditions and a set of downstream consequences, with the pairing of entry conditions being determined largely by the social coordinating role of the relevant normative status. Dispensing with norms in favor of normative statuses enables us to more clearly bridge the gap between fact and norm. My account of normative statuses is deeply indebted to Brandom's (1994) inferentialist approach to norms and normativity—though I do not consider myself any sort of inferentialist.

8 My approach to normative statuses owes a great deal to Brandom 1994.

9 No doubt, we sometimes seek, as it were, external validation of our merely internal warrants. For example, we may seek to (objectively) verify a theory by appeal to evidence from the world below. Or we may seek external ratification of norms proffered up to rational others with whom we do not yet stand in full normative solidarity.

10 Taylor (in progress).

11 Culminated competent reflective endorsement is thus my candidate for solving what Bratman (2004) calls the problem of agential authority. What inner states, Bratman asks, are such that "when they guide, *you* govern?" To the extent that I understand the question, I answer that the "agent" governs when she is guided by states that are the outcome of culminated competent reflective endorsement. This statement is not a conceptual-analytic claim about what the ordinary concept of an agent comes to. Nor is it itself a normative claim about what aspects of an agent "deserve" authority over others. Our claim is, rather, nothing more or less than a descriptive and explanatory claim in deep speculative psychology. Pretend, again, that it is a statement made from the perspective of Martian anthropologist who is engaged in no normative inquiry about what is to be, be done, or be believed. Rather my Martian anthropologist purports to describe the deep psychological structure of human agency of human rational self-management and thereby to answer the question "What in nature is (human) rational agency?"

12 Though my account of reflective endorsement bears a certain superficial family resembles to Korsgaard's (1996) views about reflective endorsement, my views, once again, are offered in defense of descriptive and explanatory project—the project of naturalizing normativity—with which Korsgaard would have no truck. Correlatively, Korsgaard's views about the significance of reflective endorsement are offered in service of a normative project—the project of answering what she calls the normative question. I am not addressing the normative question here.

13 See Korsgaard 1996 for the *locus classicus* of this sort of argument.

14 By now it should not be necessary to caution that by calling a belief or desire "worthy," we theorists intend to confer no normative status on the relevant belief or desire. Only the subject confers such a status through her culminated, competent reflective endorsement. We merely characterize her as conferring a status. We may characterize her as conferring a certain status on her own beliefs and/or desires without ourselves endorsing the relevant status. Again, *our* aims are descriptive and explanatory, not normative. To be sure, there is a question that our subject might put to herself "What status OUGHT

I to confer upon this state?" That is a question which often arises in what I below call the dialectic of ratification—particularly when one rational being *challenges* a status conferral made by another. Our task is not to engage in the dialectic of ratification with those whose conferrals compete. It is rather to characterize what in nature that competition consists in and the natural principles that ultimately govern that competition.

15 Stability under reflection plays a role in my account analogous to the role played by stable plans and intentions in Bratman 2000. Bratman thinks stable plans and intentions play a decisive role in answering the question of what he calls "agentive" authority. Relatedly, Blackburn (1998) evidently thinks that knowledge is roughly a matter of beliefs that are stable under the pressure of further evidence and inquiry. There is something right about this thought. Indeed, I defend a similar claim in Taylor (in progress). Unlike Blackburn, I see no tension whatsoever between a thoroughgoing realism and making stability under inquiry be the hallmark of that which we are pleased to honor with the title "knowledge."

16 By this stage of our argument, it may perhaps go without saying that the claim that my own "authority" is the one authority that binds is intended as a descriptive/explanatory claim and not as a normative claim. But because misunderstanding of this point seems so ready to hand, I pause to clarify one more time. The use of the word 'authority,' after all is, perhaps, bound to invite misinterpretation. But here is a way to block the temptation to construe claims about the "authority" that binds as normative claims. First, think of phrases like "my authority" as a shorthand way of talking about the bare psychological power to confer status. The claim that bindingness is rooted in nothing but the exercise of this bare psychological power is meant, in the first instance, only to demarcate the special status-conferring power I have with respect to myself and no one else. Think of normative authority not as something absolute and fixed, but as something relative and contestable. In the first instance, my "authority" is only an authority over and for myself. That just means that when I confer a normative status, I "speak," in the first instance, only for myself. Similarly, your authority is, in the first instance, only an authority over and for you. But we may each also try, in various ways, to extend our authority by having our authority taken up by the other as an authority for the other as well. If you think of "authority" in this way as limited, relative, and potentially extendable through what I below call the dialectic of ratification, you may not succumb to the temptation to look for some ultimate, intrinsically normative ground upon which our inner authority rests. Normative status rests on no external or transcendental ground. Things have normative status solely because we take them to have normative status. By taking things to have a status we thereby make them have a status. This approach clearly purports to deflate binding. Binding turns out to be rooted in nothing but concrete psychological undertakings, grounded in nothing but our own inner attitudes. Those who believe that there are minded independent, irreducible normative "facts" of the matter will no doubt be unsatisfied. And I do not pretend to have offered a knock down argument *against* such views. My aim in this essay is not refute the lesser theory, but to put forth my own better theory and to test it by its consequences.

17 In a somewhat different context, Nietzsche sees the mistake clearly: "Life *shall* be loved, because—! Man *shall* advance himself and his neighbor, *because*—! What names these Shalls and Becauses receive and may yet receive in the future! In order that what happens necessarily and always, spontaneously without any purpose, may henceforth appear to be done for some purpose

and strike man as rational and an ultimate commandment, the ethical teacher comes on stage, as the teacher of the purpose of existence; and to this end he invents a second, different existence and unhinges by means of his new mechanics the old ordinary existence" (*Gay Science*, Book 1, §1.)

There is a lot that needs untangling in this pregnant passage. But with Nietzsche's judgment that what he calls a "second existence" is an illusory invention, I fully concur. We should reject the very idea there is or must be an ultimate normative ground to human existence, an ultimate ground lying outside our merely human valuings, but endowed, nonetheless, with the power to command, approve, or disapprove those valuings. Our valuings remain always and only valuings of our own constituting. They rest on no normative authority save our own.

References

Blackburn, Simon. 1993. *Essays in Quasi-Realism*. Oxford: Oxford University Press.

Blackburn, Simon. 1998. *Ruling Passions: A Theory of Practical Reasoning*. Oxford: Oxford University Press.

Boghossian, Paul. 2006. *Fear of Knowledge: Against Relativism and Constructivism*. Oxford: Oxford University Press.

Brandom, Robert. 1994. *Making It Explicit*. Cambridge, MA: Harvard University Press.

Bratman, Michael. 2000. "Valuing and the Will." *Philosophical Perspectives 14: Action and Freedom*. Reprinted in Bratman, 2007.

Bratman, Michael. 2004. "Three Theories of Self-Governance." In *Philosophical Topics* edited by John Fischer, 32: 1 and 2: 21–46. Reprinted in Bratman, 2007.

Bratman, Michael. 2007. *Structures of Agency: Essays*. Oxford: Oxford University Press.

Castañeda, Hector-Neri. 1975. *Thinking and Doing: The Philosophical Foundations of Institutions*. Dordrecht and Boston: D. Reidel Pub. Co.

Gibbard, Allan. 1990. *Wise Choices, Apt Feelings*. Cambridge, MA: Harvard University Press.

Gibbard, Allan. 2003. *Thinking How to Live*. Cambridge, MA: Harvard University Press.

Harman, Gilbert. 1996. "Moral Relativism." In *Moral Relativism and Moral Objectivity*, edited by Gilbert Harman and Judith J. Thomson, 3–64. Cambridge, MA: Blackwell Publishers.

Harman, Gilbert. 2000. "Moral Relativism Defended." In *Explaining Value: And Other Essays in Moral Philosophy*, edited by Gilbert Harman, 3–19. Oxford: Clarendon Press.

Korsgaard, Christine. 1996. *Sources of Normativity*. Cambridge: Cambridge University Press.

Millikan, Ruth. 1984. *Language, Thought, and other Biological Categories*. Cambridge, MA: MIT Press.

Moore, George E. 1903. *Principia Ethica*. Cambridge: Cambridge University Press.

Ratzinger, Cardinal Joseph and Pera, Marcello. 2006. *Without Roots: The West, Relativism, Christianity, Islam*. Translated by Michael F. Moore. New York: Basic Books.

Rorty, Richard. 1991. *Objectivity, Relativism, and Truth: Philosophical Papers, Volume I.* Cambridge and New York: Cambridge University Press.
Rorty, Richard. 2007. *Philosophy as Cultural Politics: Philosophical Papers.* Cambridge: Cambridge University Press.
Taylor, Kenneth A. 2003. *Reference and the Rational Mind.* Stanford: CSLI Publications.
Taylor, Kenneth A. In progress. *Referring to the World: An Opinioned Introduction to the Theory of Reference.* Oxford: Oxford University Press.
Wong, David B. 1984. *Moral Relativity.* Berkeley: University of California Press.

Part III

Some Applications

11 Science and the Value of Objectivity

David Macarthur

In the history of modern philosophy, the idea that natural science is value-free has played an important role in providing a "scientific" basis for metaphysical realism. The sorts of facts natural science discovers are thought to provide a conception of *mind-independent* reality. According to this metaphysical vision, natural scientific facts play the role of truth-makers for a correspondence theory of truth, as well as supplying privileged objects and properties for legitimate referring expressions within science. Value-neutral facts are the touch-stone of objectivity, features of the world that are as they are quite independently of the mental life of human subjects. The detective's "Just the facts ma'am" and the debater's derogatory exclamation "But that's just a value judgment!" are evidence that the same fact/value dichotomy is also ingrained in the wider culture.

1 The Idea of Objective Epistemic Values

In a recent article on scientific objectivity the author begins,

> Scientific objectivity . . . expresses the idea that the claims, methods and results of science are not, or should not be influenced by particular perspectives, value commitments, community bias or personal interests, to name a few relevant factors.[1]
>
> (Reiss and Sprenge 2014)

What does it mean to say that (natural) science is value-free?[2] Obviously, this is not true of the *activity* of doing science. Values are involved in having the motivation or opportunity (say, the funding or access to the appropriate research equipment) to pursue a certain line of inquiry, the selection of data, and methods of inquiry, as well as the use to be made of the facts once they are discovered. But the thought that science is value-free is unaffected by these observations. According to the metaphysical fact/value dichotomy, underlying value-laden scientific activities are value-neutral facts. Science is in the business of theorizing about these facts: say, discovering correlations amongst them; and, on that basis, hypothesizing causal patterns and laws. Jack Smart, a representative of this tradition of thought, puts it this way,

> One person may go in for classical mechanics, another for electromagnetism. Their decisions would involve value judgments, but the law of conservation of momentum and Maxwell's equations themselves contain no value judgments. A similar mistake occurs often and especially in the writings of social scientists and historians. It is sometimes said that history is value laden because a historian has to select his facts from an immense mass of information. That does not imply that the facts so selected are value laden. There is no contradiction in saying that value judgments are relevant to what value neutral facts we want to unearth
>
> (Smart 1999, 433)

The question, then, is whether value-neutral facts are indeed unearthed by science?

Hilary Putnam has argued convincingly for a negative answer to this question on the grounds that the sorts of facts science investigates presuppose *epistemic* (or rational) values. Call this Putnam's Point. When scientists propose and test hypotheses they apply such values as Occam's Razor (i.e., do not multiply entities beyond necessity in explanation), simplicity, reasonableness, coherence, logical consistency, elegance, and so on. We can also consider that when a scientist appeals to inference to the best explanation the notion of "best" here is an epistemic value based on a judgment about what is more explanatory than what.

One response to this argument is to concede Putnam's Point but to employ the notion of an *objective* fact. The idea is that talk of "value-free" facts is to be understood as shorthand for objective facts governed by objective epistemic values. Epistemic values might be objective if there is a consensus within the scientific community about what counts as instantiating these values. Objective facts would then be value-free in so far as they are untainted by subjectivity. It is important to see that this does not rule out objective facts about subjects. For instance, there can be objective facts about a certain person's pain on a certain occasion—e.g., it is caused by a particular lesion or illness—without this truth about this person's subjective state being in any way dependent on anyone's beliefs about it, including the person's own beliefs.[3]

2 Absolute vs. Positional Conceptions of Objectivity

One reason this view can seem initially compelling is the attraction of an absolute conception of objectivity. One widely influential version of this idea, owed to Bernard Williams, is a conception of the world as it is *anyway*—a conception of the world that is "to the maximum degree independent of our perspective and its peculiarities" (Williams 1985, 139). The pre-eminent epistemic worth of the natural sciences is, on this view, a matter of striving for "absolute knowledge of how things are"—even if we are never in a position to know that we have such knowledge when we do (Williams 1978, 302). An ethical fact such as "Slavery is evil" does not count as absolute knowledge in the relevant sense. While both ethical and scientific thought present themselves as routes of discovery of a realm of truths, only science is a mode of sensitivity to features of the (real) world; that is, only "science has some chance of being more or less what it seems, a systematized theoretical account of how the world really is, while ethical thought has no chance of being everything it seems" (Williams 1985, 135). Ethics, unlike science, is *not* reflectively related to the absolute conception in such a way as to sustain the idea that it engages with a distinctive ethical reality.

According to Williams, the absolute conception is a conception of the world "*sub specie aeternitatis*" (Williams 1985, 111). He explains,

> This notion of an absolute conception can serve to make effective a distinction between "the world as it is independent of our experience" and "the world as it seems to us." It does this by understanding "the world as it seems to us" as "the world as it seems peculiarly to us"; *the absolute conception will, correspondingly, be a conception of the world that might be arrived at by any investigators, even if they were very different from us.* What counts as a relevant difference from us, and indeed what for various levels of description will count as "us," will, again, be explained on the basis of the conception itself; we shall be able to explain, for instance, why one kind of observer can make observations that another kind cannot make. It is centrally important that these ideas relate to science, not to all kinds of knowledge. We can *know* things whose content is perspectival: we can know that grass is green, for instance, though *green*, for certain, and probably *grass* are concepts that would not be available to every competent observer of the world and would not figure in the absolute conception . . . *the contrast with value should be expressed not in terms of knowledge but of science.* The aim is to outline the possibility of a convergence characteristic of science, one that could meaningfully be said to be a convergence on how things (anyway) are.
>
> (Williams 1985, 139, emphases mine)

Three things are particularly worthy of note in this passage: (1) the fact/value distinction is not a difference between knowledge and something less than that. There can be evaluative knowledge—say, ethical knowledge— according to Williams, just as there can be scientific knowledge. (2) The fact/value distinction is best understood, therefore, not as a distinction between two kinds of knowledge but rather between the absolute facts revealed by *science* (or, on Williams's reductive vision, those revealed by physics alone),[4] and the perspectival facts with which ethics deals. (3) The absolute facts of science reveal "how things (anyway) are" in a way that is "independent of our experience" using concepts that *any* alien investigator could converge on if they investigated successfully for a sufficient length of time—where the explanation of convergence is that that is how things anyway are.

We can accept that the way ethical and scientific discourse bears on the world is different. Unlike science, ethics need not be understood to require its own distinctive subject-matter. However, why must we accept Williams's fantastical idea according to which scientific inquiry—at least at its most fundamental level—is conceived of as a *transparent* representation of the world, as if how things are anyway simply impresses itself on the scientific inquirer?[5] In thinking in this way Williams is relying on the idea that an abstract inquirer, or a consensus of such, when employing 'absolute' concepts that are not grounded in human sensibility (unlike, say, color vision) or culture (unlike, say, an idea of what is rude) or practice (unlike, say, the human practices of science themselves), could reach a consensus about certain beliefs which is best explained in terms of things being as those beliefs represent them as being. Williams pictures this as the world simply impressing itself on the mind of the inquirer.

Williams is overlooking that the idea of a scientific consensus involves not only the beliefs achieved at the end of a local inquiry but consensus about the *way* they are arrived at including agreement about the application of epistemic values (e.g., simplicity, Occam's Razor, reasonableness, coherence) to the various hypotheses that the inquirers take to be candidates for explaining the relevant phenomena. This involves:

(1) *Objectivity of data*: inquiry in the natural sciences is characterized by a prior specification of the vocabulary in which the data is described. A suitable vocabulary must be equally employable by all trained inquirers on the basis of public criteria of application. In this way concepts for measurable quantities (e.g., position, velocity) are favored over concepts for qualitative or evaluative properties (e.g., color, taste, goodness).
(2) *Objectivity of scientific rationality*: the theoretical outcomes of science are the result of applying a certain method of rational inquiry according to rational norms and values accepted by working scientists.[6]

If the objectivity of science is ultimately a matter of method, Williams's object-based model is inapposite. The idea of there being alternative "perspectives" on the same "reality" is modeled on the way in which there are alternative locations from which to perceptually witness the same physical object, perhaps with the aid of technology (e.g., microscopes, telescopes)—an example being a mountain seen from different vantage points around it. On this object-based conception of objectivity, different appearances from difference locations are all explicable as revealing aspects of a unique reality or "object," in this case a particular mountain, understood as something over and above the appearances it affords. The local perspectives thus contribute to a perspective-transcendent conception of the mountain.

Objectivity is not, then, absolute, but what we might call "positional." Amartya Sen explains it thus: "The positional view of objectivity takes note of the parametric dependence of observations, beliefs and decisions on features of the person in question" (Sen 1993, 145). Since we are dealing with the community of scientists, it would be better to qualify this statement in terms of features of the person *or persons* in question. Science may transcend the perspectivity of individual observers but it remains perspectival (or positional) in a cognitive sense. The inquirer is located somewhere in history, assessing data relative to her epistemic "position" by means of certain techniques of observation, data collection, and testing (e.g., experiments) in light of a distinctive style of scientific reasoning.[7] This is a non-objectual conception of objectivity based on the applicability of a self-correcting method involving certain norms or virtues of evidence and argument.

A key feature of the positional objectivity of science is that it relies on a conception of an inquirer as no-one in particular, but always located somewhere in epistemic space. Thomas Nagel has argued:

> A view or form of thought is more objective than another if it relies less on the specifics of the individual's makeup and position in the world, or on the character of the particular type of creature that he is. (Nagel 1986, 5)

Both Williams and Nagel appeal to this thought to suggest that there is an absolute conception of objectivity, what Nagel calls "the view from nowhere" and what Williams calls "the absolute conception of reality." The idea that we can abstract from local perspectives without limit leads to the incoherent idea of a view without a viewer, a thought without a thinker—or a thinker whose thought is magically transparent to reality. A description of the world in terms of physical magnitudes is one that abstracts away from local concepts including such subjective matters as *beauty*, as well as matters of local custom such as *being polite*, and concepts that involve an appeal to human visual consciousness such as *green*. But applying the concepts of physics, say, is still something one does from a certain epistemic position, even if that is not the position of anyone in particular, e.g., consider different astronomers making observations from various telescopes of the same celestial body at the same time.

3 Objectivity is a Value

Although Putnam does not say as much, it seems clear that objectivity is another epistemic value, a positive value in terms of which we assess subjective concepts or beliefs or views—subjectivity typically being understood as a disvalue. The important point is that the objective/subjective distinction is not a unique or fixed contrast but a range of evaluative contrasts which vary depending on what we are counting as "subjective," hence "objective." It is helpful to invoke the concept/conception distinction to explain this. Our *concept* of objectivity is of a standard against which we judge subjectivity or plural subjectivities.[8] Different *conceptions* of objectivity are different ways of giving content to this standard depending on what notion of subjectivity is in question. A judgment of the color of a red ball made by a person with red-green color blindness is less objective than a judgment of a person whose color vision is normal as determined by standard tests. A judgment of the relative lengths of two lines in the Müller-Lyer illusion made solely on the basis of apparent length is less objective than a judgment based on measuring the line lengths—notwithstanding that this is a stable illusion which others could corroborate, and on that basis counts as objective relative to an observer who fails to see the difference.

Since objectivity comes in degrees, it would be better to speak of *relative* objectivity or what is *more or less* objective since there is no such thing as outright or absolute objectivity.[9] That is, there is no God's eye view. The value of objectivity is the value of establishing an *invariance* within a subject-dependent plurality—the kind of invariance depending on the form of objectivity, hence subjectivity, in question.[10] In science objectivity is based on the importance of *invariance* concerning the data and experimental results of inquiry, which must be equally available or repeatable or verifiable by others if they are to win general approval. It is a matter of establishing the cognitive position of an ideal inquirer that can be occupied by different actual inquirers. The ideal inquirer pursues answers to her questions by appeal to a certain selection of what is to count as 'data,' perhaps on the basis of conducting experiments involving scientific measuring instruments—e.g., electron microscopes, telescopes—where both the observational data and experimental results are couched in language the conditions of the employment of which are available to all trained and competent inquirers. This imposes a significant restriction on the available vocabulary that a scientific inquirer is entitled to employ. Clearly, it will not involve evaluative concepts (moral, aesthetic, political, etc.) or concepts without clear conditions of identity (e.g., belief). But that is not to say that evaluative concepts like objectivity play no role in the presuppositional background of the practice of *doing* science.

A second form of invariance is a matter of there being norms of scientific rationality (including the epistemic values that Putnam has highlighted) and a certain measure of agreement about the application of these norms. The ideal position of an inquirer is occupied by no-one in particular but it is, nonetheless, a certain position for all that, locatable at a certain time in history and characterizable in terms of a certain methodology. It is most certainly not to be equated with an absolute view from nowhere.

4 Valuing Objectivity

To understand that the objectivity of science is a value is to acknowledge our role in *valuing* objectivity in science, that the genuineness of the activity of science is judged in terms of striving for and achieving a certain level of objectivity. To endorse the vision of a view from nowhere is to suppose that the correct application of concepts in the sciences is dictated by reality itself as if inquirers play no role at all in determining whether circumstances within their purview *merit* being so described. The idea that reality itself could dictate the concepts in terms of which we describe it is one that has long been criticized by neo-pragmatists: for example, Rorty criticizes the idea of there being "sentence-shaped" facts (i.e., states of affairs) to play the role of truth-makers for a correspondence theory of truth (Rorty 1998, 36); and Putnam criticizes the very idea of metaphysically privileged "objects" describable by appeal to nature's own vocabulary (Putnam 1992, 122). To think this way is to forget that science is a practice which operates according to an evolving rational method of inquiry conducted by us according to certain interests and purposes of ours.[11]

We can make this point from another direction by noting that a distinction between 'objective' and 'subjective' in some domain is not something given—as if by the world itself—but something *instituted* by our practices of establishing a certain invariance that matters to us. Unfortunately, the inchoate realization that the objectivity that science achieves is a value has led some thinkers to a skepticism that denies that there is *any* objectivity worthy of the name. Rorty is one such skeptic. From the point of view of the plural positional account of objectivity defended here, Rorty makes the fatal misstep of identifying objectivity with the view from nowhere, the God's-eye-view of the metaphysical tradition beginning with Plato. He writes,

> pragmatists would like to drop the idea that humans are responsible to a non-human power. We hope for a culture in which questions about the "objectivity of value" or the "rationality of science" would seem equally unintelligible. Pragmatists would like to replace the desire for objectivity—the desire to be in touch with a reality which is more than some community with which we identify ourselves—with the desire for solidarity with that community.
>
> (Rorty 1991, 39)

It is only because Rorty understands objectivity in absolute or metaphysical terms as a matter of "standing in immediate relation to a non-human reality" (Rorty 1991, 21)—something he elsewhere labels "Objective" with a capital "O" (Rorty 1982, xvi)—that he sees the dialectical situation as one of having to choose between objectivity and solidarity. On the pluralist account we are considering, solidarity, understood as a commitment to intersubjective agreement, is itself clearly understandable as a form of objectivity.[12]

At this point it is worth observing that we have come full circle. The objective facts that science discovers were supposed to be those that presuppose objective epistemic values that, as Williams and Nagel show, there is a temptation to think have no taint of subjectivity. They appear to hold to the myth that these values simply impress themselves upon us under certain conditions wholly independently of us. The absoluteness of absolute objectivity is imagined to require this. But now we see that we can only make sense of *instituting* objective values in human practices by appeal to intersubjective agreements which are not independent of us. But what does this mean exactly? The relevant form of agreement is not, as Rorty sometimes suggests, an agreement in opinions about what is justified or not. The objectivity of science is not a matter of achieving a majority of votes in a poll of acknowledged experts. At a minimum there must be *some* agreement amongst scientific experts about what the appropriate epistemic values are and *some* agreement about their past applications.

5 The "Taint" of Subjectivity

We have seen that the idea of something being objective or relatively objective depends on there being an invariance in contrast to a variation of something subjective or relatively subjective. But how the distinction between objective and subjective is instituted in some area is not itself something settled wholly independently of subjects and subjectivity. That is the moral of the view from nowhere. In contrasting objective thing kinds such as *galaxies more than a hundred light-years from earth* with social thing kinds like linguistic practices, Brandom remarks,

> The criterial classification of things into objective and social is itself a social, rather than objective or ontological, categorization of things according to whether we treat them as subject to the authority of a community or not.
>
> (Brandom 1979, 190)

The point Brandom is making can be generalized: instituting *any* particular distinction between objective and subjective with respect to some subject-matter is not done wholly independently of subjects. It depends on human practice (the "social"). So what goes wrong in the Williams/Nagel view of absolute objectivity is the thought that what settles the objectivity of things, at the least perspectival end of the polarity, is simply the world itself. This is what the metaphor of transparency is supposed to register. Inquirers under certain circumstances simply see that things are thus and so anyway, as if this was something settled in complete independence of them. However, it is not the world that classifies things into objective and subjective but us.

Of course, when we consider the objectivity of scientific rationality this dependence on us is anything but straightforward. Somehow within the communal practice of science, judgments of individual inquirers about what is the best explanation of some phenomena help to sustain the distinction between what is correctly so taken and what merely seems to be correct. What makes the dependency at issue hard to understand is that it is not a simple matter of social convention: the authority of correct reasoning is not settled by communal consensus since it makes sense to suppose that a majority are wrong; and since it seems to resist explicit formalization one loses the sense of exactly what has supposedly been agreed to.

The sort of objective normativity at issue here is well displayed by the concept of truth. The kind of agreement relevant to the institution of a distinction between true and false within our practices of forming and criticizing beliefs is part of what Wittgenstein is getting at when he writes,

> "So you are saying that human agreement decides what is true and what is false?"—It is what human beings *say* that is true and false; and they agree in the *language* they use. That is not agreement in opinions but in form of life.
>
> (Wittgenstein 1958, §241)

What is true is an objective matter that is not settled by the decision of a majority about what is collectively taken to be true. For there to be an institution of judging a way things are to be true or false at all there must be agreement in what Wittgenstein calls "form of life" which involves our life with (the logic of) language. We must agree not exactly in what is true or false ("opinions") but in what makes sense and what does not—what Wittgenstein called "possibilities of phenomena" (Wittgenstein 1958, §90). Wittgenstein's remarks about measurement also hold for judgment in general: just as "what we call 'measuring' is partly determined by a certain constancy in results of measurement" so, too, what we call 'judging' is partly determined by a certain constancy in what we take to be correct judgment (Wittgenstein 1958, §242). This does not mean there is any given thing about which we must be right, but it does imply that we cannot make sense of always being wrong. In Quine's image of Neurath's boat, all that is required is that we somehow keep afloat.

6 Objectivity and Truth

Let us consider the relation between truth and objectivity in more detail. Truth is objective in so far as it is a norm in terms of which we judge a plurality of beliefs and believers as correct or incorrect. To take something to be true is distinct from, and does not settle, whether it is true. Beliefs and believers strive for truth but there is no guarantee that they have achieved it or not. In philosophy, this is perhaps the most common form of objectivity, which we may put this way: *believing does not make it so*—whether the believing is individual or collective. Truth is objective, then, in the sense that truth is a norm of which beliefs and believers strive for but may fall short. Yet objectivity does not entail truth.

Consider the Müller-Lyer Illusion again. To see that the two lines differ in length is normal for the human visual system. Consequently, one who does not see the difference is less objective than one who does. In such a case we might call this an "objective illusion" since the two lines are in fact equal in length; they only *look* different to normal observers under normal conditions. Other examples of objective falsehood include seeing a mirage under the right conditions and seeing the sun move around an apparently stationary earth—both observer-independent matters that can be ratified by appeal to other observers.[13] What these examples show is that objectivity does not imply truth: indeed, in these cases objectivity aligns with falsity. And this fits what we know of scientific practice. The objective methods of science often produce flawed theories, false explanations, and erroneous "observations."[14]

7 A Pragmatist Pluralist Vision of Objectivity

The conception of objectivity defended in this paper has its roots in three themes of pragmatism: 1) anti-abstractionism; 2) pluralism; and 3) considering agents as inquirers within a democratic experimentalist community of inquiry.[15] On the pragmatist vision, objectivity is inevitably subject-involving in so far as it is *we*, not the world, that decide what is partial or biased or "interested." We have seen that in science the relevant notion of objectivity is plural and procedural—procedures that reflect historically situated "styles of reasoning" as Hacking has called them. This may seem to suggest a damaging relativism. What rescues it from that is that the procedure is itself open to revision: it too can be corrected for its own (perhaps implicit) partialities, biases, interests, and ideologies. If we consider these forms of subjectivity as epistemic vices, then we can see the history of epistemology as a history of unearthing and correcting for such vices. As Daston and Gallison put it,

> That epistemic virtues should be multiple and historical should be the unsurprising consequence of the largely negative mission of epistemology: they were called into being to counter equally multiple and historical epistemic vices.
>
> (Daston and Gallison 2007, 377)

The history of our interest in objectivity is really the history of our interest in avoiding various identifiable forms of subjectivity that introduce an unwanted variability into our thinking. Science may become more objective, in the sense that we learn about and control for more and more subjective distorting factors over time, without its being the case that there is an upper limit to this process. We have to keep in mind that our interest in establishing invariant results in science does not entail that we cannot vary, by improving, our idea of what counts as invariant.

Notes

1 Julian Reiss and Jan Sprenger 2014, "Scientific Objectivity," Ed Zalta (ed.) *Stanford Encyclopedia of Philosophy*, URL = <http://plato.stanford.edu/archives/sum2016/entries/scientific-objectivity/>.

2 Henceforth I will, for convenience, often just use the term "science" for "natural science." The question of the social sciences and their differences from the natural sciences can be set aside for present purposes.

3 Of course, objective facts about one's own subjectivity create a special problem for ontological claims of mind-independence. For metaphysical realists the mind-independence claim applies in the first instance to non-psychological reality: what exists (or is real) would be so independently of the thoughts, feelings, experiences, etc., of any sentient being. This, of course, is a partial way of thinking about reality not simply because an important part of what exists are minds and their states (thoughts, experiences, etc.) but because all sorts of artifacts including art, architecture, music, language, and furniture are not strictly mind-independent. For one thing, these objects are causally dependent upon minds for their existence, though not in the way that states of mind are.

4 Williams (1986, 111) remarks, "The concepts [of the absolute conception] are, roughly, the concepts of physics."

5 McDowell remarks, "This is to picture science as a mode of inquiry in which the facts can directly imprint themselves on our minds, without need of mediation by anything as historically conditioned and open to dispute as canons of good and bad scientific argument" (1986, 380).

6 I take it that the products of successful science inherit their objectivity from the objectivity of the scientific method by which they were discovered.

7 For this use of "style of reasoning" see Ian Hacking, "Language, Truth and Reason," in Hacking (2002).

8 As Hilary Putnam sees it, objectivity makes available a notion of better and worse answers, or interpretations, or norms, etc. See, e.g., Putnam (1990), especially Chapters 8–11.

9 Thomas Nagel (1979, 196) is right to the extent that he treats our ordinary talk of objectivity and subjectivity as "shorthand" for "a polarity" of more of less objective/subjective outlooks.
10 As Robert Nozick (2001, 79) remarks, "A property or relationship is objective when it is invariant under the appropriate transformations."
11 McDowell too criticizes this "idea of a transparent application of rationality" (1986, 381).
12 Despite his official skepticism about objectivity, Rorty sometimes acknowledges that solidarity is a form of objectivity. For example, he occasionally describes his move as the "wish to reduce objectivity to solidarity" (Rorty 1991, 22). Of course, both the metaphysical vision of absolute objectivity and Rorty's reductionism fail to see that objectivity is a flexible and plural notion contrasting with different conceptions of subjectivity.
13 These examples are discussed by Sen (1993, 133).
14 Arthur Fine (1998, 18) remarks, "Indeed, from the history of the natural sciences, we know that over the long run the best scientific practices turn out wrong answer after wrong answer."
15 For further discussion, see Macarthur 2016.

References

Brandom, Robert. 1979. "Freedom and Constraint by Norms", *American Philosophical Quarterly*, 16(3): 187–96.

Daston, Lorraine and Gallison, Peter. 2007. *Objectivity*. New York: Zone Books.

Fine, Arthur. 1998. "The Viewpoint of No-one in Particular", *Proceedings and Addresses of the American Philosophical Association*, 72(2): 7–20.

Hacking, Ian. 2002. *Historical Ontology*. Cambridge, MA: Harvard University Press.

Macarthur, David. 2016. "A Kant-inspired Conception of Pragmatism as Democratic Experimentalism." In *Pragmatism, Kant & Transcendental Philosophy*, edited by Gabriele Gava and Robert Stern, 67–84. London: Routledge.

McDowell, John. 1986. "Notice: Ethics and the Limits of Philosophy", *Mind*, 97(379): 377–86.

Nagel, Thomas. 1979. *Moral Questions*. Cambridge: Cambridge University Press.

Nagel, Thomas. 1986. *The View from Nowhere*. Oxford: Oxford University Press.

Nozick, Robert. 2001. *Invariances: The Structure of the Objective World*. Cambridge, MA: Harvard University Press.

Putnam, Hilary. 1990. *Realism with a Human Face*. Edited by James Conant. Cambridge, MA: Harvard University Press.

Putnam, Hilary. 1992. *Renewing Philosophy*. Cambridge, MA: Harvard University Press.

Rorty, Richard. 1982. *Consequences of Pragmatism*. Minneapolis: University of Minnesota Press.

Rorty, Richard. 1991. *Objectivity and Truth: Philosophical Papers Vol. 1*. Cambridge: Cambridge University Press.

Rorty, Richard. 1998. *Truth and Progress: Philosophical Papers Vol. 3*. Cambridge: Cambridge University Press.

Sen, Amartya. 1993. "Positional Objectivity", *Philosophy and Public Affairs*, 22(2): 126–145.
Smart, John J. C. 1999. "Ruth Anna Putnam and the Fact/Value Distinction", *Philosophy* 74(289): 431–7.
Williams, Bernard. 1978. *Descartes: The Project of Pure Inquiry*. London: Penguin.
Williams, Bernard. 1985. *Ethics and the Limits of Philosophy*. London: Fontana.
Wittgenstein, Ludwig. 1958. *Philosophical Investigations*. Translated by G. E. M. Anscombe. Oxford: Blackwell.

12 The Environment and the Background of Human Life

Nature, Facts, and Values

Piergiorgio Donatelli

1 The Environment and the History of Culture

From where is it that our concern for the natural environment comes—the concern which is expressed in the fear that magnificent landscapes, majestic trees, snowy peaks, ecosystems, animal species and local climates will all disappear forever? Advances in knowledge and technology have spurred on reflection on these matters, as also have certain shifts in morality. Ominous signs of the human impact on the planet over the last half-century have in fact become inescapable. It has become evident to what extent the growth of developed societies entails the destruction of the environment, through the use of polluting technologies, fossil fuels, and life-styles based on cars, meat-eating, heating and cooling systems, and so forth. At the same time moral concepts have changed. There is greater awareness of the value of nature, in all its forms, which is increasingly seen as something precious to be protected. In this awareness different sentiments and attitudes come together, both nature-oriented—an appreciation of nature's independence with regard to human intervention—and human-oriented—an aesthetic appreciation of natural landscape, an interest in the equilibrium of ecosystems in which the human species can prosper, breathing clean air, for example.

In our civilization (unlike others, where different considerations are called for)[1] perception of the value of natural environments is a relatively recent acquisition. I would like to propose a framework for the reading of the processes of transformation of beliefs in this field that will serve in the exposition of the position I will later defend, and I will begin with the account offered by Eugene Hargrove (1989). With Hargrove, we may observe that an interest in the value of nature in relation to our conduct arises at the moment when, in various ways, nature begins to be seen as something on which we may have an effect, an effect that may be tested and evaluated in terms of human values: of beauty, of pleasure, of the meanings we find in it. In this light, essentialist perspectives do not encourage such an attitude to nature. For example, nature considered in terms of essential forms, as in certain important currents in Ancient

Greek thought, does not allow for consideration of the fragile and varying character of nature, nor does it allow for an appreciation of the value inhering, not in hidden essences, but in contingent and sensible forms; but then, neither does the scientific conception of nature, as this arises in the modern era with physics, with its concentration on primary qualities and its disinterest in secondary qualities, and therefore in the natural world as it is experienced through the senses.

Conversely, an ethical sensibility arose through the modalities of attention that are cultivated in sciences such as biology, botany, and geology, for which neither the rationalism of pre-modern science, nor the reductionism of the new sciences were congenial (Hargrove 1989, ch. 1). These sciences are based on the observation of nature as it is commonly experienced, and, particularly in geology and natural history, this could not leave out of account the role of time, with its historically contingent and singular events. Such modalities of observation and explanation make available a notion of nature situated at the level of the common experience of the senses and of human conduct. The nature that presents itself to these sciences is that which manifests as beautiful and interesting, or ugly and terrible, and which is at the same time fragile, because it is subject to the modifications human beings impose on it. What is opened up here is a natural world that lives at the same level as our perceptions and actions, and thus forms part of our human interests. Another important source of such interests, stressed by Hargrove (1989, ch. 3), is the aesthetic. The taste for informal gardens, for landscape painting, for poetry, and with it the romantic notion of nature, leads to a form of attention that focuses on its fragile and contingent forms; but the sciences themselves, such as botany and geology, encourage an aesthetic consideration of nature as something interesting and beautiful, but also fragile and malleable, and therefore in need of our care.

These developments in science and aesthetics intersect, and may conflict with, developments in philosophy. Hargrove observes that the dominant philosophical conception in modernity, the skeptical and Cartesian tradition, has not helped to consolidate an interest in the natural environment. In the general framework he outlines, the moral and human interest in the environment, in the broadest sense, requires a concept of nature that occupies the same level as that of our ordinary perceptions and attitudes: a concept of nature which is not substituted by a reductionist conceptual space that has no room for our sensations and attitudes, but rather one in which nature appears as a family of circumstances and beings which converse directly with us, that speak of beauty or of fear or disquiet, a space that offers us images with which to describe our human world and in which to try and read our lives. At the same time, in order to be an authentic object of human and moral interest, nature must not simply be a fictional or merely sentimental object, not just an occasion for experiences of private vision or entertainment. We need to be able to trust the fact that our ordinary experience reveals to us the way nature is, because it is on these grounds that we may develop an interest in it and a concern for its affairs. The experience of nature can be the basis of our moral commitment only if it is taken seriously, and not treated as a construct of the imagination or an expression of our states of mind. We need some form of realism to nurture a certain trust in our experience when it reveals to us a natural world full of interest, if we wish to act on this basis. As Thoreau writes: "Talk of mysteries!—Think of our life in nature,—daily to be shown matter, to come in contact with it,—rocks, trees, wind on our cheeks! the solid earth!" (Thoreau 1988, 95).[2]

But this realism is made difficult by the philosophical currents indicated by Hargrove. We are now familiar with the accounts of twentieth-century modernity—such as those advanced by Charles Taylor (1995), Bernard Williams (1995), John McDowell (1994), and in a different way Michel Foucault (2005)—which maintain that crucial lines in modern philosophy have largely rendered a concept of nature conceived of as a hospitable place for humans inoperable—a nature in the light of which we can understand human interests and values. On the contrary, it has left us with a concept of nature that is functional only to the interests of the sciences and to their modalities of observation and research, leaving aside whatever is the object of a broader and richer range of faculties and interests, among which are human moral and aesthetic interests in the widest sense.

2 A Place for Nature in Ethics

This theme has been dealt with by the great advocates of the value of the natural environments in the last centuries, such as John Stuart Mill and Immanuel Kant. We can read Mill's whole system as an attempt to take on the one hand the methods of science, and their relevance, outside of science itself, to society and democracy; and to put these together with, on the other hand, different modes of relations with human beings, animals, and natural objects (plants and landscapes, say), which require the development of other faculties than those cultivated in scientific research, faculties that have to do with sentiments and the imagination. Kant, to move back a century, treats the beautiful or the sublime in nature as one of the pillars on which the reconstruction of human culture is built, of what he calls the transcendental character of humanity. We in fact entertain an intellectual interest in beauty because this reveals nature as a hospitable place for human beings, reflecting our own constitution, or because in the sublime it is revealed as inhospitable, dangerous, and scary, sentiments which throw us back again to our transcendental constitutions as human beings.

For Mill and for Kant, in different ways, the experience of beauty in nature (in these different forms) is profoundly meaningful, because it provides a notion of nature—which is not that on which science (or morality) operate—that allows humanity to stand out in its character, and this in its turn has moral significance. In this sense it is important for both to establish a contrast between the spontaneous activity of nature and the realm of human interests and purposes. Nonetheless, nature as it is revealed in aesthetic experience is difficult to connect to the world out there: at the root of this difficulty there is a crucial fact that has to do with the transformation brought by modern science and its various ramifications. Modern science introduces a concept of nature that is emptied of secondary qualities and of all of such features which converse directly with human interests. Environmental ethics is situated on the difficult terrain represented by a type of realism that places at its center nature as it is experienced by us ordinarily (that is, not through scientific specialization) through beauty, and as it is revealed in poetry, in novels, in painting.

We can see therefore that dominant currents in modern culture have not made appreciation of nature easy, and environmental ethics still lives in the wake of this difficulty. But this perspective must be integrated with another consideration which is that the conception of nature in the ancient world—which portrays natural events as a realm of signs that speak the language of humans in diverse and often obscure ways—offers another and different form of resistance to environmental ethics. Pre-modern nature seen as the grand scenery of human lives is not thought of as something on which humans may intervene, inflicting wounds and diminishing its value. Thus, it is only in modernity that such an awareness has been acquired, deriving both from what we know and what we can do (both aspects, knowledge and technology, having grown exponentially in recent decades); and this is also at the root of our ethical interest in nature.

If we wish, we can portray the history of scientific, and therefore moral and political, modernity as the story of liberation from an attitude of fearful respect for nature, both in terms of our environment and in terms of our own biological nature. The philosophical tradition of the Enlightenment, that runs, for example, from Hume's *On Suicide*, to Mill's *Nature*, exposes this attitude as definitively superstitious; but now we are discovering the value of an attitude of respect for nature that is able to inspire the imposition of limits to our intervention. The concerns of contemporary environmental ethics thus allow us to better read the contrast between the attitudes of antiquity and of modernity. In pre-modern civilization an attitude of awed respect for nature did not allow for an awareness of the possibility of radical human intervention, nor therefore of its risks and its potential. Modern civilization has for the most part extirpated this attitude of respect for nature, bringing to the fore the possibility of human intervention; but in so doing it has lost the sense that the environment that surrounds us is the source of meanings that are woven into the tissue of sentiments on which our attitudes of attention and care and our moral sensibility are based. The issues of environmental ethics place before us the need to account for a richer, more complex and more diversified attitude to nature, whose concerns will include both theoretical knowledge and technological goals (having to do with our well-being, justice, and so on), but also a broader human concern for the value of nature as a sphere that is independent of us. The difficulty of moral realism in relation to nature is therefore also the difficulty of identifying a dimension of appropriate experience in which this attitude to nature is situated.

3 The Sources of our Interest in Nature

If we consider the themes of environmental ethics in this light, the traditional ethical approaches explored in this discipline may look unsatisfactory. The most common approaches represent environmental problems, and reasonably so, as problems having to do with our own interests. There is no doubt that the problems created by the transformation of agricultural techniques, by the anthropization of the territory, and by the dramatic fallout from climate change with its potential to destroy the entire planetary ecosystem, are all ones which affect our interests directly and can thus be described in the various theories from this perspective[3]. The natural environment is the source of goods that are fundamental to our lives as well as of emotional well-being that has to do with our feelings and imagination. What we do to the environment has an effect on our interests and our welfare. A conception, like that of utilitarianism, that puts our welfare at the center of our moral concerns, will have no trouble in including environmental questions as worthy of moral consideration. What is more, if, according to a sentio-centric perspective (which utilitarianism may advance) the criterion of worthiness of moral consideration is extended to all beings able to feel pleasure and pain, we will have to account for the effects of our actions on the environment in relation to a circle of beings that goes beyond the human to include animals and other forms of life that are threatened by human intervention. Conversely, Kantian and contractualist ethical conceptions have greater difficulty in accounting for our duties regarding the natural environment. These are the characteristic difficulties concerning the "frontiers of justice" with regard to beings that are not classifiable as agents able to make a contract, to whom we owe respect as rational agents like ourselves (see Nussbaum 2007 for her criticism of this). But these positions too have indirect ways of accounting for "duties" toward the environment.

It is my contention that these approaches can explain one of the sources of *our* moral concern for nature. It is certainly reasonable to maintain that the environment is a matter of concern for us, and as a basic means for the realization of our human ends it merits moral consideration. Various problems attach to following this line of argument, but not to the point of invalidating the approach, which certainly remains comprehensible. We are concerned about the environment because we are concerned about ourselves, about our interests and projects, and our community; and perhaps also the wider community of living beings having sensory perception of the effects of our actions. We may even include the communities who will live in the future. But here I am suggesting that this is not the only source of our moral consideration of the natural environment. My position is that in and through the concerns we have with regard to modes of cultivation of the earth, of breeding animals, of anthropizing the whole planet, of destroying uncontaminated landscapes and so on, another type of consideration of the value of nature is shown—one that is different from what is found in the mainstream ethical conceptions.

Natural environments have a value that resides, for example, in their beauty, or in the fact that they surprise or overwhelm us. These feelings reveal the existence, full of meaning, of entities that are for the most part independent of us but at the same time tell us something profound about ourselves. Both features are significant. If we discovered them to be artificial—artificial beyond a certain threshold such that the sense of their independence from us is destroyed—they lose all interest for us (a point made by both Kant and Mill). This independence is a matter of degree. The beauty of a landscape is linked to the human work which shaped it, and yet it can transmit a sense of the life of the plants, animals, and odors arising from the earth, all of which makes for a world that is independent of us and full of interest. We cannot specify in advance the threshold beyond which the sense of the independence of nature is lost, but there is a threshold which accounts for our attitudes in this area.

At the same time, nature attracts or repels us, provoking a considerable variety of reactions, because it tells us something about ourselves; whatever its different forms, different landscapes, different forms of animal life (the beauty of wild horses running, or the ferocity of lions in the savannah), it tells us also of our own animal nature. I would like to put forward the idea that in our experience of nature as something mostly independent from us aspects of ourselves are revealed that are not seen in the evaluative and deliberative experiences in which we formulate goals and try to realize them. Though, of course, in proposing goals and trying to realize them nature has a hand. Our proposals and attempts at their realization occur within natural environments, using natural materials, or else those manufactured from other natural materials; but in this type of activity we concentrate on what we want and what we can do, and the sense of nature as an independent material that resists our intervention does not come to us as something meaningful or profound to do with ourselves, but rather as a fact of which to take note. Here our interest, addressed as it is to what we want and how to obtain it, has to do with non-moral evaluation and deliberation as much as moral. In other circumstances, however, the independence of nature and its resistance to our will strikes us as a deep and meaningful fact, a fact that reveals the broader circumstances that define who we are and the basis of our interests.

4 Attitudes to Nature

I would now like to list a series of examples of feelings and attitudes to the natural environment that throw light on the characteristics here indicated. I would like to make use of some of the examples given by Wiggins (2000), which I will read in the light of the perspective I am outlining, integrating them with it[4].

A) The sense of beauty and harmony offered by landscapes and natural scenery. This recalls Kant's idea of natural beauty. In effect his analysis is very useful here. Kant shows that the experience of beauty consists in awareness of our feeling at home in nature, the feeling that nature is our place, it is made for us. His analysis helps to explain a set of attitudes in which we appreciate nature as a benign and friendly place to which we are grateful. In this type of sentiment the harmony we see in nature tells us something about our own sense of harmony and of order, one that gives structure to our sense of things. Connected to this sense of harmony there is also a sense of love for individual organisms. Iris Murdoch has developed the idea that beauty in nature (though she sees this as secondary to beauty in art, reflecting a characteristic shift that separates us from the eighteenth century) teaches us to free ourselves of our self-centeredness and educate ourselves to look in depth at reality outside of ourselves and our needs. As she writes: "we take a self-forgetful pleasure in the sheer alien pointless independent existence of animals, birds, stones and trees" (Murdoch 1997, 370–71).

B) A different case is what frightens us, what freezes the blood. We can think of various examples, and among these there is the sense of disquiet that the loneliness of nature can provoke, which may be linked to a feeling of nature as alive and watching us, as taking our measure. Such feelings may incline towards, or may reject, a sense of magic or superstition, but in either case they call up images of living nature to express the unease or anguish aroused in a lonely place where nature dominates and overwhelms us. John Wisdom, commenting on Wordsworth's *Tintern Abbey*, writes that: "Wordsworth's feeling *is* the feeling that the world is haunted, that something watches in the hills and manages the stars" (Wisdom 1953, 164). What I'm talking about is something many of us have experienced, even just in a deserted city park in summer, an uneasy feeling that finds a powerful and anguished expression in a film like Peter Weir's *Picnic at Hanging Rock*.

A related though different feeling may be experienced through contact with animals when we see them as alien beings whom we look at from beyond an abyss that cannot be crossed. In the movie *Grizzly Man* Werner

Herzog presents and comments on the footage made by the explorer and environmentalist Timothy Treadwell, right up to his tragic end when he and his girlfriend are eaten by a bear in front of his own camera. Herzog shows us the non-human ferocity of the grizzly, which seems to inhabit a form of life so alien to our own. As Herzog says in his voice-over: "What haunts me is that in all the faces of all the bears that Treadwell ever filmed I discover no kinship, no understanding, no mercy, I see only the overwhelming indifference of nature." The eyes of the bears express only "a half-bored interest in food," Herzog says, and this interest in food comes across as something completely different from our eating; it is a purely animal interest that has nothing to do with us. Here animal life is alien to us and freezes our blood. In both these very different cases nature appears as foreign to us. A certain duplicity in nature emerges here; its familiar character can transmute at any moment into something unfamiliar and alien in which we no longer find ourselves. But this character, too, tells us something about ourselves, about the fact that we live in a world which is on the one hand our own scenario, and on the other a mysterious place that doesn't belong to us at all.

C) Yet another attitude is that of admiration for what is great and wonderful and arouses a sense of awe: the endless unfurling of mountain peaks, the stormy sea. Here we are in the area of the concept of the sublime that was of so much importance in the eighteenth century. Once again Kant helps us offer an analysis of the way in which nature strikes us in this family of examples. On Kant's analysis, in the feeling of the sublime the vastness or the power of nature annihilates us or terrifies us because it seems to make the efforts of the imagination or desire seem worthless; but by reaction this reveals to us the infinite meaning of our nature, of our superiority over nature. The greatness and power of nature makes us at the same time small and great: we become great in the light of the vastness of nature; light is thrown onto us by this contrast. In Kant's elaboration of it the contrast may seem to illuminate the greatness of human beings in a way that looks indulgent (Iris Murdoch was very critical of it), but the contrast also allows us to see the other side—nature depicted as something great and terrible.

D) Other attitudes again are those addressed to nature, still as a sphere that is independent of us, but where this sense of independence takes the form of a challenge to a frightening adversary, worthy of our admiration and respect. A family of representations of this could be drawn from extreme sports, that is, in relation to competitive activities where the nature that is challenged can be either one's own or external, or both. In all sports there is a certain element of challenge to one's own nature, of putting it to the test. These attitudes may be enhanced when the challenge is addressed to the natural environment, as with rock climbing, where the adversary is the mighty mountain that wins our respect. Or we might think of the challenge to violent nature, like sailing in stormy seas, or even in hunting where this is practiced in a way that arouses admiration and respect for the hunted animal. Here we have feelings that recall those that Kant categorized as of the dynamic sublime—the sublime that admires and is repulsed by the power of nature, though in this case nature is not only an object of contemplation, but of challenge, of competition.

E) One final example I want to mention is that of the care of nature in its fragility and its need of our attention. Traditional agriculture and animal husbandry are pervaded by this attitude, which is compatible with crop-growing and with the breeding of animals for work in the fields or for meat.

In all these examples nature emerges not as a mere object of knowledge, nor as part of the factual considerations in the light of which we elaborate our moral considerations; but rather as feelings imbued with an aesthetic dimension, but at the same time not merely as regards taste. What they have to do with is rather the dimension that Mill calls *aesthetic* (in various places, e.g., in his essay *Bentham*), that is to say, the dimension of admiration that he distinguishes both from the sphere of moral approval and from that of human sympathy. These reveal nature as something independent of us, not indefinitely malleable, something hard that offers the background against which we move as human beings with our interests of various kinds. In these different circumstances nature does not appear to us as something to take account of in our deliberations, to get to know, or simply from which to take pleasure or other particular sensations. Instead, in these circumstances nature appears to us as a solid background to our movements and our existence, a background that reveals itself in all these various ways. These attitudes and feelings toward nature show that the world out there matters to us, as the place in which we move with our interests and purposes—a world of nature, of animals—of human beings as well, even though I have offered few examples of how our one nature, our own body, can strike us in a similar variety of ways[5]. In such situations we appreciate nature as that which offers a context for our activities, our purposes, our *human* interest in things. The appreciation of nature in these cases also reveals our recognition of the bases of our human interest in things.

This does not mean that these bases cannot change, bringing about a turn in the system of our interests. But at any one moment we have to do with some basis, in some sense natural, of our interests, and there are circumstances in which we express our recognition of this basis in the various forms I have suggested: respect, love, admiration, wonder, fear, anguish.

5 The Metaethics of Our Interest in Nature

I now wish to examine briefly the metaethics of this view of nature in its various forms, and look at a question frequently raised in environmental ethics, that of anthropocentrism. The way in which I have presented the family of attitudes to nature under consideration takes a clear position on this question. In the light of these attitudes nature has a value from a human point of view. The only way to appreciate this value is to cultivate sentiments of this type, to make oneself receptive to these dimensions of sensibility and experience. Nature has value from the point of view of these dimensions of experience. The source of value is therefore the human point of view. In this light we can criticize conceptions which argue that such a value is discernable from the point of view of the universe, as proposed by Henry Sidgwick and G.E. Moore. But such a perspective may also be encouraged by the philosophical idea of offering a characterization of human nature in its various dimensions, and, therefore, also of values, from the viewpoint of the evolutionary sciences, in as far as these projects tend to assume that science offers a point of view from nowhere: an idea maintained for example by Bernard Williams under the notion of the "absolute conception of reality" (which he himself criticized elsewhere and later in his work).[6] This notion may be seen to gain formidable strength among those who see science as the only authoritative intellectual enterprise available. Within these limits, therefore, the position I have suggested here is an anthropocentric one.

Having established this point, however, we can also recognize that the value that these attitudes individuate is in no way a human interest. In this respect we may follow the precious indications given by Bernard Williams (1995), further developed by David Wiggins (2000). Williams considers two main forms of attitude to nature: the sense of gratitude and of peace that it may evoke in us, and that of challenge and of terror. Just as we have become aware of the ever greater power of control that we have over nature, the Promethean attitude of fear has become more important because this reveals "the thought that nature is independent of us, something not made, and not adequately controlled" (Williams 1995, 239). Williams here takes up Kant's point from his own perspective, offering some of the examples listed above. What interests me here is the metaethical point he makes. He writes: "it is not these feelings in themselves that matter. Rather, they embody a value which we have good reason, in terms of our sense of what is worthwhile in human life, to preserve, and to follow, to the extent that we can, in our dealings with nature" (Williams 1995, 239). Thus, the source of value in nature is our human point of view: the scale of measurement, as Wiggins writes, is the "human scale of values" as manifest in feelings and attitudes of this type. The value that we attribute to nature in the examples under examination depends on the fact that nature *matters* to us: but the point to consider is the *way* it matters. As Wiggins writes: "The human scale of value is by no means exclusively a scale of human values" (Wiggins 2000, 16). In these examples, the values that have to do with nature matter to us precisely in that nature is independent of us, friend or foe or whatever else; our appreciation is addressed to non-human values.

The main descriptions offered by ethical theories, which see their task as illustrating the structure of *human* interests, leave out of account the attitudes I have listed here, which are also those to which Williams and Wiggins refer in their defense of environmental reasons. It is precisely because these are *not* to do with human interests that this is so. Such attitudes elude in the first place utilitarian description (contemporary utilitarianism, that is; Mill on the other hand includes these attitudes as crucial sources of value in order to account for *his* notion of utility). Here nature is not revealed in its utility for human beings. These feelings unveil a value that is independent of whatever it is we need from nature and instead reveal the complex framework of our lives. Even the emotional well-being that utilitarians count among the forms of utility is not what is at stake here. Nature is not attributed value simply because it has the capacity to evoke particular experiences and feelings in us. On the contrary, the authentic dimension of these feelings is not that of being sought for the pleasure they give, but inasmuch as they are part of the activities that allow us to experience the feelings themselves. Here Kant's precise descriptions of the types of feelings involved in (what we may call) aesthetic experience are invaluable. He does not wish to identify in the experience of the beautiful and the sublime the kind of agreeableness we might seek intentionally. Within the sensibilities illustrated by the attitudes I have listed we have our experience of the value of nature, and this resides in features that are appreciated not because they give us pleasure or because they serve our ends, but because they demand our respect, admiration, fear, reverence, and so forth.

The scale of value is human but the values are not human. We feel respect, admiration, fear (all features typical of relations among humans) even for scenes in nature. Therefore, shall we say, there is still a certain anthropomorphism here, which is not connected with interests but rather attributes to nature as a whole the range of our feelings. We face a stormy sea as we face an adversary, we challenge our own nature climbing glaciers in the mountains, we admire the endlessness of mountain peaks, and we feel welcome and protected in a landscape full of grass and flowers, of clean air, with snowy peaks on the horizon. In these cases nature speaks to us and we speak to her through this range of feelings. These are the confines of the anthropocentrism and anthropomorphism I am describing. But what we appreciate is nature as independent from us and not our own interests, or occasions for obtaining pleasant or exotic sensations.

A similar form of reasoning can be applied to the other main ethical theories. Contractualism and Kantianism don't seem to have a direct interest in nature. (Again I am speaking of contemporary theory, Kant made some vital contribution to these matters especially in his work on aesthetic judgment). As Christine Korsgaard puts it, at the center of the moral life there is the autonomy which imposes its form on the world, and does not receive any (Korsgaard 1996, 5) On this view there is no way to account for a sort of value which I have described, with Wiggins, as foreign to human values. In this type of approach, the source of value coincides with its object, that is to say, with the rational agent. It remains true nonetheless that the object of lawmaking is part of nature: it is our desires, our self-love, that constitute the material on which the moral law operates. On the one hand such desires, such natural goods, as Korsgaard calls them (2005), are the appropriate object of moral actions only in so far as what governs them is the form of the law—within the limits in which they are ratified by the moral law (in other words, as natural human goods). On the other hand, we can compare our natural goods with those of other beings such as animals, even though they are not moral goods from their point of view. Following this line of thinking we could perhaps come to imagine some kind of natural good that we have in common not only with animals but also with other organisms like plants (Korsgaard 2005, 106). In this perspective we can determine indirect duties towards other beings with which we share those natural goods which, in our own case as rational beings, are the direct object of the moral law. In the moment in which we legislate morally, conferring moral form on our self-love, we recognize a form of good (self-love: the natural good of an organism) that we can identify also in other forms of life, and make it the object of indirect duties. It is clear that this idea of Korsgaard's, whatever its merit in relation to the interpretation—and continuation—of Kant's philosophy, in any case establishes the good of animals and maybe plants by extension from human interests; while in the attitudes I am discussing though the scale is human, it is not a scale of human values.

If we go back to the historical Kant, it is more useful to leave aside the *Critique of Practical Reason*, in that if we stay with the moral Kant there is no immediate room to account for this other type of argument. Nonetheless, for Kant moral reasoning is only one of the uses of reason, and there are others. Human reason is marked also by uses such as the aesthetic, in which a certain experience of nature as beautiful or sublime illuminates the character of human reason itself. In this light, our capacity to appreciate the value of nature as something independent from us, a type of experience that Kant wishes to separate completely from the moral, nonetheless illuminates the character of reason, which, in its practical use guides us in our moral life. Reading this backwards, therefore, aesthetic experience illuminates also the nature of reason that guides us morally.

The inadequacy of ethical theories such as utilitarianism and Kantianism to find room for the kind of interest in nature I have described here also applies to objectivist neo-Aristotelian views, according to which something is a good if it is necessary for the flourishing of a human individual. We have seen how in fact the attitudes considered above do not bring nature within the sphere of our interests, even if these are construed in the Aristotelian sense. But there is a fundamental objection I would raise against a teleological approach. In fact, this offers a lesson that might appear congenial to my perspective, given that a teleological conception of organisms extends practical evaluation beyond the human borders. The good of an organism has to do with the flourishing of the type of organism, or form of life or species, to which the single individual belongs (Foot 2001). In this perspective, for example, light and water are among the natural goods of the geranium on my windowsill. Along this line of thinking we can certainly extend the theory of goodness, though it would need to be developed to ascertain its limits. For example, in the classic version, the human being who uses certain faculties is placed on top and this allows the creation of a hierarchy of natural goods. The natural good of organisms like human beings might not include the consideration of the good of organisms different from them, such as non-human animals and plants. Instead, in the hands of an author like Nussbaum, this framework is reshaped in the direction of an appreciation of the pluralism of forms of life such that the conception of which goods require our intervention is reoriented (Nussbaum 2007).

But the fundamental objection is another one. The perspective I am developing wishes to advance a humanistic (or anthropocentric) defense of the non-human value of nature. Whereas, if we follow the line of a teleological approach, we are required to imagine the extension of practical necessity to all organisms as if this were offered from the point of view of nowhere, from which we describe the nature (the essence) of the organisms. For each organism, as a member of a species, we can identify the goods it needs, water and sun for the geranium, for example. The goods and the practical needs of human beings are accounted for in the same way, considering the human being as a member of a species. (The picture may get more complicated if it is maintained that the reflection on human goods is carried out from a human point of view, and not from the point of view of nowhere. In which case this would not be a metaphysical reflection on the nature of human beings but a practical reflection conducted within the practical rational faculty. We should consider though that such a reflection is in any case conceived of an exercise in practical rationality, which is in its turn treated as one of the characteristics that identify the human species. In the last analysis, therefore, the credentials of the reflection carried out from within the practical dimension derive from the fact that this exemplifies a characteristic that we get hold of metaphysically as one of the aspects that constitute human beings as that kind of species.) But, in the attitudes I mentioned above, the non-human value of nature appears through the exercise of an attention and a human capacity of response that are simply not seen if we imagine we can raise ourselves above the partiality of our reactions and feelings in order to describe the essence of organisms. This point is made clear if we turn again to Kant. Within the experience of beauty, on his analysis, we come to realize that nature has precisely the form appropriate to being governed by the intellect, it shows this special purpose. But the purpose individuated in the experience of beauty (the free play of the faculties) does not pertain to things, but to our mode of seeing nature as if it possessed an order that makes it appropriate to being governed by the intellect (*Critique of the Power of Judgment*, 5: 184). The pleasure experienced satisfies a need for order and expresses the agreeable sense of being at home in nature. We would not grasp this value in nature if, in standing in and before it, we read there an intrinsic order, a purpose in whose light needs and goods can be discerned, as occurs in this sort of Aristotelian thinking. In conclusion, therefore, if we follow the line of a teleological approach, we encounter both the benefits and the difficulties characteristic of this perspective, but we will clearly not be able to account for the kinds of attitudes under examination here.

6 Nature as the Background to Human Activities

I would like to end by offering a positive characterization of this type of value of nature. As we have seen in the examples of attitudes given, value appears in the form of considering nature as a whole, as something independent from us, as a scene that we cannot reconstruct in terms of what interests us, what we need, what pleases us. In this light we can understand the problems of Mill and Kant discussed above, in as far as they wanted to vindicate the hard reality, independent from us, of that which is revealed in the aesthetic experience of nature, but also to separate the sense of this reality from the object of science. At the same time, this notion of nature as a whole speaks to our deepest concerns in the different ways we have seen. Nature shows itself to us as a scene of beauty that inspires a love for single things and a respectful attention for the existence of each being; as a mysterious scene that gives depth to our very existence; as a place that evokes a sense of nobility and grandeur; as an adversary that endows our challenges and life itself with meaning; as an independent organism needful of care. Now we can see in these sorts of attitudes (and in others that could be listed in which nature takes on this role of interlocutor in relation to a range of feelings and human responses) a way of recognizing nature as the background that creates the appropriate context for human activities.

The value we attribute to nature is the value which we recognize in what characterizes the scene that makes sense of our activities and of ourselves. For example, we appreciate nature as the context in which there are such things as running, challenging the resistance of the ground and our own fatigue. In a world in which all this disappeared, this thing called "running" would also disappear. Or, to take an example closer to reality, in the situation in which running in sporting contexts employs increasingly modified terrains, and in which the physical capacities of the athlete are radically changed (by various means), the activity of "running" changes and is transformed into something very different.[7] With regard to the attitudes I have listed, and to this complex approach, the point is that we appreciate nature as a background that opposes resistance to us, allowing for characteristic human activities such as walking and running to make their mark, but also it allows for activities like walking in natural places that may hold surprises for us and appeal to our sense of adventure. Adventure and its pleasures are also connected to human activities such as walking that presuppose a wide world which is for the most part not under our control, a world that can surprise us, that is multi-layered and has depth. The point is, however, naturally a general one and has to do with the sense in which the human activities that we value, that we criticize, and so forth, have their meaning against a background which consists of many things. But among these things there is this: that there is a natural environment in which they take place. Human activities are shaped, and thus we may say also *limited*, by this background, and nature has value also in that it contributes to the form of this background.

As we have seen, in these attitudes to nature, to natural environments, to living organisms and so forth, what is intended is something not completely manufactured by us. We have to leave this limit or borderline between natural and artificial vague and imprecise, but we cannot let it disappear. The sense of a limit between the natural (what we have not produced) and the artificial (what we have produced) is what activates the attitudes I have referred to; these are directed to environments and organisms that are independent from us, that we have not produced, and that we cannot control completely. These are not notions that can be exclusively clarified by science: they form part of our everyday concepts; they can be reviewed in various ways but they are not empty. The possibility of us creating new forms of life has an effect on these attitudes and these concepts, but it does not abolish them. If in the long run this does happen, it will mean that the scenery of our thinking and our feelings will have changed so radically as to make such attitudes incomprehensible, the mere legacy of a past which is dead for us; in such a case, moral reflection would be something totally different from what it is now. For the moment, our activities of various kinds, including that in which we exercise our moral sensibility, are what they are because we assume that there exists something grand and not made by ourselves, which is this planet (within a cosmos) and an entire system of forms of life and environmental conditions. As Williams holds, these are *our* ideas. We could put it like this: they are cultural constructions of what is natural, on which the explanations offered by that particular enterprise which is science have an effect, but they are not entirely governed by these.

My contention is that these ideas are important as they have to do with the shape of the complex scenery of our thought and our way of life, in which the various reasonings that belong to specific areas have a place, such as the reasons related to good sporting practice, or the reasons we formulate in reflecting on our intervention in the natural environment, and so on. These human ideas envisage a distinction between what is not entirely made by us and what instead is our product, and incorporate a notion of threshold in relation to our interventions, beyond which we do not recognize something as natural, in the sense that it is a source of the kinds of attitudes listed above. The attitudes I have discussed and the type of moral sensibility they illustrate show the importance that *present* facts and circumstances—that could change—have for us. These do not indicate the definitive measure of their value, but they certainly indicate a value that they have for us.

7 Conclusion

In this chapter I have considered and vindicated the importance of an attitude towards the natural environment understood as the wider and, to a certain degree, independent background of our practices. The independence and the resistance that this background poses to our interests and desires respond to a sense of fundamental importance and depth.[8] The relevance of the considerations advanced in the present text for the theme of the alleged distinction of facts and values should now be clear. The independence and the distance established by animals and environments is a kind of fact which articulates our overall conceptual dimension: it is in fact the background which makes sense of the importance and of the depth of our inhabiting the world, of our encounters with others, and of the many activities discussed in the text. It is a fact which *concerns us*: we can account for it only by deploying the overall scheme of human values (in the sense in which Wiggins uses this expression). The reality of the environment, of animals, and nature is not detached from the overall pattern of values, from the sense of importance and depth it expresses. We can unfold the measure of this reality only by unfolding the measure of what matters to us.[9]

Notes

1 See, e.g., Jamieson 2001, part I.
2 Cited in Hargrove 1989, 100.
3 See Jamieson 2001; Light and Rolston III 2003; Gardiner, Caney, Jamieson, and Shue 2010.
4 An unpublished lecture by Cora Diamond, "The Problem of Impiety," was also in the back of my mind when I made this list of examples.
5 On this, see Donatelli 2015a.
6 Williams 1978. For a revision, see Williams 2006, especially the homonymous essay in the collection. The extended exchange between Williams and Hilary Putnam on this issue is not always very helpful though.

7 Arguments of this kind are to be found in Sandel 2007; I discuss them in Donatelli 2015a.

8 In this text I did not develop another way of unfolding the significance of this kind of realism, which shows instead how the independence of natural environments might activate relations and encounters of many kinds: the independence and the distance that animals, environments, and ecosystems establish with us represent in fact also the basis of the coexistence with them. On this, see Donatelli 2015b.

9 I wish to thank Catherine Bearfield for the translation of an earlier version of this text and Sarin Marchetti for the editing.

References

Diamond, Cora. "The Problem of Impiety." Unpublished ms.

Donatelli, Piergiorgio. 2015a. *Manières d'être humain. Une autre philosophie morale*. Paris: Vrin.

Donatelli, Piergiorgio. 2015b. "Forms of Life, Forms of Reality", *Nordic Wittgenstein Review*, 4: 43–62.

Foot, Philippa. 2001. *Natural Goodness*. Oxford: Clarendon Press.

Foucault, Michael. 2005. *The Hermeneutics of the Subject*. Edited by Frédéric Gros and Arnold I. Davidson, translated by Graham Burchell. New York: Palgrave Macmillan.

Gardiner, Stephen, Caney, Simon, Jamieson, Dale, and Shue, Henry. (Eds). 2010. *Climate Ethics. Essential Readings*. Oxford: Oxford University Press.

Hargrove, Eugene C.1989. *Foundations of Environmental Ethics*. Englewood Cliffs, NJ: Prentice Hall.

Jamieson, Dale. (Ed.). 2001. *A Companion to Environmental Philosophy*. Oxford: Blackwell.

Korsgaard, Christine. 1996. *The Sources of Normativity*. Cambridge: Cambridge University Press.

Korsgaard, Christine. 2005. "Fellow Creatures: Kantian Ethics and Our Duties to Animals." In *The Tanner Lectures on Human Values*, Vol. 25/6, edited by Grethe B. Peterson. Salt Lake City: University of Utah Press.

Light, Andrew, and Rolston III, Holmes. 2003. *Environmental Ethics. An Anthology*. Oxford: Blackwell.

McDowell, John. 1994. *Mind and World*. Cambridge: Harvard University Press.

Murdoch, Iris. 1997. "The Sovereignty of Good Over other Concepts." In *Existentialists and Mystics: Writings on Philosophy and Literature*. London: Allen Lane/Penguin.

Nussbaum, Martha Craven. 2007. *Frontiers of Justice*. Cambridge, MA: Harvard University Press.

Sandel, Michael J. 2007. *The Case against Perfection: Ethics in the Age of Genetic Engineering*. Cambridge, MA: Harvard University Press.

Taylor, Charles. 1995. "Heidegger, Language, and Ecology." In *Philosophical Arguments*. Cambridge, MA: Harvard University Press.

Thoreau, Henry David. 1988. *The Maine Woods*. London: Penguin.

Wiggins, David. 2000. "Nature, Respect for Nature, and the Human Scale of Values", *Proceedings of the Aristotelian Society*, 100: 1–32.

Williams, Bernard. 1978. *Descartes: The Project of Pure Enquiry*. Harmondsworth: Penguin.
Williams, Bernard. 1995. "Must a Concern for the Environment Be Centred on Human Beings?" In *Making Sense of Humanity and Other Philosophical Papers (1982–1993)*. Cambridge: Cambridge University Press.
Williams, Bernard. 2006. *Philosophy as a Humanistic Discipline*. Edited by A.W. Moore. Princeton: Princeton University Press.
Wisdom, John. 1953. "Gods." In *Philosophy and Psychoanalysis*. Berkeley: University of California Press.

13 Fact/Value Complexes in Law and Judicial Decision

Douglas Lind

Law stands in a complicated relationship with philosophy's perennial fact/value dichotomy. On the one hand, law formally separates facts from values. Courts in both the common law and civil law traditions commonly distinguish questions of fact from questions of law. Many courts and administrative agencies structure their opinions into distinct sections for 'findings of fact' and 'conclusions of law.' Jury trials impart to juries the role of fact-finders while denying them any say whatsoever over the controlling legal rules or principles. Appellate courts in the United States generally employ a highly deferential standard of review for a trial court's factual determinations whereas they review matters of law *de novo*. These conventions of legal practice suggest that law embraces a sharp divide between facts and legal values or norms.

Yet on the other hand, law in many respects blurs facts and values. The enactment of a law, while an expression of social value and policy, also sets down a legal fact. Beyond separating questions of law from those of fact, courts sometimes tag issues as raising 'mixed questions of law and fact.' In jurisprudence, the noble tradition of natural law built a teleological conception of law, employing Aristotelian practical reason, which identified its essence and fundamental principles according to fitness to human well-being. From this, the natural law jurists, like their modern counterparts in normative jurisprudence, came to measure the force of precepts found in the lowly factual enactments of human positive law in terms of their conformance to the value-rich dictates of the posited higher law.

The relationship between law and the fact/value dichotomy is thus quite ambiguous. Adding to the ambiguity is the relative indifference shown the dichotomy by legal scholars. A scarce few legal thinkers have ever paid it any mind. Two stand out. In metaphysical jurisprudence—the study of the nature of law—most noteworthy is the neo-Kantian legal philosopher, Hans Kelsen. From his earliest works Kelsen wrote of the tension between 'is' and 'ought,' and emphasizing that the validity of a legal norm depends on how it is arrived at, not on whether it corresponds with some assumed substantive value (Kelsen 1992, 15, 56–61).

His *General Theory of Norms* includes an extended discussion of what he considered to be the "incontrovertible difference" between 'is' and 'ought' (Kelsen 1991, 63). Abiding by that "irreducible duality" with care, Kelsen insisted that all norms in his theory of positive law, including the basic norm, could be derived from other norms or ascriptions of value (Kelsen 1991, 63). Further, he reproved natural law theory on the ground that it rests on a failed attempt by Aquinas to deny this duality between 'is' and 'ought' (see Kelsen 1991, 71–2). This line of criticism was followed by Julius Stone, who likewise chided natural law jurisprudence for breaching the fact/value divide by, in effect, committing the naturalistic fallacy (Stone 1965, 212–18). John Finnis retorted that Kelsen and Stone misunderstood natural law theory, arguing that natural law, especially in the Thomistic tradition, draws its principles from practical reason, not by inference from facts about human nature (see Finnis 1980, 26–48).

Beyond Kelsen's firm affirmation of the fact/value divide and this brief exchange over the grounding of natural law, the fact/value distinction has received little attention from those inquiring into the nature of law. Nonetheless, there is a sense in which it strikes a responsive nerve elsewhere in jurisprudential thought. For the dichotomy goes to the heart of the most enduring problem of jurisprudence: defining the function, scope, and limits of judicial power. Since the dawn of Western law, the role and function of the judge has troubled politicians, populaces, and jurists alike. The fear is that the power to sit in judgment of one's fellow humans and allocate the benefits and burdens of the law is a power so fraught with opportunity for abuse that it's wielding must be rigidly limited and restrained. The belief is that the necessary restraint is best achieved by making the content of the law determinate and fixed antecedent to the exercise of judicial power, and firmly limiting that power to an assessment of whether the facts of a concrete case fit the law. That is, the values of the law must be judge-proof, while the function of the judge must be fact-limited.

The effort to achieve such an uncompromising separation of fact and value in adjudication has significantly impacted the development of Western law. History shows that the effort has earned at best a checkerboard of faint success. Nonetheless, the resolute drive to limit judicial decision-making to finding facts and squaring them against *the law as it is*, while reserving to legislatures or others all say as to questions of value or policy, and as to what *the law ought to be*, continues unabated to this day. This effort to so winnow facts from values in law and judicial decision-making takes us to the second prominent instance where legal philosophers have confronted the fact/value distinction. Around the turn of the twentieth century, jurisprudential movements gained favor in both Europe and the United States that challenged the compartmentalized picture of law and judicial practice. The free law movement in Europe along with the loosely affiliated jurists known as the American legal realists argued that the practice of adjudication cannot sustain a formal divide between facts and values. Among these legal thinkers, the American realist Karl Llewellyn most directly confronted the fact/value dichotomy, averring that empirical study of law discloses the entanglement of 'is' and 'ought' in adjudication (see Llewellyn 1931, 1236–7, 1252–5).

In this chapter I reconsider and build upon the work of those jurists of a century ago who sought to dispel the image of judicial decision-making as a highly restricted, ministerial activity. I ratify their empirical findings that judges in practice regularly engage with the law's values, influence its policies, and make and modify its content. Yet I go beyond their descriptive accounts of fact/value entanglement to propose a way of understanding legal judgment that necessarily collapses the fact/value dichotomy. I argue that all attempts to compartmentalize the law rest on a misunderstanding of the nature of legal precepts and their application in judicial decision-making. For precepts—a legal system's rules, principles, doctrines, standards, etc.—necessarily *fuse* facts and values. They are conceptual composites best understood as *fact/value complexes*. This chapter aims to develop such an understanding of legal precepts and their judicial application. My argument comes in three main sections. First, I provide some background on the efforts across the ages of jurisprudential history to deny the interconnection of facts and values in judicial practice. Second, I consider the free law and legal realist response. Finally, I argue for understanding legal precepts as fact/value complexes. This understanding complements the empirical work of the free law and legal realist jurists. Yet it offers an analytically stronger claim as to the fact/value interconnection in law, thereby allowing a more complete rebuttal of the shadowy dream that facts and values can be rigidly separated in law and judicial decision.

1

Efforts to restrain courts from exercising any authority over the substantive content of law date from at least the middle years of the Roman Republic. The Romans adopted two tactics in pursuit of such restraint. First, they tried enacting systematic codifications of the law. Second, they entrusted legal scholars with power to authoritatively interpret it.

In the fifth century BCE, the Roman commission known as the decemvirate promulgated the Law of the Twelve Tables, a statutory codification of existing custom. Pressure for this fixed, written legal code had come from the plebeians who felt the laws were being unfairly administered to favor the patricians. The plebeians directed most of their discontent at two classes of officials who exercised judicial or quasi-judicial power: the *iudici*, or judges, and the consuls. The plebeians especially resented the consuls. The highest political officeholders in Rome at the time, the consuls enjoyed a range of functional authority, including judicial power.

In exercising that power, they did not consider themselves bound by the law. The Law of the Twelve Tables was an attempt to wrest interpretive control away from the consuls as well as ordinary judges, by making the law so clear and definite as to render interpretation unnecessary. To an extent, it succeeded. The Twelve Tables became the basic foundation of Roman law. Cicero would write in the first century BCE that the Twelve Tables were so central to Roman civilization that school children were taught to recite them from memory (Cicero 1928, II.9). Yet the centuries obscured their definitiveness. By Cicero's time, the Roman legal system had become quite flexible and interpretative discretion by consuls and ordinary judges had become, once again, quite far-reaching. This caused the Romans to turn to the second measure for arresting judicial discretion: the placing of interpretive authority in the hands of legal scholars, the jurisconsults, whose interpretive commentaries on the law came to be treated as binding on judges. The most famous was Gaius, an otherwise obscure jurist whose *Institutes*, an elementary legal textbook, became the primary authoritative source on Roman law for over four centuries. In about 533 Gaius's treatise was superseded by the Emperor Justinian's *Corpus Juris Civilis*. Publication of the *Corpus* marked a return to codification as the principal Roman method for restraining judicial power. Intended as a complete and comprehensive codification of Roman law, the *Corpus* included a compilation of all imperial enactments, an encyclopedia of selected jurisconsult writings, and a student textbook covering the basic principles of Roman law. The whole of the *Corpus*, including the textbook, was issued as an imperial statute with the full force of law. Justinian prohibited Roman lawyers and judges from referencing any other legal texts, including earlier works of Roman jurists, except for those expressly incorporated into the *Corpus*. Gaius's treatise survived as it provided the model for the student textbook, the *Institutes*.

The *Corpus Juris Civilis* is the cornerstone of the civil law or Roman law tradition. While its influence waned with the decline of Rome, it later revived so profoundly as to become the lodestar for the development of law throughout most of Europe and many countries beyond (see Stein 1999, 38–68). Its impact on Western law to this day cannot be overstated. This includes its dual approach to restraining judicial freedom of decision: codification and reliance on juristic commentary. Codification in theory obviates the need to interpret law. A complete, comprehensive, and definitionally exhaustive legal code would not require judges to interpret a legal precept or take any measure of its underlying values or policy when applying it to the facts of concrete cases. The practice of giving authoritative voice to juristic writings supplements codification by giving cover to a legal code insofar as it falls short of complete, comprehensive clarity. Separately and in combination, the two approaches strive to make law a logically closed system of axioms and corollaries—a system of precepts so settled and determinate that the exercise of judicial power would require only a logical, mechanical assessment of how the operative facts of a case square with the law.

The influence of these twin measures to confine the scope of the judicial function reached its height during the eighteenth and nineteenth centuries. Those centuries saw a major codification movement sweep across continental Europe. The French *Code Napoléon* was promulgated under Napoleon I in 1804. Drafted by a commission of four prominent jurists and modeled on Justinian's *Corpus Juris Civilis*, the *Code Napoléon* culminated a French codification effort begun during the reign of Louis XIV in the late seventeenth century. In good part, the French interest in codification came in response to the fact that French law had become by the seventeenth century largely the product of a strongly independent judiciary. The *Code Napoléon* directly sought to squelch that judicial influence over the law's values and norms. It provided: "In the decision of cases submitted to them, judges are forbidden to lay down general rules of conduct" (*French Civil Code* 1804, art. 5).

Similar concerns about insufficiently checked judicial power induced Frederick the Great of Prussia to oversee the creation of a civil code that took more than half a century to complete and in the end included over 17,000 articles. Published in 1794, the *Allgemeines Landrecht für die Preussischen Staaten* ("Universal Territorial Law for the Prussian States") was designed to make Prussian law "judge-proof" (Merryman 1985, 39). It provided detailed solutions for every legal dispute its writers could foresee. Judges were directed to abide by the solutions and refrain from independently interpreting the code provisions. If a case presented a fact situation unaddressed by the codified solutions, the judge was instructed to pass the matter on to a special Statutes Commission. Judges who bypassed the Statutes Commission and tried to interpret the code themselves were subject to punishment (see Robinson et al. 1985, 422). By the end of the nineteenth century, civil codes had been enacted in several more countries across Europe, as well as in a number of countries elsewhere that had been impacted by European imperialism. Most bore the influence of the *Code Napoléon.*

Concurrent with the codification movement, European jurisprudence from the seventeenth through the nineteenth centuries tended to diminish the judicial function. Judges were routinely characterized as relatively powerless, passive officeholders confined to observing code provisions while giving deference to juristic commentary. German philosopher Christian Thomasius stated matter-of-factly in 1688: "There are, therefore, two parts of judicial jurisprudence: *the interpretation of laws*, which you could for the sake of distinguishing it term the jurisprudence of professors and doctors of law; and the *application of laws*, which is the jurisprudence of *advocates and judges*" (Thomasius 2011, 85–6). In his *A Methodological System of Universal Law* (1741), Dutch jurist Johann Gottlieb Heineccius wrote that it is "the office of a judge to apply laws to facts or actions" (Heineccius 2008, 473). In France, the Baron de Montesquieu in 1748 depreciated the judicial function as "in some measure next to nothing. . . . [T]he national judges are no more than the mouth that pronounces the words of the law, mere passive beings, incapable of moderating either its force or rigor" (Montesquieu 1949, 156, 159). Italian legal philosopher Cesare Beccaria, in his highly influential *On Crimes and Punishments* (1764), considered it a legislative prerogative to declare the meaning of the law, while judicial decision-making is a programmed exercise in logical deduction:

> For every crime that comes before him, a judge is required to complete a perfect syllogism in which the major premise must be the general law; the minor, the action that conforms or does not conform to the law; and the conclusion, acquittal or punishment. If the judge were constrained, or if he desired to frame even a single additional syllogism, the door would thereby be opened to uncertainty.
>
> (Beccaria 1963, 15)

The attitude toward law and judicial decision articulated by Thomasius, Heineccius, Montesquieu, and Beccaria continued to influence jurisprudential thought through much of the nineteenth century, coming later to be called by its critics Legal Formalism. While those legal thinkers wrote specifically about legal systems in the Roman civil law tradition, formalist jurisprudence was not confined to civil law jurisdictions. Jurisprudential efforts to deny that judges make law and to circumscribe judicial decision-making within the parameters of logical deduction became prominent during the eighteenth and nineteenth centuries in the common law world as well. Sir Matthew Hale wrote in 1713 that:

> The Decisions of Courts of Justice, tho' by Virtue of the Laws of this Realm they do bind, . . . do not make a Law properly so called, (for that only the King and Parliament can do); yet they have a great Weight and Authority in Expounding, Declaring and Publishing what the Law of this Kingdom is, . . . and tho' such Decisions are less than a Law, yet they are a greater Evidence thereof than the Opinion of any private Persons, as such, whatsoever.
>
> (Hale 1971, 45)

William Blackstone concurred, famously arguing in his *Commentaries on the Laws of England* that the common law consists of general rules and customs "declared in the decisions of the courts of justice," but not made therein (Blackstone 1979, I 73). Judges are the "depositary of the laws," Blackstone maintained, and their decisions offer "the principal and most authoritative evidence that can be given, of . . . [all that] shall form a part of the common law" (Blackstone 1979, I 69). But by oath judges are "sworn to determine . . . according to the known laws and customs of the land" (Blackstone 1979, I 69). They are "not delegated to pronounce a new law;" (Blackstone 1979, I 69) nor is it within judicial power to "depart[ ] from" or "alter" existing law (Blackstone 1979, I 70–1). Accordingly, while judicial decisions provide in general the most reliable evidence of the law, Blackstone insisted "that *the law*, and the *opinion of the judge* are not always convertible terms, or one and the same thing; since it sometimes may happen that the judge may *mistake* the law" (Blackstone 1979, I 71).

Hale and Blackstone's position that the decisions of common law judges are only *evidence* of preexisting law drew rebuke from Jeremy Bentham. To Bentham, it was imperative to admit that judges *make* law. Yet he took no solace in that fact. Characterizing the judge as "the under-workman" of the legislator (Bentham 1823, 398), Bentham cautioned that common law judges must be prohibited from tinkering with the dictates of positive law. He wrote:

> Should there be a Judge who enlightened by genius, stimulated by honest zeal to the work of reformation, . . . should seek with his sole hand to expunge the effusions of traditionary imbecility, and write down in their room the dictates of pure and native Justice—Ah! . . . that amendment from the Judgment Seat is confusion, that partial amendment brought at the expence of universal certainty is but . . . universal evil.
>
> (Bentham 1774, 223–4)

Similar sentiments gained favor in the United States. As the new American republic began to aggregate its own body of legal materials in the nineteenth century, legal scholars sought to collate that material and regiment judicial practice into a distinctly American 'science of jurisprudence.' Like the modern civil law jurists, the American legal scholars championed a jurisprudence predicated on definitional clarity and completeness, formalistic methodology, and judicial restraint. They endeavored to define and articulate the "essential meaning" of the fundamental principles of law with such "precision" as to ensure that those principles can "be measured and enforced under a general system, in which nothing is to be left to the dangerous discretion of judges or jurors" (G.D.A.P. 1861, 142–4). In the words of one American formalist: "The great end of a refined jurisprudence is to circumscribe the sphere of judicial discretion within the narrowest limits practicable; and requires that the rules of conduct, by which the responsibilities of the citizen are to be tried, shall, to the very utmost, be clearly defined and plainly proclaimed" (G.D.A.P. 1861, 144). Speaking of the general principle of negligence, this jurist continued: "And the high aim of the philosophic jurist will ever be to perfect a standard of duty, of wide range, plainly marked, which he can set up firmly between the bench and the jury box, for the guidance and control of both" (G.D.A.P. 1861, 144).

James V. Campbell, a Michigan Supreme Court Justice and law professor at the University of Michigan, lamented in 1880 how the administration of justice in the United States had become inefficient due to "the indigestibility of our monstrous bulk of law material" which was being cited as authority in spades without adequate understanding of the underlying general principles (Campbell 1880, 257–8). Campbell argued that to forestall this trend, the "only practical refuge is among the text-writers" (Campbell 1880, 271). He called for jurists to write bantam text-books modeled on Justinian's *Institutes* (Campbell 1880, 284–5). These were to condense and synthesize the general principles and doctrines of the various branches of law into "complete formulas" that would be "authority everywhere, and universally useful" (Campbell 1880, 273). On Campbell's account, the texts would "contain every thing essential" and "include every exception and qualification" (Campbell 1880, 283), thereby ensuring that judges would not veer into "legislating instead of expounding the law" (Campbell 1880, 269).

While Campbell envisioned a formulaic ideal where "every formula . . . [would] convey [by] a clear and complete meaning . . . a perfect enunciation of a principle" (Campbell 1880, 283), he did not think that logic was the only tool judges would need to apply the principles. He found "the strict methods of the logicians" incapable of determining the outcomes in legal cases, for the "rigid logicians do not make allowance enough for the variable, but inevitably present, element of human nature, and they do not furnish the best rules for use in legal controversies" (Campbell 1880, 286). Other American formalists disagreed. Everett V. Abbot, for example, portrayed judicial decision-making in strict syllogistic terms. Reminiscent of Beccaria, he wrote:

> Thus the juridical procedure takes a form which in its lowest terms is a syllogism, wherein the major premise is the predication of a juridical principle, the minor premise is a predication that the case at bar comes within its terms as an instance of it, and the conclusion is the joinder of the two in the final judgment of the court. The ascertainment of the major premise is the province of the jurist through the process of logical reasoning; the ascertainment of the minor premise is the province of the court through its process of investigating facts; and the conclusion—that is, the judgment—follows, or should follow, inevitably upon these two.
>
> (Abbot 1897, 482)

Like the efforts among civil law jurists to limit the judicial function to deductions from value predicates embedded in codified sets of principles and rules, the American common law formalists thus labored to make the law fixed and determinate antecedent to judicial decision-making. Both formalistic movements sought to lessen the judicial office by safeguarding the law's values and norms from judicial influence. Both created a formalistic bulwark meant to confine judges to finding facts and applying the law thereto, and to prevent them from exercising any discretion over the controlling principles of law. The movements differed only as to the source of those principles and of the law's ultimate values.

The civil law approach sought to ensconce the law's basic values and precepts in written codes, such as the *Code Napoléon* and the Prussian *Allgemeines Landrecht*, that reflected the sovereign will augmented by the wisdom of notable jurists. In principle, this approach reduced all law to the fixed definitions and detailed specifications found in the codes, while restricting the judicial function to logical deductions from those codes. It thus aimed to forestall all active adjudicative creativity—even to the extent, as a practical matter, of obviating judicial interpretation outright.

The common law, by contrast, would seem to reserve a more robust role for judges to exert influence over the law's content and basic values. Yet the evidence or declaratory theory of precedent championed by Hale and Blackstone depicted common law adjudication as a modest and relatively ministerial enterprise, clearly not a process of active lawmaking. The American formalists shared that perspective. They denied that courts make law. And they protested against any judicial role in articulating fundamental jural values or norms. Thus, while the decisions of common law judges declare and apply law, formalists such as Campbell and Abbot argued that the law's fundamental principles and basic norms originate *antecedent* to and *outside* judicial practice. They did not suggest that judicial interpretation is wholly unnecessary. But they did seek to reduce adjudication to a highly constrained, mechanical task, thereby just as fully separating the law's values and norms from the judicial function as the civil law jurists.

2

By the late nineteenth century and into the twentieth, legal formalism came under sustained attack both in Europe and the United States. In Europe, a number of legal philosophers rallied in favor of a relatively far-reaching "judicial freedom of decision" (see Ehrlich 1917, 79; Gény 1917, 2). These free law jurists reproved the "excessive formalism" that came from restricting judicial interpretation of law to logical deductions from civil codes and other formal written sources of law (Gmelin 1917, 85). They argued that judicial decision-making, by its very nature, is a creative enterprise that necessarily calls upon judges to affect the content of law and engage with its norms and values. Some in the free law movement emphasized the subjective, emotive aspect of judicial decision-making. Johann Georg Gmelin, an Austrian judge, cautioned that, "True justice cannot be found in cold, logical ratiocinations, nor by a wisdom garnered in dusty books," for the judicial function requires that the judge act "not with his head merely but also with his heart" (Gmelin 1917, 88). Others called attention to the normative content of all law. The prominent Austrian legal philosopher Eugen Ehrlich argued that all legal precepts, including the rules of construction meant to limit the freedom of judges, unavoidably incorporate value judgments into judicial decision-making—at a minimum the value judgment that every case should bring about a "fair result" (Ehrlich 1917, 65). Because of this Ehrlich favored a "dynamic conception" of judicial practice (Ehrlich 1917, 78). He argued that courts should treat the operative facts of the cases before them as "coefficients of social tendencies," and that they should regard legal rules and principles as living, adaptable products of socio-historical development (Ehrlich 1917, 79).

French philosopher François Gény agreed that the judicial function goes beyond performing logical deductions from the formal sources of law and that courts of necessity work with and influence the law's fundamental values. He advanced a quite broad notion of judicial discretion—"free decision on the basis of scientific investigation" (Gény 1917, 5). Gény maintained that the formal sources of positive law—statutes and custom—provide the most reliable authority and "best guide" for determining the law's meaning and resolving legal controversies (Gény 1917, 2). Where those formal sources are clear and unambiguous, courts must give them their logical import. But where the formal sources lack clarity or are incomplete, Gény argued that "free scientific investigation must supplement them and find the rule which governs the relations" at issue (Gény 1917, 21). This supplemental responsibility entrusts courts with an expansive, legislative-type discretion to conduct "free research" into the nature of law and the values underlying legal rules (Gény 1917, 6). It enjoins courts to search for the "objective factors to which all problems of positive law must look for their solution" (Gény 1917, 6).

While Gény by no means thought it appropriate for courts to fashion legal precepts at will, he did see judicial freedom of decision as more far-reaching than the logical application of existing written law to the facts of concrete cases. He reasoned: "Although the facts themselves will in part suggest the rule, yet its real origin must be sought in a superior order of ideas which can be discovered only by an effort of reasoning" (Gény 1917, 16). Gény conditioned this judicial inquiry into the 'superior order of ideas' on the limiting principle that judge-made legal precepts must "spring directly from the nature of things" (Gény 1917, 11). There, in the 'nature of things' he discerned two ultimate values or ideals: *justice* and *social utility*. These two ideals comprise, on Gény's account, the "objective factors" that together form the "ultimate standard" or "ends" toward which all adjudication should aim. He wrote:

> I believe that . . . every body of laws should tend toward realizing, in the life of humanity, on the one hand an ideal of justice, on the other an ideal of utility, meaning by the latter expression that which by common opinion is considered as promoting the welfare of the greater number. The two great guide-posts of our investigation should be, therefore, justice and general utility.
>
> (Gény 1917, 14)

Gény conceded that by themselves the ideals of justice and utility are "nothing but empty forms" (Gény 1917, 14). Yet he did not consider them impotent; nor did he see them as working in isolation. For he further posited three secondary principles: "(1) The principle of autonomous will; (2) the principle of public order or superior interest; (3) the principle of equilibrium of private interests" (Gény 1917, 21). To Gény, these secondary principles "animate [the] whole system of law" and give the 'empty' ideals of justice and utility substantive content (Gény 1917, 21). In combination, the two abstract ideals and the three animating principles comprise the essence of Gény's scientific method of free judicial research. In the final analysis, they reduce judicial decision to a single evaluative standard: a utilitarian balancing of interests.

Similar to the European free law movement, legal formalism came under attack starting in the late nineteenth century in the United States. Like the free law jurists, the American legal thinkers took a variety of approaches. As early as 1881 Oliver Wendell Holmes, Jr., famously mused, "The life of the law has not been logic: it has been experience" (Holmes 1881, 1). More than any other jurist, Holmes with this pithy aphorism laid bare the errant simplicity of legal formalism. Some years later he proposed a more complete judicial philosophy. His starting point was to assume that every valid legal precept is justified by the values or ends that underlie it. Judges accordingly are responsible for relating every rule of law they apply "accurately and definitely to an end which it subserves" (Holmes 1897, 457). Holmes admonished judges for failing to fully articulate and disclose those ends. For his part, analysis into the law's fundamental values resulted in his advancing a utilitarian-styled philosophy of adjudication similar to Gény.

That is, in resolving controversies at law, Holmes urged judges to concede the limits of logic, history, and precedent in favor of advancing the well-being of society as measured by economics and statistics. Every judge, he argued, should "seek an understanding of economics" (Holmes 1897, 474). For every case should turn ultimately on economically-determined "considerations of social advantage" (Holmes 1897, 467). Every rule of law offers social advantages and comes with social costs.

Judges must weigh the benefits and costs presented by competing rules for decision. Fulfilling that duty, Holmes maintained, necessarily transfers to judges a substantial lawmaking function. That function extends to scrutinizing legislation, as Holmes called upon judges "to consider and weigh the ends of legislation, the means of attaining them, and the cost" (Holmes 1897, 474). It also includes inquiring into constitutional meaning, even when measuring the social advantages of alternative interpretations allows judges to consider principles found outside the text of the instrument itself (see Holmes 1897, 467–8). For ultimately, on Holmes' account, the judicial function entrusts to judges substantial discretion over the law's fundamental values and ends, such that they ought to "consider more definitely and explicitly the social advantage on which the rule they lay down must be justified" (Holmes 1897, 468).

A very different and more blistering criticism of legal formalism came from the loosely affiliated American jurists known as the legal realists. Unlike Holmes, the legal realists did not propose a normative standard of judicial decision. They offered a wide ranging rebuff of legal formalism whereby they sought to discredit all attempts to restrain the judicial function. Most notable among the legal realists is Karl Llewellyn.

Llewellyn was an ecumenical legal critic. He remonstrated against not only legal formalism, but about nearly every jurisprudential tradition and most aspects of the American legal system. Few legal traditions or practices escaped his often cynical pen. To Llewellyn, the ascent of legal formalism in the United States was attributable to the influence business interests came to exert over the law. He argued that by the latter part of the nineteenth century "the Business Man [had taken] hold of the ideology of America" (Llewellyn 1934, 208). The captains of industry had successfully equated the national welfare with *laissez-faire* capitalism, appropriated for business ends "the best brains of the bar" (Llewellyn 1934, 209), and twisted the content and practice of law into a "bourgeois, business, and buccaneer ideology" (Llewellyn 1934, 211). It was in service of this ideology that, in Llewellyn's view, legal formalism emerged.

Llewellyn thus did not regard legal formalism as a principled effort to separate by function the branches of government or protect the values and legal norms of a culture from judicial manipulation. Rather, he saw it as a convenient stratagem to hold the law captive while the robber barons plundered. The "business captains needed a stable footing in the law," Llewellyn reasoned, a jurisprudence that could provide "for clarity and certainty, for a firm foundation . . . for a *solid* something" that would render the law "sufficiently straitjacketed in out-moded moulds not to catch up too fast with novel predatory practices" (Llewellyn 1934, 208, 209). Legal formalism accommodated. Its "mechanized" veneer gave the appearance of principled restraint and judicial fidelity to the law's fundamental values (Llewellyn 1934, 209). Yet the reality was that its model of judicial decision amounted to nothing more than "the judges' notions of the way things should be" (Llewellyn 1934, 209). That is to say, in Llewellyn's view, judicial decision-making had become tilted injudiciously toward favoring the shared interests of business, bench, and bar, such that the true logic of formalist adjudication was that "decisions which we *like* are 'sound,' and therefore precedent. But decisions which we do *not* like are 'unsound,' and therefore to be disregarded" (Llewellyn 1934).

Llewellyn responded to this troubling characterization of the administration of law in the United States with a systematic program of reform. Aspiring to craft "an empirical *science* of law" (Llewellyn 1930, 444), he urged that law be studied according to its effects, that it be conceived of as a means toward achieving social ends, and that judicial decision-making be assessed consistently at the descriptive level (see Llewellyn 1931, 1236, 1246–50, 1255, Llewellyn 1934, 212, 447–8, 454, 209). His achievements are best measured, like those of the other legal realists, in terms of how he raised new questions and influenced the critical study of law and legal institutions. He brought into question the stability and certainty of the law by characterizing it as inherently a process of movement, change, and adaptation. He expressed distrust and skepticism toward traditional legal concepts and rules as fair descriptions of how courts actually decide cases. For there is a disjunction, he argued, between "paper rules" ("what the books . . . say 'the law' is") and "real rules" ("the practices of the courts") (Llewellyn 1930, 448); and accordingly, there is need for "the careful severance of rules *for* doing (precepts) from rules *of* doing (practices)" (Llewellyn 1931, 1237). Understood as a practice, judicial decision-making, he maintained, cannot be reduced to a formal process of logical deduction. For value judgments unavoidably become intertwined with factual findings in judicial deliberation. Exposing that entanglement became one of Llewellyn's principal goals. Its pursuit helped him effect what he considered legal realism's "most fascinating result:" the disambiguation of 'is' and 'ought' in legal discourse (Llewellyn 1931, 1252).

As his critique of legal formalism and the American legal system makes clear, Llewellyn found facts and values—'is' and 'ought'—unavoidably intertwined in law and judicial decision. Part of the evil he perceived in legal formalism was its theoretical pretense that judicial decision-making is logical, objective, and value-neutral, while in practice it abetted judicial practices that truckled to the interests of the propertied and powerful. Llewellyn confronted this coarse reality of American law by advocating the "*temporary* divorce of Is and Ought for purposes of study" (Llewellyn 1931, 1236). While value judgments abound in judicial decision-making, he argued that "during the study of what courts are doing . . . [it is critical] to disregard the question what they ought to do" (Llewellyn 1931, 1236). That is, Llewellyn was convinced that improving the judicial administration of law and making sound recommendations as to how law *ought* to be requires first "knowing objectively, as far as possible, what that part of law is now doing" (Llewellyn 1931, 1237). Hence the need to temporarily "disentangle Is and Ought" (Llewellyn 1931, 1252).

Llewellyn thought that the temporary decoupling of fact and value in the study of judicial practice would accomplish two ends. First, it would expose the hidden or unacknowledged values that surreptitiously were guiding judicial decisions. For his part, Llewellyn, an expert in the law of contracts and commercial law, put forth positions in economic and social philosophy that were at odds with many of the accepted positions of business and the bar. Like other realists, he sought to reform the law by replacing the presupposed values of the dominant economic order with the alternative values he championed (see Llewellyn 1931, 1237).

Second, Llewellyn was convinced that by revealing the extent to which facts and values are unavoidably intertwined in judicial decision, decoupling them would validate their entanglement. Once the entanglement was legitimized, courts could embrace the normative side of their practice and engage in value judgments in open court, rather than furtively. To Llewellyn, this would vitalize adjudicative practice by underscoring "the need for courts to face squarely the policy questions in their cases" (Llewellyn 1931, 1254) and by "demonstrat[ing] that the field of free play for Ought in appellate courts is vastly wider than traditional Ought-bound thinking ever had made clear" (Llewellyn 1931, 1252). Indeed, on Llewellyn's account, the field of free play for the normative in adjudication even extended to courts' factual findings. Contrary to the established view that the ascertaining of facts in a case at law involves value-free, objective inquiry, Llewellyn argued that "the classification of raw facts is largely an arbitrary process, raw facts having in most doubtful cases the possibility of ready classification along a number of lines" (Llewellyn 1931, 1253). Moreover, he also included in that field of free normative play the hallowed common law practice of following precedent. Without besmirching that practice, Llewellyn argued that courts enjoy vast discretion in the interpretation of precedent. It is a discretion, he claimed, that cannot be reduced to logic or definitional clarity, for "only policy considerations and the facing of policy considerations can justify 'interpreting' (making, shaping, drawing conclusions from) the relevant body of precedent in one way or in another" (Llewellyn 1931).

3

Karl Llewellyn and others in the legal realist and free law movements thus significantly eroded the perceived separation of fact and value in law and judicial decision. They showed that the efforts, dating back to the Roman Republic, to remove value judgments from judicial practice through precisely written codes, juristic authority, and formalistic procedures were largely unavailing. Still, the drive to restrain judges from making policy determinations or exercising authority over the law's substantive values or content remains robust to this day. 'Judicial restraint' sounds a rally-cry of remarkably resilient appeal. In the words of one recent legal writer: "judicial restraint is the quintessential rule by which judicial bodies throughout the world should govern themselves" (Mendelsohn 2010, 57). The refrain of restraint, in fact, seems only to have grown louder since the time when the free law and legal realist jurists sought to squelch it. Not uncommonly, courts in the United States especially receive rebuke under the charge that they "have tended to run amok, have become makers rather than simply interpreters of the law, and have come to exercise a degree of authority . . . that is unparalleled" in Western legal history (Wood 1997, 49).

The fact/value dichotomy thus retains substantial influence in law. It continues to strike a responsive chord for many as the operative principle of judicial legitimacy. Why did the efforts of the free law and legal realist jurists to discredit it apparently achieve so little? Part of the reason is likely that the dichotomy readily aligns with the ideal of the modern nation-state as a government of separated powers. That governmental ideal includes a legislative branch as well as a judicial branch. The neatest demarcation of functional powers is to lodge all lawmaking power in the legislative branch and to define the judicial function so as not to overlap or encroach upon the legislative. The fact/value dichotomy matches that ideal.

Further, the critiques of legal formalism and the fact/value divide offered by jurists such as Gény and Holmes may well have been compromised by self-inflicted wounds. Those two both began their analyses of the formalist method by describing adjudication and showing it to be a more vital practice than formalism allowed. But they then converted their analyses to the advocacy of consequentialist normative theories. This drew attention away from their highly profitable critiques of formalism. And it likely diminished the influence of those critiques since their respective utilitarian-styled jurisprudences long ago fell into disrepute.

A final reason why the free law and realist jurists fell short in dispelling the compartmentalized view of law is that, to the extent they did not move on to normative jurisprudence, their work was firmly limited to the empirical. Llewellyn, for example, set out to create an empirical science of law. He fully committed to the empirical method, insisting that he aimed to "*consistently* keep description on the descriptive level" (Llewellyn 1931, 1250). This commitment to an empirical study of what courts *do in practice* is highly worthy and produced a good number of valuable insights into adjudication. It produced ample evidence that in judicial practice facts and values intertwine. Yet it failed to provide any grounding for that entanglement. That is, while the realist and free law empirical studies are quite persuasive in showing that there is no fact/value dichotomy in judicial decision-making, those studies do not compel the conclusion that the practice of adjudication could not be otherwise.

Hence, they do not discourage proponents of judicial restraint from arguing that courts exceed their proper function when they make value and policy judgments. Just because courts do, as a matter of empirical fact, entangle facts and values does not mean that they should do so or must continue along that path.

This suggests that something more is needed to diminish the influence of the fact/value dichotomy in the context of law and judicial decision-making. Empirical evidence of fact/value entanglement as produced by the free law and legal realist jurists is necessary, but insufficient. I argue in the remainder of this chapter for augmenting that evidence with inquiry into the nature of legal propositions. For legal precepts—the propositional rules, standards, doctrines, principles, and maxims used by courts to resolve legal controversies—express, by their nature, hybrid propositions: part fact assertions, part value predicates. They are best understood as composite concepts—*fact/value complexes*. Understanding legal precepts as fact/value complexes provides a theoretical base of support for the empirical evidence that adjudication interconnects facts and values. It is a base, I suggest, that shows that the interconnection is necessary and that courts could not act otherwise.

4

I owe the term 'fact/value complex' to Julius Stone. In the 1960s Stone introduced it to refer to certain legal precepts, such as the constitutional protection of due process and the negligence standard of reasonable foreseeability, which require courts to combine factual inquiry and normative evaluation (Stone 1966, 737–8 and 1968, 264). Stone gave the concept of fact/value complexes his most extensive treatment in a 1985 law journal article on the reasonable foreseeability standard. That standard, a hallmark of modern negligence law, stipulates that a defendant's duty of care extends to those damages or injuries that a reasonable person, under the circumstances, would have foreseen at the time the action complained of took place. Stone argued that this standard presupposes that every judicial "finding of 'reasonable foreseeability' (or the contrary) entails value- (or 'policy-' or 'justice-') judgments" (Stone 1985, 542), for "a court correctly applying such a precept is necessarily involved not only in a finding of facts, but in a preliminary evaluation of the facts, involving *inter alia* the contexts and the socio-ethical merits of the parties' positions, before it can give legal judgment" (Stone 1985, 539). In this sense, the principle of reasonable foreseeability models what Stone meant by a fact/value complex. For in determining whether a defendant acted negligently, it calls upon a court to not merely conduct a thorough inquiry into the facts, but to conceive of them as "facts-as-evaluated-by-the-normative-standard-of-reasonableness" (Stone 1985, 564). In this, the standard "permits (and indeed requires) the court to take account of a very wide, if not unlimited, range of policy considerations in determining what damage the defendant *ought* reasonably to have foreseen as likely to flow from his careless acts or omissions" (Stone 1985, 564). The standard of reasonable foreseeability, that is, necessarily intertwines inquiry into facts and considerations of policy and value.

While Stone celebrated the indeterminacy and the "leeways of choice" that fact/value complexes like reasonable foreseeability offer courts (Stone 1985, 565), he cautioned that not all legal precepts accord such decisional freedom. Only those legal precepts that take the form of 'principles' comprise fact/value complexes. Principles, he argued, are framed in general terms and tend "to have a range of built-in flexibility" which renders them readily adaptable to new and changing circumstances (Stone 1985, 541). This generality and inherent flexibility forges their judicial application toward "some predication of facts, but also *requiring moral evaluation of these facts by the decision-maker*" (Stone 1985, 539). Legal 'rules,' Stone maintained, are different. 'Rules' are set forth in greater specificity than 'principles' and lend themselves to more determinate interpretation and application. A rule, that is, is a legal "precept predicated on a detailed set of facts" (Stone 1985, 549). It is "a precept detailed enough to be applied in each particular case without resort to normative evaluations such as 'justice' or 'reasonableness' " (Stone 1985, 542).

Stone used the Australian High Court case *Hacksaw v. Shaw* (1984) to illustrate the difference between rules and principles. The issue in *Hacksaw* was whether a landowner was liable in tort for injuries he inflicted on a trespasser. Shaw owned a farm. He stored gasoline there for running his farm machinery. Since thieves were regularly pilfering the gas at night, Shaw decided to lay in ambush one night to confront any prowlers. A car drove onto his farm with headlights off. Hacksaw was a passenger. The driver of the car alighted and began to siphon gas. Shaw fired two warning shots from a rifle into the car, unintentionally striking Hacksaw in the front seat.

Hacksaw subsequently brought suit seeking compensation for the injuries she sustained. The case required a judicial choice between two competing standards of duty for determining negligence. On one side stood a special common law rule that would impose on Shaw, as an occupier of land, only a limited duty to Hacksaw, as a trespasser, not to set a trap or inflict intentional harm. On the other side was the general negligence duty of care that would hold Shaw responsible for firing his rifle into the car if he reasonably should have foreseen that there could be a passenger inside. The Australian High Court chose the latter. Their choice, according to Stone, highlights the difference between a principle founded on a "normative fact-value complex"—here the reasonable foreseeability standard—and a rule, calling only for factual inquiry (Stone 1985, 556 n.66). Had the High Court chosen to follow the rule of limited duty to trespassers, the fact that Hacksaw was a trespasser would alone have been decisive. Application of the rule in light of that single fact would have been mechanical and inflexible. But in choosing the standard of reasonable foreseeability, the Court broadened its inquiry to include not only the factual circumstances of the trespass, but also the normative question of what Shaw ought reasonably to have foreseen. The fact/value complex of reasonable foreseeability, that is, called for both a richer factual inquiry into the circumstances of the trespass and for an evaluative assessment of reasonable foreseeability (Stone 1985, 555–7).

Stone's notion of fact/value complexes has the potential of revealing the unavoidable interconnection between facts and values in adjudication. It calls to mind Charles Stevenson's claim that certain ethical judgments possess a "hybrid character" (Stevenson 1963, 221). Stevenson observed that specific judgments such as 'he is courageous' contain both a factual component and an evaluative component. The judgment roughly translates to "he habitually makes a stand against danger in a manner that is admirable" (Stevenson 1963, 221). The subject component of this judgment offers a partial factual description of an agent. It carries no evaluative force. But the predicate does. 'Admirable' expresses evaluative approval. Hence, to state 'he is courageous' reports certain facts about an agent that necessarily implicate evaluative considerations. The concept 'courageous' inextricably fuses these factual and evaluative components (see Stevenson 1963, 221).

While I borrow the term 'fact/value complex' from Julius Stone, I use it differently. For as employed by Stone, the notion falls short of providing a conclusive account of how legal precepts interconnect facts and values. It falls short in three respects, being at once too narrow and too broad, and in not fully appreciating how inextricably facts and values are melded in these complex propositions. In addressing these points, I make three claims about the nature of fact/value complexes: (1) the extension of the class fact/value complexes is vast and not limited to legal principles stated in highly general terms; (2) the range of discretion judges have in applying fact/value complexes is not broad and unlimited, but a conservative freedom to respect a legal precept's evaluative component within the system of ordered discourse that is judicial practice; and (3) fact/value complexes do not merely entangle, but *fuse* their factual and evaluative components.

(1) The extension of the class fact/value complexes is vast and not limited to legal principles stated in highly general terms.

Stone's characterization of fact/value complexes is too narrow in that he defines it as extending only to those legal precepts that are 'principles' as distinct from 'rules.' The first problem with this characterization is that there is no clear distinction in law between rules and principles. Stone articulates the difference he sees by stating that principles (1) are framed in general terms, (2) possess built-in flexibility that accommodates adaptation to new factual circumstances, and (3) necessarily require moral evaluation of the facts. By contrast, he characterizes rules as (1) being stated in detailed specificity, (2) requiring determinate and relatively fixed interpretation and application, and (3) not requiring normative evaluation (Stone 1985, 539, 541–2, 549). Stone's discussion of *Hacksaw v. Shaw* was meant to illustrate these points of contrast. The trouble, however, is that these distinctions are not borne out in judicial practice.

Stone's two paradigm principles are the negligence standard of reasonable foreseeability and the constitutional guarantee of due process of law. Contrary to his position, courts do not heed Stone's distinction between principles and rules in referring to those two precepts. Case law variously refers to the reasonable foreseeability standard sometimes as a principle (*Sorrells* 1993, 322, *Krasnecky* 2002, 1289, *Bertram* 2006, 225), other times as a rule (*Clohessy* 1996, 56, *Atherton* 1979, 637, *Swartz* 2002, 1098). Ambiguity surrounds due process as well. Every application of the constitutional due process guarantee requires first identifying the precise due process right at issue, for that determines the applicable test. Courts often refer to the germane test as a principle (*Rogers* 1960, 541; *Zaber* 2010, 640; *Hope Clinic for Women, Ltd.* 2013, 49), yet not uncommonly as a rule (*State Plant Bd.* 1959, 408; *Grine* 2004, 27) or standard (*Hennigan* 1987, 364, *Chen* 2004, 25). In one recent decision, the United States Supreme Court referred to the due process right at issue as implicating a *principle*, a *rule*, a *framework*, a *standard*, and a *maxim* over the course of two short paragraphs, seemingly treating the five terms as interchangeable (Williams 2016, 5–6). In judicial practice, there is just not the clear demarcation that Stone suggests between a legal rule and a legal principle. Hence, his attempt to define 'fact/value complex' as extending to principles but not rules rests on a false disjunction.

Further, and more importantly, even if there were a clear line separating legal principles from rules, the integration of facts and values in fact/value complexes occurs in legal precepts of the latter category as well as the former. In fact, it is unclear whether there are any legal precepts that are not fact/value complexes. There certainly are some legal precepts that are framed in more general terms than others. Provisions guaranteeing individual rights such as due process of law or the "right to respect for private and family life" guaranteed under Article 8 of the European Convention on Human Rights are notoriously 'open-ended' (Council of Europe 1988, art. 8). Because of the general terms they employ, these rights provisions quite understandably call for normative evaluation in their application. Yet greater or lesser degree of generality in the phrasing of legal precepts is not a reliable measure for identifying some as fact/value complexes and others not. Nor is a precept's inherent flexibility a valid criterion.

Consider the common law rule against perpetuities. The rule stipulates that a future interest in real property must vest within a certain number of years—21 years following the death of the last 'life-in-being' at the time the interest was created. This rule would seem, on Stone's account, to lie far afield from the realm of fact/value complexes. It is termed a 'rule'; it is framed in specific terms; it is infamously inflexible; and as a rule whose application requires intense factual scrutiny of property transactions and the counting of years, it would appear unaffected by normative considerations. Yet the rule's infamy lies in the judicial tumult it has spawned regarding the social policies and values it puts in conflict.

The English House of Lords created the rule against perpetuities in 1682. Henry Howard, 22nd Earl of Arundel, had left a plan of testamentary inheritance that included future interests for his as yet unborn (and unplanned) grandchildren. Their Honors invalidated the plan because it offended a vital public policy favoring the marketability of real property (see Duke of Norfolk's Case 1682). The rule against perpetuities has proven very difficult to apply. Part of the difficulty comes from its factual complexity. The rule literally requires measuring lives and calculating possible contingencies. It is violated if there is *any possibility* that a future interest will not vest within the permitted period of years (Gray 2002, 191). The trouble the rule fosters also comes in part from a struggle over its extension. The rule arose from a singular concern about intergenerational family gift transactions. The English courts, intent on limiting restraints on the alienation of real property, sought to curtail landowners' efforts to control their land in perpetuity. In the late nineteenth century, however, courts in England and the United States started applying it to commercial transactions. This extension has been roundly criticized and has proven to be a source of judicial headache (see *Pathmark Stores, Inc.* 1995, 1193, *Atlantic Richfield Co.* 2014, 1185–9).

Yet the greatest source of difficulty in applying the rule against perpetuities comes from the public policies and principles of social morality it places in juxtaposition. The rule's underlying purpose is well-understood—"to avoid fettering real property with future interests dependent upon contingencies unduly remote . . . , thus working an indirect restraint upon alienation, which is regarded at common law as a public evil" (*Weber* 1936, 808). This purpose, however, often comes in conflict with the high value the law places on interpreting wills and trusts so as to honor and not thwart decedents' donative intent (Gray 2002, 601). It also frequently threatens to derail otherwise economically and socially beneficial commercial transactions and work against the general policy that the law disfavors the destruction of contracts due to uncertainty (see *Malad, Inc.* 2008, 626; *Coulter & Smith, Ltd.* 1996, 1262). As a result, courts struggle to apply the rule in ways that avoid anomalous results. Some courts sidestep the conflicts by stating that they are "not permitted to relax the rule against perpetuities" and are duty-bound to recognize that it "contains no exceptions" (*Haggerty* 1958, 420). Others convulse at the rule's often harsh consequences and at how obdurately it can crush competing values. Hence not uncommonly, courts shun applying it in a "rigid or remorseless manner" (*Wong* 1963, 534). Many interpret the factual scenarios and instruments at issue before them so as to avoid finding the rule violated, if at all possible (see *Plummer* 1926, 655–61; *Gore* 1994, 430; Gray 2002, 601). In recent years, courts in some jurisdictions have begun mollifying the rule by focusing on the *reasonableness* of a contingent future interest, rather than the temporal *remoteness* of its vesting (see *Mattern* 1963, 320, *Atlantic Richfield Co.* 2014, 1187). These judicial struggles do not exemplify judicial overreaching. They do not amount to illicit judicial tampering in the realm of policies and values. Rather, they manifest the nature of a fact/value complex: that judicial application of a legal precept to the facts of an individual case unavoidably obligates the court to engage with the values and norms that the precept puts in play.

This melding of facts and values in judicial decision-making exhibited by the rule against perpetuities occurs in every branch of law. This includes 'rules' in the law of negligence, such as the rule of limited liability to trespassers that the High Court of Australia bypassed in *Hacksaw v. Shaw*. That rule rests just as fully on considerations of social policy as the reasonable foreseeability standard that the Court followed and that Stone highlights as a paradigm fact/value complex. While I hesitate to make the universal claim that *every* legal precept is a fact/value complex, I am at a loss to think of any that are not. The early pragmatist Edgar A. Singer argued, in the context of science, that "no categorical account of Nature, *i.e.*, no image of Nature 'in the concrete,' can be given which does not embody a series of [social value] choices" (Singer 1924, 204). This holds equally true in law. For laws do not come into being in a value-free context. Law is an instrument of social control. It aims to facilitate cooperative interaction, provide for fair distribution of social benefits and burdens, and establish means for resolving conflict and ironing out disagreement without resort to force. One of the central means used to achieve these goals is the enactment of laws, every one of which aims to implement some immediate social policy as it reflects, indirectly and by (often unconscious) implication, a series of value presuppositions. To engage judicially with facts in law thus is to engage with the values, norms, and policies that those facts embody and which, in turn, make them *legally relevant facts*.

Consistent with Charles Stevenson's suggestion regarding the realm of ethics, the expression of such fact/value engagements takes the form of "hybrid" propositions: fact/value complexes. What characterizes a legal precept as a fact/value complex thus is not its generality or its flexibility, but its hybrid propositional form. Fact/value complexes are part fact assertion, part value predicate. The factual component of a fact/value complex is only legally relevant because of its association with the evaluative component. That factual component cannot be contemplated by a court without taking the evaluative predicate into account as well. The dream dating back to ancient Rome to strictly confine the judicial function to engaging with facts without touching the content of law or its values therefore defies fulfillment.

(2) The range of discretion judges have in applying fact/value complexes is not broad and unlimited, but a conservative freedom to respect a legal precept's evaluative component within the system of ordered discourse that is judicial practice.

I have argued that courts cannot, as a practical matter, engage with legally relevant facts without also engaging with the values and policy-infused content of the law that make those facts relevant. From this it does not follow, however, that judges enjoy broad, unlimited discretion over choosing which policies to embrace or which values to invoke.

This takes us to the second sense in which I break from Stone's account of fact/value complexes. His account is overly broad. Speaking of the reasonable foreseeability negligence standard, Stone wrote, "All relevant policy factors . . . can and should, in the present view, be dealt with in . . . determinations of reasonable foreseeability. The fact-value complex of 'reasonable foreseeability' is sufficiently indeterminate to accommodate all these factors" (Stone 1985, 552). The question is what policy factors are relevant. Stone answered that question by suggesting that the reasonable foreseeability standard "permits (and indeed requires) the court to take account of a very wide, if not unlimited, range of policy considerations in determining what damage the defendant *ought* reasonably to have foreseen as likely to flow from his careless acts or omissions" (Stone 1985, 564).

This characterization of the range of discretion judges enjoy in applying fact/value complexes is overbroad. Judges may not tramp at will through the foggy mists of values and norms. For judicial decision-making is a craft-bound practice wherein facts and values interconnect in a system of ordered discourse. That system both entrusts judges with substantial freedom of decision while confining their exercise of that freedom within certain practice-oriented conditions of excellence. Judges enjoy a degree of deliberative freedom not unlike that advanced by some in the free law and legal realist movements. Adjudication is, as Eugen Ehrlich argued, a "dynamic" practice that requires judges to view the facts in contested cases as "coefficients" of the value-preferences and social tendencies reflected in legal precepts (Ehrlich 1917, 65). Per François Gény, judicial practice obligates courts to "supplement" the formal sources of written law to the extent those sources are incomplete or ambiguous (Gény 1917, 21). This supplemental responsibility, as Karl Llewellyn so aptly described, extends to judges a wide "field of free play for Ought" wherein they cannot avoid confronting, without shirking their duty, the policy considerations and value judgments that the law embodies (Llewellyn 1931, 1252).

Yet while the judicial 'field of free play for Ought' is quite wide, it is not unlimited. Its width is bounded by the functional values and conditions of excellence that underlie judicial practice. As the American jurist Benjamin Cardozo once remarked, "The judge, even when he is free, is still not wholly free. He is not to innovate at pleasure" (Cardozo 1921, 141). Judicial freedom of decision—the freedom to innovate and to exercise creative choice over matters of value and policy, and even as to findings of fact—extends only to filling gaps or resolving indeterminacies in the formal sources of law. The manifest intent of those sources, the meaning of legal precepts insofar as settled and unambiguous, and the stated policies and values aligned with them demand judicial respect. Far from enjoying, as Stone would have it, a "wide, if not unlimited" discretion over matters of policy and value (Stone 1985, 564), the freedom of judges to venture beyond the values embodied in the formal sources is highly circumscribed. Judges certainly can contemplate and engage with the policies and values contained in the evaluative component of a fact/value complex. Indeed, they *must* do so. For otherwise they are not giving the composite fact/value precept full consideration. But the values they engage must be *found* values. Judges cannot import their own values; nor can they engage with social values other than those the precept itself embodies. For as stated by Cardozo, no judge is "a knight-errant roaming at will in pursuit of his own ideal of beauty or of goodness" (Cardozo 1921, 141).

(3) *Fact/value complexes do not merely entangle, but fuse their factual and evaluative components.*

In closing I want to briefly raise one final point of departure between my sense of fact/value complexes and Stone's. The enduring effort to compartmentalize law so as to restrain judicial freedom rests on a false notion of legal facts. Those who strive to separate fact and value in adjudication have always held a view of legal facts reminiscent of the logical positivist conception of 'fact.' The logical positivists conceived of facts as things that could be reported or confirmed in brute or raw form by mere observation, generally through sensory experience. As Hilary Putnam once noted, this conception of 'fact' allowed them to draw "a sharp and clear distinction between value terms and descriptive terms" (Putnam 2002, 24). Those who try to mark a sharp boundary between facts and values in law and judicial decision share this sense of fact.

In law, however, there are no raw facts. Indeed, I question whether facts discovered or discerned in any central field of cognitive inquiry are genuinely raw or brute. Here we can again benefit from the insights

Edgar A. Singer offered nearly a century ago as to scientific inquiry. Singer argued that in the context of doing science, "the individual [is] faced with a group of facts; but not of bare facts, for insofar as these have even enough meaning to be pointed out as facts they bear the traces of description with all that this implies of past choices" (Singer 1924, 221–2). Similarly, any facts that hold enough meaning to be treated as *adjudicative facts* bear the traces of legal relevance with all that this implies of prior social value and policy choices. No adjudicative fact, that is, can be raw, for it is only by being found to be legally relevant by a court acting within the ordered discourse practice of adjudication that a fact comes to carry legal meaning. Legal facts thus are two steps removed from bruteness: substantively, by being connected with some legal precept and the values or policy that it embodies; and procedurally, through a judicial finding affirming that connection.

Since every fact/value complex includes a factual assertion, and since, as has just been shown, all factual assertions in law make claims about facts that are legally relevant, the factual component of every fact/value complex is necessarily interconnected with the value component. Because of this, Stone's account of fact/value complexes requires a slight amendment. Stone correctly emphasized that fact/value complexes *require* courts to make value judgments (Stone 1985, 542, 551). But he implied that those required evaluative judgments were in some sense separate from the court's factual findings. Hence, his notion of fact/value complexes implies an *entanglement* of fact and value. That is not strong enough.

'Entanglement' does not adequately reflect the strength of the interconnection between facts and values in fact/value complexes. For the verb 'entangle' suggests a weaving together of things initially separate and at least logically possible of receiving separate treatment in the future. Fact/value complexes are hybrid propositions of a more robust order. Consider again Charles Stevenson's example of the moral judgment, "He is courageous." This hybrid ethical proposition contains two components, one factual, the other evaluative. But it asserts a single judgment. It is not two entangled judgments of the sort Stone suggests for legal fact/value complexes.

Following Stevenson, I argue that fact/value complexes be understood as judgments that *fuse* their fact and value components. They are single judgments involving the application of a legal precept to the operative legally relevant facts of a concrete case. As such, fact/value complexes exhibit a truly hybrid propositional form: neither reducible to fact nor value, but an inextricably composite fact/value judgment. For as Singer argued, to say that "every judgment capable of truth or error *involves* a question of fact" does not mean "that it *reduces* to a question of fact" (Singer 1924, 212). To this I would add that because all judicial application of facts in the context of fact/value complexes involves a question of value or policy does not reduce adjudication to a value judgment. But it does reveal that the practice of judicial decision-making unavoidably requires judges to engage with the substantive content and underlying values and policies embodied in legal precepts. Further, it validates as logically necessary the empirical descriptions of judicial practice presented so accurately by the free law and legal realist jurists. And finally, it shows that the quixotic efforts dating from ancient Rome to compartmentalize the law and make its content and values judge-proof has met with faint success for good reason. For to think that judges can contemplate and apply legal precepts without engaging both components of a fact/value complex is to tilt not just at windmills but against the wind.

References

Abbot, Everett V. 1897. "Keener on Quasi-Contracts II", *Harvard Law Review*, 10: 479–512.

Allgemeines Landrecht für die Preussischen Staaten. 1794.

Atherton v. Devine, 602 P.2d 634 (Okla. 1979).

Atlantic Richfield Co. v. Whiting Oil and Gas Corp., 320 P.3d 1179 (Colo. 2014).

Beccaria, Cesare. [1764] 1963. *On Crimes and Punishments*. Translated by Henry Paolucci. Indianapolis: Bobbs-Merrill.

Bentham, Jeremy. [c. 1774] 1977. "A Comment on the Commentaries." In *The Collected Works*, edited by J. H. Burns and H. L. A. Hart. Oxford: Clarendon Press.

Bentham, Jeremy. [1823] 1977. "A Fragment on Government." In *The Collected Works*, edited by J.H. Burns and H.L.A. Hart. Oxford: Clarendon Press.

Bertram v. Malheur County, 129 P.3d 222 (Or. App. 2006).

Blackstone, William. [1765] 1979. *Commentaries on the Laws of England*. Vol. 1. Chicago: University of Chicago Press.

Campbell, James V. 1880. "Materials of Jurisprudence," *American Law Review* 14: 257–88.

Cardozo, Benjamin. 1921. *The Nature of the Judicial Process*. New Haven, CT: Yale University Press.

Chen v. Stewart, 2004 UT 82, 100 P.3d 1177 (Utah 2004).

Cicero. 1928. *De Legibus*. Translated by C.W. Keyes. Cambridge, MA: Harvard University Press.

Clohessy v. Bachelor, 237 Conn. 31, 675 A.2d 852 (1996).

Coulter & Smith, Ltd. v. Russell, 925 P.2d 1258 (Utah App. 1996).

Council of Europe. 1988. "Protocol to the Convention for the Protection of Human Rights and Fundamental Freedoms (European Convention on Human Rights) as Amended by Protocol No. 11." *Council of Europe Treaty Series 155*. Strasbourg: Council of Europe.

Duke of Norfolk's Case (1682) 3 Ch. Cas. 1, 22 Eng. Rep. 931.

Ehrlich, Eugen. 1917. "Judicial Freedom of Decision: Its Principles and Objects." In *Science of Legal Method: Selected Essays by Various Authors*, translated by Ernest Bruncken and Layton B. Register, 47–84. Boston: Boston Book Co.

Finnis, John. 1980. *Natural Law and Natural Rights*. Oxford: Clarendon Press.

French Civil Code. 1804.

G. D. A. P. 1861. "The Doctrine of Negligence", *American Law Register*, 9: 129–44.

Gény, François. 1917. "Judicial Freedom of Decision: Its Necessity and Method." In *Science of Legal Method: Selected Essays by Various Authors*, translated by Ernest Bruncken and Layton B. Register, 1–46. Boston: Boston Book Co.

Gmelin, Johann Georg. 1917. "Dialecticism and Technicality: The Need of Sociological Method." In *Science of Legal Method: Selected Essays by Various Authors*, translated by Ernest Bruncken and Layton B. Register, 85–145. Boston: Boston Book Co.

Gore v. Beren, 254 Kan. 418, 867 P.2d 330 (1994).

Gray, John Chipman. [1942] 2002. *The Rule against Perpetuities* (Fourth edition). Edited by Roland Gray. Washington, D.C.: Beard Group Inc.

Grine v. Sylvania Schools Bd. of Educ., 2004 Ohio 6904 (Ohio App. 2004).

Hacksaw v. Shaw [1984] 155 C.L.R. 614.

Haggerty v. City of Oakland, 161 Cal.App.2d 407, 326 P.2d 957 (1958).

Hale, Sir Matthew. [1713] 1971. *The History of the Common Law* (Third edition). Edited by Charles M. Gray. Chicago: University of Chicago Press.

Heineccius, Johann Gottlieb. [1741] 2008. *A Methodological System of Universal Law*. Translated by George Turnbull. Edited by Thomas Ahnert and Peter Schröder. Indianapolis: Liberty Fund.

Hennigan v. State, 746 P.2d 360 (Wyo. 1987).

Holmes, Oliver Wendell, Jr. 1881. *The Common Law*. Boston: Little, Brown and Co.

Holmes, Oliver Wendell, Jr. 1897. "The Path of the Law,", *Harvard Law Review*, 10: 457–78.

Hope Clinic for Women, Ltd. v. Flores, 2013 IL 112673, 991 N.E.2d 745 (Ill. 2013).

Kelsen, Hans. 1991. *General Theory of Norms*. Translated by Michael Hartney. Oxford: Oxford University Press.

Kelsen, Hans. [1934] 1992. *Introduction to the Problems of Legal Theory*. Translated by Bonnie Litschewski Paulson and Stanley L. Paulson. Oxford: Clarendon Press.

Krasnecky v. Meffen, 777 N.E.2d 1286 (Mass. App. 2002).

Llewellyn, Karl N. 1930. "A Realistic Jurisprudence—The Next Step", *Columbia Law Review*, 30: 431–65.

Llewellyn, Karl N. 1931. "Some Realism about Realism—Responding to Dean Pound", *Harvard Law Review*, 44: 1222–64.

Llewellyn, Karl N. 1934. "On Philosophy in American Law", *University of Pennsylvania Law Review*, 82: 205–12.

Malad, Inc. v. Miller, 199 P.3d 623 (Ariz. App. 2008).

Mattern v. Herzog, 367 S.W.2d 312 (Tex. 1963).

Mendelsohn, Allan I. 2010. "Judicial Restraint in International Law", *Federal Lawyer*, 57: 52–55.

Merryman, John Henry. 1985. *The Civil Law Tradition: An Introduction to the Legal Systems of Western Europe and Latin America*. Stanford, CA: Stanford University Press.

Montesquieu, Baron de. [1748] 1949. *The Spirit of the Laws*. Translated by Thomas Nugent. New York: Hafner Press.

Pathmark Stores, Inc. v. 3821 Associates, L.P., 663 A.2d 1189 (Del. Ch. 1995).

Plummer v. Roberts, 315 Mo. 627, 287 S.W. 316 (1926).
Putnam, Hilary. 2002. *The Collapse of the Fact/Value Dichotomy and Other Essays*. Cambridge, MA: Harvard University Press.
Robinson, O. F., Fergus, T. D., and Gordon, W. M. 1985. *An Introduction to European Legal History*. Abingdon, UK: Professional Books Ltd.
Rogers v. Richmond, 365 U.S. 534 (1960).
Singer, Edgar Arthur, Jr. 1924. *Mind as Behavior and Studies in Empirical Idealism*. Columbus, OH: R.G. Adams & Co.
Sorrells v. M.Y.B. Hospitality Ventures, 435 S.E.2d 320, 334 N.C. 669 (1993).
State Plant Bd. v. Smith, 110 So.2d 401 (Fla. 1959).
Stein, Peter. 1999. *Roman Law in European History*. Cambridge: Cambridge University Press.
Stevenson, Charles. 1963. *Facts and Values: Studies in Ethical Analysis*. New Haven, CT: Yale University Press.
Stone, Julius. 1965. *Human Law and Human Justice*. London: Stevens & Sons Ltd.
Stone, Julius. 1966. *Social Dimensions of Law and Justice*. Stanford, CA: Stanford University Press.
Stone, Julius. 1968. *Legal System and Lawyers' Reasonings*. Sydney: Maitland Publications Pty Ltd.
Stone, Julius. 1985. "The Way of a Judge with a Principle—A Tribute to Professor Chaim Perelman," *Northern Kentucky Law Review*, 12: 537–66.
Swartz v. McNabb, 830 So.2d 1093 (La. App. 2002).
Thomasius, Christian. [1688] 2011. *Institutes of Divine Jurisprudence*. Edited by Thomas Ahnert. Indianapolis: Liberty Fund.
Weber v. Texas Co., 83 F.2d 807 (5th Cir. 1936).
Williams v. Pennsylvania, 579 U.S. ___ (2016), slip opinion.
Wong v. DiGrazia, 60 Cal.2d 525, 386 P.2d 817 (1963).
Wood, Gordon S. 1997. "Comment." In *A Matter of Interpretation: Federal Courts and the Law*. Edited by Antonin Scalia. Princeton, NJ: Princeton University Press.
Zaber v. City of Dubuque, 789 N.W.2d 634 (Iowa 2010).

Contributors

Carla Bagnoli is Professor of Philosophy at the Università di Modena. Her areas of interest are Moral and Social Epistemology, Metaethics, Practical Reasoning, Action Theory, Philosophical Psychology, Philosophy of Mind, Philosophy of the Emotions, and Kantian Constructivism. She is the author of *Dilemmi morali* (2006), and *L'autorità della morale* (2007). She is also the editor of *Morality and the Emotions* (2011), *Che fare? Prospettive filosofiche sull'azione* (2013), and *Constructivism in Ethics* (2013).

Maeve Cooke is Professor of Philosophy at University College Dublin. Her philosophical interests lie in Critical Social Theory, the Philosophy of Law and Religion, Autonomy, Ideology, and Power. She is the author of *Language and Reason: A Study of Habermas's Pragmatics* (1994), and *Re-Presenting the Good Society* (2006). She also edited Jürgen Habermas's *On the Pragmatic of Communication* (1998).

Piergiorgio Donatelli is Professor of Philosophy at Sapienza Università di Roma. His areas of expertise are Ethics, the History of Moral Philosophy, Bioethics, and the philosophies of Wittgenstein, John Stuart Mill, and Michel Foucault. He is the author of *Wittgenstein e l'etica* (1998); *La filosofia morale* (2001); *Introduzione a Mill* (2007); *La vita umana in prima persona* (2012); *Etica. I classici, le teorie e le linee evolutive* (2015), *Manieres d'être humain* (2015). He also edited Cora Diamond's *L'immaginazione e la vita morale* (2006) and Cora Diamond's and James Conant's *Rileggere Wittgenstein* (2010).

Douglas Lind is Associate Professor at the University of Virginia Tech. He specializes in the Philosophy of Law, Professional Ethics, Pragmatism, and the Later Philosophy of Wittgenstein. He practiced law for a number of years, specializing in Environmental Law, as an Assistant Attorney General for the State of Missouri, and in private practice in Philadelphia, Pennsylvania. He is the co-editor of *Pragmatism, Law, and Language* (with G. Hubbs, 2013).

John McGuire teaches at the School of Politics and International Relations, University College Dublin. His research interests include Contemporary Critical Theory, Cosmopolitanism, Cynicism, and theories of social change. He is the author of *Friends of the Gods? Cosmopolitans, Cynics, and the Persistence of Platonism* (forthcoming).

David Macarthur is Associate Professor at the University of Sydney. He worked on Contemporary Pragmatism, Naturalism, Skepticism, Philosophical Quietism, Philosophy of Language, Wittgenstein, and the Philosophy of Art. He is the co-editor of *Naturalism in Question* (with M. De Caro, 2004), and *Naturalism and Normativity* (with M. De Caro, 2010). He also co-edited Hilary Putnam's *Philosophy in an Age of Science: Physics, Mathematics, and Skepticism* (with M. De Caro, 2012) and edited Hilary and Ruth Anna Putnam's *Pragmatism as a Way of Life: The Lasting Legacy of William James and John Dewey* (2016).

Giancarlo Marchetti is Associate Professor at the Università degli studi di Perugia. He wrote on Pragmatism, Epistemology, Philosophy of Language, Ethics, and the History of Modern and Contemporary Philosophy. He is the author of *Il neo-pragmatismo* (1999) and *Verità e valori. Tra pragmatismo e filosofia analitica* (2008). He also co-edited *Il Pragmatismo. Dalle origini agli sviluppi contemporanei* (with G. Maddalena and R. M. Calcaterra, 2015).

Sarin Marchetti is Postdoctoral Fellow at the Università degli Studi di Milano. He works mainly in Ethics, Metaphilosophy, Pragmatism, and the History of Analytic Philosophy. He is the author of *Ethics and Philosophical Critique in William James* (2015), and *James* (2016). He is also the co-editor of *Pragmatism and the European Traditions: Encounters with Analytic Philosophy and Phenomenology before the Great Divide* (with M. Baghramian, 2017).

Alex Miller is Professor of Philosophy at the University of Otago. His research interests are the Philosophy of Language and Mind, Metaethics, Metaphysics, and Rule-Following. He is the author of *Philosophy of Language* (1998, 2007 2nd ed.) and *An Introduction to Contemporary Metaethics* (2003, 2013 2nd ed.). He also co-edited *Rule-Following and Meaning* (with C. Wright, 2002).

Hilary Putnam was one of the most influential American philosophers, who made major contributions to the Philosophy of Mathematics, Philosophy of Physics, Philosophy of Language, Philosophy of Mind, Pragmatism, and the later Wittgenstein. His most recent books are *The Collapse of the Fact/Value Dichotomy* (2002); *Ethics Without Ontology* (2004); *Philosophy in An Age of Science* (2012); and *Naturalism, Realism, and Normativity* (2016).

Ruth Anna Putnam is Professor of Philosophy, Emerita, at Wellesley College. She wrote profusely on American Philosophy, William James, John Dewey, Epistemology, Ethics, and Political Philosophy. She is the co-author of *Pragmatism as a Way of Life: The Lasting Legacy of William James and John Dewey* (with H. Putnam, 2016) is also the editor of *The Cambridge Companion to William James* (1997).

Mauro Rossi is Professor of Philosophy at the Université du Québec à Montreal. His did work on Metaethics, Normative Ethics, Moral Psychology, the Philosophy of Mind, and the Philosophy of Economics. He is the editor of *Normativity and Normative Psychology* (2012), and is currently working on a book on *Happiness and Wellbeing* (with C. Tappolet).

Naomi Scheman is Professor of Philosophy and Gender, Women's, & Sexuality Studies, Emerita, at the University of Minnesota. Her work explores the intersections of Epistemology, Politics, and Feminist Theory. She is the author of *Engenderings: Constructions of Knowledge, Authority, and Privilege* (1993), and *Shifting Ground: Knowledge & Reality, Transgression & Trustworthiness* (2011). She co-edited *Is Academic Feminism Dead? Theory in Practice* with R. Brewer, M. L. Fellows, S.N. Garner, A. Kaminsky, and J. Pierce, (2000), and *Feminist Interpretations of Wittgenstein* with P. O'Connor, (2002).

Robert Schwartz is Distinguished Professor of Philosophy at the University of Wisconsin at Milwaukee. He published extensively on the Philosophy of Psychology, Philosophy of Science, Language Philosophy of Mathematical Cognition, and Pragmatism. He is the author of *Rethinking Pragmatism* (2012).

Christine Tappolet is Professor of Philosophy at the Université de Montreal. Her research interests include Metaethics, Normative Ethics, Moral Psychology, Normativity, and Emotion theory. She is the author of *Émotions et valeurs* (2000), *Emotions, Value, and Agency* (2016), and co-author of *Les Concepts de l'éthique. Faut-il être conséquentialiste?* (with Ruwen Ogien, 2008). She is the co-editor of *Weakness of Will and Practical Irrationality* (with Sarah Stroud, 2003), and of *The Modularity of Emotions* (with Luc Faucher, 2008).

Kenneth Taylor is Herny Waldgrave Stuart professor of Philosophy at Stanford University. His work lies at the intersection of the Philosophy of Language and the Philosophy of Mind, with an occasional foray into the History of Philosophy. His other interests include Semantics, Reference, Naturalism, and Relativism. He is the author of *Truth and Meaning: An Introduction to the Philosophy of Language* (1998), *Reference and the Rational Mind* (2003), and *Referring to the World: An Introduction to the Theory of Reference* (forthcoming).

Index

Made in the USA
Las Vegas, NV
29 October 2024

02a09cd7-d319-44c1-b464-6d2088c3475aR01